The Norman Conquest: A Zooarchaeological Perspective

Naomi Jane Sykes

BAR International Series 1656
2007

Published in 2016 by
BAR Publishing, Oxford

BAR International Series 1656

The Norman Conquest: A Zooarchaeological Perspective

ISBN 978 1 4073 0092 4

COVER IMAGE *Sheep, cattle, and pig corbels from twelfth-century Kilpeck Church, Herefordshire (Photos by R. Jones)*

BAR Publishing is the trading name of British Archaeological Reports (Oxford) Ltd. British Archaeological Reports was first incorporated in 1974 to publish the BAR Series, International and British. In 1992 Hadrian Books Ltd became part of the BAR group. This volume was originally published by Archaeopress in conjunction with British Archaeological Reports (Oxford) Ltd / Hadrian Books Ltd, the Series principal publisher, in 2007. This present volume is published by BAR Publishing, 2016.

Printed in England

BAR
PUBLISHING

BAR titles are available from:

BAR Publishing
122 Banbury Rd, Oxford, OX2 7BP, UK
EMAIL info@barpublishing.com
PHONE +44 (0)1865 310431
FAX +44 (0)1865 316916
www.barpublishing.com

Contents

Please note that the CD referred to above has now been replaced with a download available at www.barpublishing.com/additional-downloads.html

Preface and Acknowledgements

Essentially, this book is my 2001 thesis – slimmed down, tarted up and minus the typos that littered the original – otherwise the content remains largely unchanged. It has been updated only to a small degree, hence the embarrassing 1980s/1990s skew of the bibliography. It has, however, benefited from reference to several seminal works which were yet to be published when I was writing my thesis: Benoît Clavel's superb 2001 synthesis of zooarchaeological data from northern France; Cecile Callou's 2003 excellent work on rabbits; Hugh Thomas's beautiful book *The English and The Normans: Ethnic Hostility, Assimilation and Identity 1066-c.1220* and, more recently, James Barret *et al.*'s work on medieval fish.

I have endeavoured to add in other data from post-2001 zooarchaeological reports but sample sizes are still pitifully small for some of the questions I have attempted to address. As new data emerge, I hope that others may develop or disprove the theories I have proposed here, and I would certainly encourage archaeologists to take another look at the Norman period which is so deserved of re-investigation.

Because this book is, in effect, composed of other people's work, I am indebted to all of those zooarchaeologists whose results I have used – a glance at the bibliography highlights the star players. In addition to this, there are several people without whose work and help this book would simply not have been possible: Clive Gamble, Annie Grant, Terry O'Connor, Peta Sadler, Dale Serjeantson and Jean-Hervé Yvinec. My thanks go also to Umberto Albarella, Robin Bendrey, Alban Gautier, Daffyd Pierce, Kristopher Poole and Sylvia Warman for allowing me to use their work in advance of publication, and to Lorrain Higbee, Dale Serjeantson, Richard Thomas and Fay Worley for providing rabbit metrics. Umberto Albarella, Robin Bendrey and Kristopher Poole kindly read sections (or all) of the book and I am very grateful for their helpful comments. David Hinton deserves to be singled out for the support he has given me over the past few years: always reading drafts of my papers, always tossing them at me as 'rubbish', always right – you are my favourite pedant.

Mum and Dad: to have provided me with moral and, at times, financial support without ever really knowing what I do is a sign of both your kindness and eccentricity. You've received very little from me in return and this book, that I know you'll never read and which will only add to the weight of clutter in your house, is hardly the pay-back you deserve.

Richard Jones: took photos, digitised maps, endured me and proof-read the book – all thankless tasks but thank you.

List of Figures

List of Tables

Chapter 1: Introduction

Investigation of social and economic change has always been central to archaeology. As part of this, population movements have frequently been emphasised as instigators of transition. This is particularly the case in British archaeology where, as an island, migration episodes tend to be viewed as highly significant. The Norman Conquest was the last and perhaps most famous of Britain's invasions, resulting in the almost complete replacement of the Saxon elite, both lay and ecclesiastical. Because the events surrounding the Conquest are so well documented, 1066 has come to be held as a significant watershed.

Yet, despite the abundance of textual evidence concerning it, the Norman Conquest has received surprisingly little attention from archaeologists. Whilst there are numerous archaeological texts concerning the Roman (for example, Millet 1990), Saxon (Myres 1970; Webster 1986; Welch 1991) and Viking invasions (Graham-Campbell 1990; Richards 1992), only one major book (Rowley 1997) has attempted to examine the Norman Conquest from an archaeological perspective. Furthermore, Rowley's book placed more emphasis on the historical than archaeological evidence, arguing that the archaeological record is a poor medium for detecting Norman impact, and re-iterating the widely held belief that, without guidance from history, the Conquest would remain archaeologically invisible.

While I accept that archaeological evidence for Conquest-related change may be difficult to discern, it seems possible that the current dearth of data may be due to a lack of investigation, rather than an actual absence. To date, both the range and depth of archaeological analysis has been limited. Although it is commonly cited (for example Hurst 1976, 343; Rowley 1997, 26) that artefacts do not change from the pre- to post-Conquest period this subject is, in truth, extremely under-researched. And, when detailed analyses have been undertaken, they have produced promising results (for example Brown 1997a, 1997b; Cotter 1997; MacGregor 1991). It is on this basis that this book sets out to undertake a detailed zooarchaeological analysis of the Norman Conquest, whereby data are considered by site-type to detect subtle temporal variations, if present, in human-animal relationships.

It is only in recent years that sufficient information, from all types of site, has become available to allow such a study. Before this, zooarchaeologists were generally able to consider animal management only at a site level. Some inter-site comparisons (Clutton-Brock 1976 and Grant 1988a) and regional syntheses (Levitan 1987a, 1987b; Crabtree 1994; Huntley and Stalibrass 1995) have been undertaken but most have been spatially or temporally restricted, focusing on either the pre- or post-Conquest period. One notable exception is Noddle's (1975a) study of eight English medieval sites, which together spanned the Saxo-Norman transition. Noddle also considered continental data, which she recognised as vital for discerning potential foreign influences. Similar examinations of foreign assemblages have since been achieved by O'Connor (1989a) and Crabtree (1994). There remains, however, substantial scope for the multi-period analysis of regional/national animal economies and their comparison with continental data.

The aim of this book is to show that zooarchaeological and historical data can be used together profitably to provide a new perspective on the Normans and their conquest of England. In order to accomplish this, the Norman Conquest is examined at the macro, meso and micro scale, which can be translated as the Norman Empire, Saxo-Norman England and specific Saxo-Norman sites, respectively.

1.1: The Norman Empire

Although in Britain the Normans are most famous for their Conquest of 1066, Figure 1 shows that by the late eleventh century the Normans were active not only in northern Europe but also in southern Italy, Sicily and the Levant. Throughout the twelfth century, links between these different regions, especially between England and Sicily, became increasingly close, resulting in considerable political and cultural exchange (Cassady 1986; Loud 2003). To understand the impact of 1066, England cannot, therefore, be viewed in isolation. Here, data from sites across the Norman Empire are considered with the hope of detecting the introduction of foreign elements or customs into post-Conquest England.

At the same time, because the Normans have been hailed as catalysts, rather than importers, of change (Rowley 1999, 13), post-Conquest England may demonstrate social or economic shifts that cannot be linked to the imposition of foreign traditions. It is for this reason that the situations in both pre- and post-Conquest England are also examined.

1.2: Saxo-Norman England

Since at least the 1860s scholars have debated whether the Norman Conquest caused significant change to Anglo-Saxon society (for example Round 1895; Stenton 1908; Brown 1973, 1984) or if continuity prevailed (notably Freeman 1867-1879; Stenton 1943; Barlow 1966): the history of the debate concerning the Conquest has been summarised by Chibnall (1999). Originally it was considered that 1066 resulted in the introduction of a pre-existing 'Norman Package' containing, amongst other things, feudalism, motte-and-bailey castles, Romanesque architecture and deer parks. It is now recognised that the arrival of these elements was less clear-

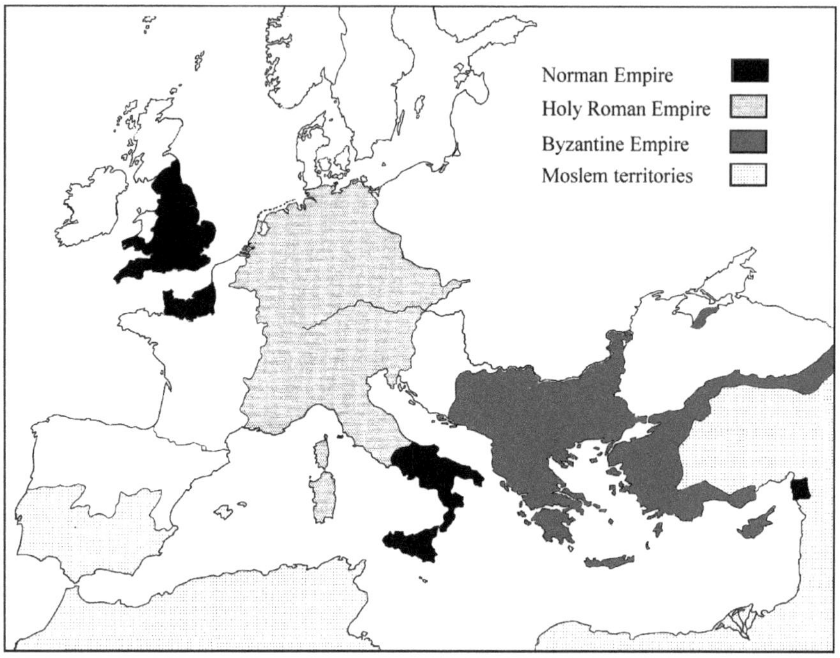

Figure 1: The Norman Empire (after Davis 1976)

cut: many were already present in Anglo-Saxon England and should not, therefore, be viewed as Norman achievements (for a discussion of feudalism see Campbell 1991; for the origins of castles see Davison 1969 and Thompson 1991; for architecture see Fernie 1982, and for parks see Liddiard 2000, 2003 and Sykes in press). Although most modern academics (such as Chibnall 1986; Campbell 1991; Golding 1994; Williams 1997) now accept basic continuity and reject the concept of a 'Norman package', the idea that the Conquest impacted dramatically on British wildlife – with the introduction of fallow deer (*Dama dama*), rabbits (*Oryctolagus cuniculus*) and pheasants (*Phasianus colchicus*) – is still prevalent amongst historians, zooarchaeologists and ecologists (Henderson 1997; Rackham 1997; Rowley 1997, 131 and 1999, 104; Yalden 1999, 157; Bartlett 2000, 672; Langbein and Chapman 2003). Widespread support has also been retained for the idea that the Normans altered hunting traditions, both through the imposition of Forest Law and the introduction of the *chasse par force de chien* – the classic form of medieval hunting whose terminology was heavily Gallicised (Gilbert 1979, 58). The possibility that the Normans established new animal species or hunting traditions in post-Conquest England is as significant as whether they introduced castles or Romanesque architecture but it is a question that has received little attention. This book seeks to redress the balance.

The investigation of hunting traditions is particularly pertinent within the context of Norman studies as it has the potential to inform not only on patterns of wild animal exploitation but also on socio-political structure, the two being very much linked (Wickham 1994, 161). Although it is now acknowledged that the Normans did not introduce feudalism, the extent to which the Conquest reinforced social inequality remains uncertain: studies of hunting practices may help to clarify the situation. The development of social hierarchy is here also considered through the examination of consumption practices, since ethnographic studies have demonstrated that cuisine is strongly linked to social structure (Goody 1982).

Inextricably woven into the social and political fabric of Saxon and Norman England was the economic set up. Rowley (1984, 21) has stated that 'agricultural and industrial processes were unaffected by the Conquest' and certainly historical evidence provides little indication of economic change. However, most late eleventh- and twelfth-century documents refer to the aristocracy and the Conquest itself, rather than the activities of the lower classes. It seems highly possible that changes in animal husbandry or exploitation could have occurred whilst escaping the attention of the medieval writer. Even in the more recent Post-medieval period there seems to be some discrepancy between the historical evidence and the 'true' situation. For example, the so-called Agricultural Revolution has been dated to the eighteenth century by some historians (Orwin 1949; Trevelyan 1957; Ritvo 1987) whilst others ascribe the transition to the sixteenth or seventeenth century (Trow-Smith 1957; Kerridge 1967). If historians have found it difficult to date such a significant phenomenon, which occurred in a period when writings on agriculture were common, it seems unlikely that subtle changes in Saxo-Norman husbandry practices would be historically detectable. Consequentially, zooarchaeology may hold a key to establishing whether Conquest-related shifts in the rural economy occurred. Investigations of animal husbandry, management and product distribution systems are, therefore, central themes of this book. Animal-based industries, such as butchery, bone working and tanning, are also examined to inform on the development of specialist trades.

Medieval Ireland provides a useful comparison for the situation in Saxo-Norman England. The Anglo-Norman Conquest of Ireland occurred in 1169 and McCormick (1991) has studied its effects on the country's wild and domestic fauna. Although the Conquest of Ireland took place more than one hundred years after 1066, the patterns of social change, such as the dispossession of the elite, were the same as those experienced in England. Historical evidence for rural life in pre- and post-Conquest Ireland is far superior to that for England (see for instance Kelly 1997) and by considering this information in conjunction with the zooarchaeological data, McCormick has developed a detailed picture of Conquest-related change. Here, McCormick's findings are used as a comparative dataset against which the English evidence is viewed.

Examining synthesised evidence at a national or regional scale obviously results in a loss of definition concerning site-based patterning; hence there is a need to consider the data at a more detailed level.

1.3: The Site Level

Any examination of the Conquest at a meso and macro scale must be borne out of detailed studies at the micro, or rather site, level. Such analyses provide not only the basic data for studies of economic and industrial change but also give resolution to issues of social or cultural practice.

The issue of Norman identity has been a topic of considerable debate. Eleventh- and twelfth-century documents suggest that the Normans perceived themselves as a people, or '*Gens*', regardless of their geographical location (Loud 1982; Bliese 1991; Potts 1995; Shopkow 1997) but archaeological absence of a distinct material culture led some scholars to question whether the Normans did indeed possess a common identity (Davis 1976; Rowley 1999: 13). It is now widely recognised that, although material culture is frequently invoked in expressions of ethnicity, there is not a fixed one-to-one relationship between ethnic groups and artefact typology (Jones 1997). Instead symbolic assertions of ethnicity are fluid, situational and generally enmeshed with a group's existing cultural practices and social, economic and political structure. Archaeological identification of ethnicity requires, therefore, the adoption of a contextual and historical approach that considers changing patterns in the production, distribution and consumption of material culture.

To date, zooarchaeological investigations of ethnicity have lagged behind other specialist fields of archaeology. Up to now they have generally been founded in period- or site-specific studies of diet, notably the analysis of species presence/absence, or butchery patterns (for instance Langenwalter 1980; Ijzereef 1989; Scott 1996). Indeed, these are sources of evidence that shall also be examined here: butchery marks in the hope of detecting cultural practices (as for instance Sackett 1986; Kenyon 1997) and diet to determine whether the Normans sought to project a pious image (for example Potts 1995; Shopkow 1997) through their cuisine. If zooarchaeological evidence for Norman identity is to be recognised, however, it is necessary to consider the entire production-distribution-consumption process, in other words the 'social life' of animals and their products (Appadurai 1986; Hamilakis 2000). Food production, procurement and distribution can be as symbolic as the act of consumption itself. This is particularly true in the case of hunting, an activity that in farming societies – where the exploitation of wild resources is unnecessary for survival – is often employed to negotiate power and authority (Hamilakis 2003). The Norman 'love of hunting' has also come to be viewed as a marker of their identity. Whether this was actually the case is considered by studying the way in which wild animals were incorporated into patterns of display.

1.4: The Data-Set

A large quantity and range of zooarchaeological data from England and the Continent has been synthesised for the purpose of this book. The details of these are as follows.

1.4i: English Sites

Although this book concentrates on the Saxo-Norman transition, the need to demonstrate that changes, if present, were Conquest-related rather than reflections of long-term trends dictated the decision to consider sites dating to between the fifth and fourteenth centuries. The main hindrance to this approach was the problem of dating – for instance the similarity of late pre-Conquest and early post-Conquest artefacts has given rise to the label 'Saxo-Norman', which covers an approximate date range of AD 950-1200. Despite this, it was possible to categorise most assemblages into the groups presented in Table 1. In total, data from 289 British assemblages were utilised: the sites from which they are derived are listed in Table 2 and their geographic locations are shown in Figure 2. Most were collated from published reports but a number of assemblages were examined or reanalysed personally. Several sites – Steyning, Boltophs and Burpham (all in Sussex) – were re-examined specifically to check purported identifications of fallow deer or rabbit specimens. Other sites – the Cheddar Palaces (Somerset), Portchester Castle (Hampshire) and Pevensey Castle (Sussex) were examined in detail on the basis of their superior dating and multi-period status. For some research questions, sites outside these date groups were considered and they are also shown on the site list and location map.

Period	Date Range	No of Assemblages
Early/Mid-Saxon	5th to mid-9th	69
Late Saxon	mid-9th to mid-11th	62
Norman	mid-11th to mid-12th	70
High Medieval	mid-12th to mid-14th	88

Table 1: Correlation between period labels and date ranges (centuries) with the number of assemblages considered for each.

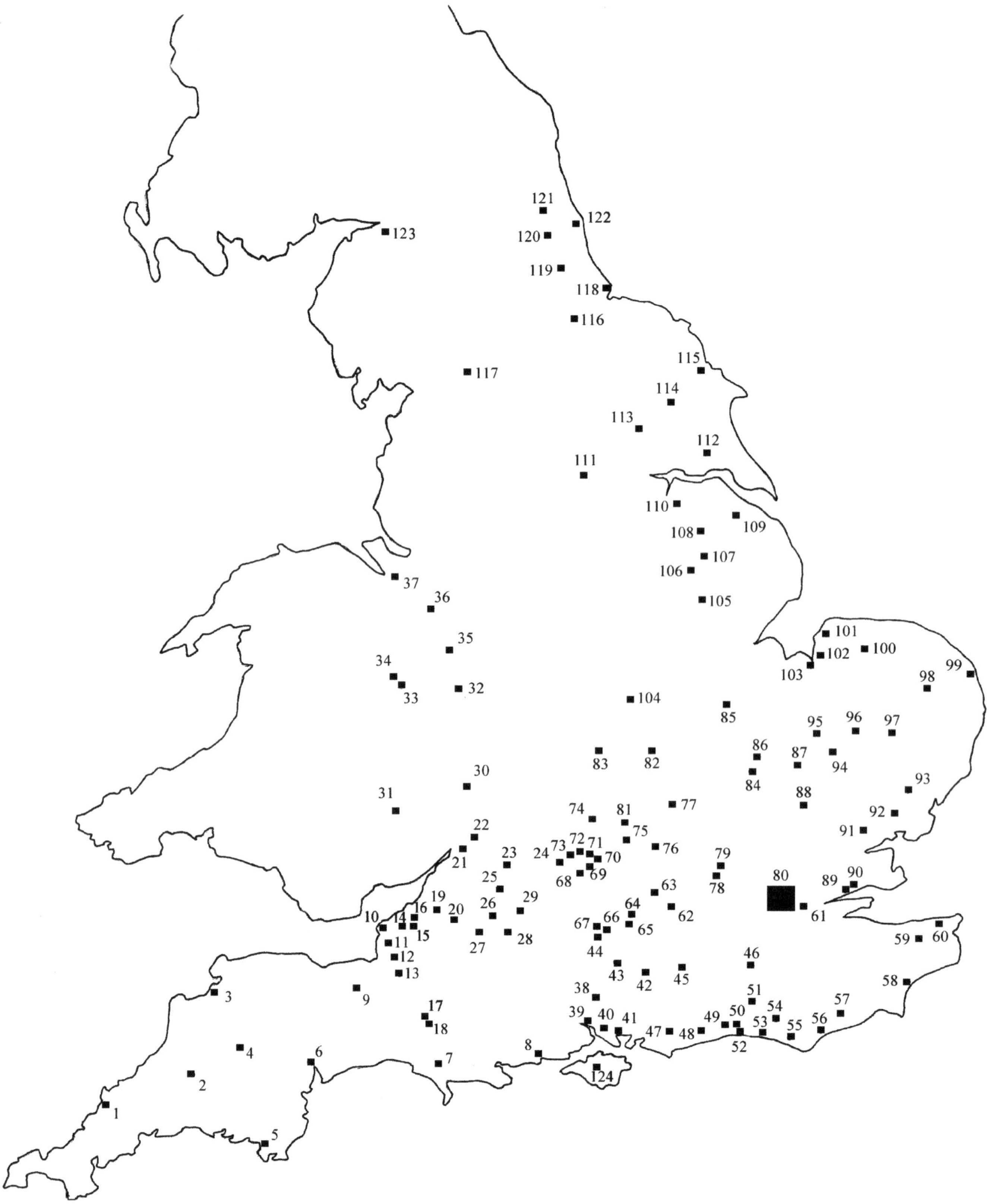

Figure 2: Location of English sites considered in this study (see Table 2 for name and reference)

Table 2: Sites considered in this study (see Figure 2 for location)

No	Site	Reference
	Cornwall	
1	Mawgan Porth	Clutton-Brock 1976; 1977
2	Launceston Castle	Albarella & Davis 1996
	Devon	
3	Westward Ho!	Levitan & Locker 1987
4	Okehampton Castle	Maltby 1982
5	Bantham	Coy 1981a
6	Exeter	Maltby 1979
6	High Street, Exeter	Maltby cited in Levitan 1987a
6	Trickhay, Exeter	Maltby cited in Levitan 1987a
	Dorset	
7	Poundbury	Buckland-Wright 1987
8	Christchurch	Coy 1983
	Somerset	
9	Benham's Garage, Taunton	Levitan 1984b
9	Priory Barn, Taunton	Levitan 1984b
10	Brean Down	Levitan 1990
11	Cheddar Palaces	Higgs & Greenwood 1979
12	Wells Museum	White n.d
13	The Mound	Darvill & Coy 1985
14	Eckweek	Davis 1991a
15	Cadbury Congresbury	Noddle 1970
16	Cleeve Abbey	Locker 1989
17	Ilchester	Levitan 1982
18	Catsgore	Everton 1982
	Avon	
19	Bristol Castle	Noddle cited in Levitan 1987a
19	Mary-Le-Port	Noddle 1985b
20	Bath	Grant 1979
	Gloucestershire	
21	Upton	Noddle 1969
21	Westgate, Gloucester	Maltby 1979b
22	Winchcombe, Gloucester	Levitan 1985
23	Barnsley Park	Noddle 1985c
24	Lechdale	Maltby 2003
	Wiltshire	
25	Postern Mill	Currie 1993
26	Yatesbury	Sykes n.d.
27	Trowbridge	Bourdillon 1993
28	Cadley Road, Collingbourne	Hamilton-Dyer 2001
28	Ludgershall Castle	Poole n.d.
29	Ramsbury	Coy 1980
	Hereford and Worcestershire	
30	Friars Street, Droitwich	Locker 1992
30	Upwich	Meddens 1997
31	Croft Ambrey	Whitehouse 1974
31	Deansway 1, Hereford	ABMAP
31	Hereford City	Noddle 1985
	West Midlands	
32	Dudley Castle	Thomas 2005
	Shropshire	
33	Wroxeter	Meddens 1987, 2000; Hammon pers comm.
34	Town houses, Shrewsbury	Noddle & Bramwell 1983
34	Shrewsbury Abbey	Jones 2002
	Staffordshire	
35	Stafford Castle	Sadler *pers. comm.*
	Cheshire	
36	Crown car park, Nantwich	Fisher 1986
37	Dominican Friary, Chester	Morris 1990
	Hampshire	
38	Abbots Worthy, Winchester	Coy 1987
38	Crowder Terrace, Winchester	Bourdillon 1992
38	East Suburbs, Winchester	Serjeantson 2000a
38	Henley's Garage, Winchester	Bourdillon 1992
38	New Road, Winchester	Bourdillon 1992
38	North Suburbs, Winchester	Bourdillon 1992
38	Sussex Road, Winchester	Bourdillon 1992
38	St John Street, Winchester	Serjeantson 2000a
38	Victoria Road, Winchester	Serjeantson 2000a
38	Winchester Defences	Bourdillon 1992
38	Northbrook	Bourdillon 1992
39	Hamwic	Bourdilln & Coy 1980
39	Quilters Vault	Bourdillon 1979
39	Southampton New Town	Noddle & Bramwell 1975
40	Bishops Waltham	Coy 1985
41	Portchester Castle	Grant 1975, 1976, 1977
42	Cowdery's Down	Maltby 1983
43	Brighton Hill South	Coy 1995
44	Faccombe Netherton	Sadler 1990
	Surrey	
45	Guildford Castle	Sykes *et al.* 2005
46	Bell Street	Isles *et al.* 1995
	Sussex	
47	Carne's Seat	Beech 1986
48	Burpham	Collected personally
49	Steyning	Collected personally
50	Bramber Castle	Westley 1977
50	Boltophs, Bramber	Stevens 1990
51	Friars Oak, Hassocks	Stevens 2000
52	Old Erringham	Westley 1980
53	Marlipins	Sykes 2003
54	Lewes Castle	Sykes 1997
54	Lewes Priory	Stevens 1997
54	Lewes Friary	Sykes 1997
54	North Street, Lewes	Sykes 1997
54	Clay Hill, Ringmer	Sykes n.d.
55	Bishopstone	Gebbles 1977, Ingrem n.d.
56	Pevensey Castle	Powell & Serjeantson forth.
57	Battle Abbey	Locker 1985
	Kent	
58	Sandtun	CluttonBrock 1976, Murray & Hamilton-Dyer 2001
59	Canterbury Castle	King, 1982
59	Canterbury Cathedral	Driver 1990
59	St Gregory's Priory	Powell & Serjeantson forth.
60	Monkton	Bendrey 2003
61	Pepperhill Lane, Northfleet	Charles & Ingrem 2001
	Berkshire	
62	Wraysbury	Coy 1989a
63	Lake End Road	Powell 2002
63	Lot's Hole	Powell 2002
64	Ufton Nervet	Westley 1974
65	Abbey Wharf, Reading	ABMAP
66	Thatcham	King 1962
67	Cheap Street, Newbury	Coy 1997a
67	Batholomew Street, Newbury	Coy 1997a
	Oxfordshire	
68	Wantage	Maltby 1996
69	Seacourt	Jope 1961/2
70	Audlett Drive, Oxford	Levitan 1992
70	Dean Court, Oxford	Jones 1994b
70	Barton Court Farm	Wilson 1986
70	Codrington Library, Oxford	Collected personally
70	Oxford Castle, Oxford	Marples 1976
70	Osney Abbey, Oxford	Wilson & Allison 1985
70	Queen Street, Oxford	Wilson *et al.* 1983
70	Shaken-Oak, Oxford	Jewel 1968; Brodribb 1972
70	Codrington Library, Oxford	Collected personally
70	St Aldates, Oxford	Marples 1977

Table 2 (Continued)

No	Site	Reference
70	St Ebbes, Oxford	Wilson *et al.* 1989
70	Stert Street, Oxford	Wilson 1979
70	The Hamel, Oxford	Wilson and Bramwell 1980
70	Dominican Priory, Oxford	Harman 1985a
70	Old Gaol sites, Oxford	Wilson *et al.* 1975
71	Cassington	Wilson 1962
71	Yarnton	Mulville *et al.* 2004
71	Eynsham Abbey	Ayres *et al.* 2003
72	Witney Palace	Ayres & Serjeantson 2002
73	New Wintles	Noddle 1975
74	Banbury Castle	Wilson, 1976; Gamble 1983
75	Middleton Stony	Levitan 1984a
	Buckinghamshire	
76	George Street, Aylesbury	Jones 1983
76	Walton Road, Aylesbury	Noddle 1976
76	Walton Vicarage	Sadler 1989
77	Chichley	Jones 1980
77	Wolverton Turn	Collected personally
77	Pennyland	Holmes 1993
77	Hartigans	Burnett 1993
78	Latimer	Hamilton 1971
	Hertfordshire	
79	St Albans Abbey	Crabtree 1983
79	Gorhambury	Locker 1990
	London	
80	Barking Abbey	Rackham 1994
80	Baynards Castle	Bramwell 1975
80	Billingsgate	Armitage 1980
80	Jubilee Hall	West & Rackham 1993
80	Maiden Lane	West & Rackham 1993
80	National Gallery	West & Rackham 1993
80	Peabody Buildings	West & Rackham 1993
80	St Magnus	Armitage 1979
80	Westminster Abbey	Rackham 1994
80	Whitehall	Chaplin 1977
	Northamptonshire	
81	Castle Lane, Brackley	Jones *et al.* 1985b
82	Lyveden	Grant 1971
82	Marefair	Harman 1979a
82	Saxon Palaces	Harman 1985b
82	St Peters St, Northampton	Harman 1979b
82	Woolmongers St, Northampton	Armitage 1999
82	Chalk Lane, Northampton	Coy 1981b
83	Brixworth	Coy *et al.* 1977
84	West Cotton	Albarella & Davis 1994
85	Maxey	Seddon *et al.* 1964
	Cambridgeshire	
86	Eynesbury	Sykes 2004a
87	War Ditches	White 1964
	Essex	
88	Wicken Bonhunt	Crabtree 1996
89	Mucking	Done n.d.
90	Hadleigh Castle	Ellison 1975
91	Rivenhall, Chelmsford	Luff 1993a
92	Culver Street, Colchester	Luff 1993b
	Suffolk	
93	Ipswich	Crabtree 1996; Jones & Sejeantson 1983
94	West Stow	Crabtree 1990
95	Brandon	Crabtree 1996
95	Melford Meadows	Powell & Clark 2002
	Norfolk	
96	Brandon Road, Thetford	Jones 1993
96	Redcastle Furze, Thetford	Wilson 1995
96	St Nicholas Street, Norwich	Hutton-MacDonald 1999
96	Thetford 1092	Jones 1984
96	Mill Lane, Thetford	Albarella 1999
97	Scole Dickleburgh	Baker 1998
98	Alms Lane, Norwich	Cartledge 1985
98	Castle Mall, Norwich	Alberella *et al.* 1997
98	Heigham Street, Norwich	Weinstock 2002
98	Fishergate, Norwich	Jones 1994a
98	Coslany Street, Norwich	Albarella 1997c
98	St Martins-at-palace plain	Cartledge 1988
98	White Friars, Norwich	Cartledge 1979
99	Caister-On-Sea	Harman 1993
100	North Elmham	Noddle 1980
100	Spong Hill	Bond 1995
101	Sedgeford	Clutton-Brock 1976
102	Kings Lynn	Noddle 1977
103	Castle Rising Castle	Jones *et al.* 1997
	Leicestershire	
104	Bonners Lane, Leicester	Baxter 2004
104	Little Lane, Leicester	Gidney 1991
104	Boetler's Castle	Pinter-Bellows 1997
104	St Austins Friary, Leicester	Thawley 1981
104	St Peters Lane, Leicester	Gidney 1991
	Lincolnshire	
105	Quarrington	Rackham 2003
106	Flaxengate, Lincoln	O'Connor 1982a
106	Lincoln City	Dobney *et al.* 1995
106	The Park and West Parade	Scott 1999
107	Goltho	Jones *et al.* 1987
108	Nettleton	Berg 1993
109	Riby Cross Roads	Scott 1994
110	Flixborough	Dobney & Jacques 2001
	Yorkshire	
111	Sandal Castle	Griffith *et al.* 1983
112	Eastgate, Beverley	Scott 1992
112	Lurk Lane, Beverley	Scott 1991
113	Coppergate, York	O'Connor 1989a
113	Fishergate, York	O'Connor 1991
113	Blake Street, York	Bond & O'Connor 1999
113	Tanners Row, York	O'Connor 1988
113	General Accident, York	O'Connor 1988
113	Skeldergate, York	O'Connor 1984
114	Wharram Percy	Stevens; Pinter-Bellows 1992
114	Wharram South Manor	Pinter-Bellows 2000
115	Crossgates	Rutter *et al.*1958
116	Catterick Bridge	Meddens 1990
117	Ribblehead	Rackham n.d.
	Cleeveland	
118	Hartlepool Monastery	Allison 1988
118	Morrison Hall	Gidney 1990
	County Durham	
119	Thrislington	Rackham 1989
	Northumberland	
120	Yeavering	Higgs & Jarman 1977
121	Prudhoe Castle	Davis 1987
	Tyne and Wear	
122	Closegate I and II, Newcastle	Davis 1991b
122	Jarrow	Noddle 1987
	Cumbria	
123	The Lanes	Stallibrass 1993
	Isle of Wight	
124	Carisbrooke Castle	Smith 1994

1.4ii: French Sites

Medieval zooarchaeology is a growing field in northern France and there is currently much exciting work being undertaken. However, in 1998, when carrying out research for this book, there was a dearth of accessible data: many reports were unpublished and animal bones seldom archived. I found only one site in Normandy for which the bones had been retained, that of Vatteville-la-Rue – an eleventh to sixteenth-century castle located on the banks of the river Seine. My analysis of the animal remains from this site forms the foundation of the French dataset considered in this study. It was, however, insufficient as a comparative sample for the British data and additional information was sought from neighbouring departments. Substantial amounts of data were available from Picardie, where medieval zooarchaeology has been established as a specialism for more than twenty years. Numerous reports have been produced by the likes of Jean-Claude Leblay, Sébastien Lepetz and Jean-Hervé Yvinec, and a synthetic volume on the zooarchaeology of the region has recently been published (Clavel 2001). Members of the Laboratoire d'Archéozoologie in Compiègne kindly supplied me with much of my French data, including some unpublished information. Although the dataset does not, therefore, relate purely to Normandy, it provides the best available opportunity for recognising French influence. All of the French sites considered in this book are shown in Table 3 and Figure 4.

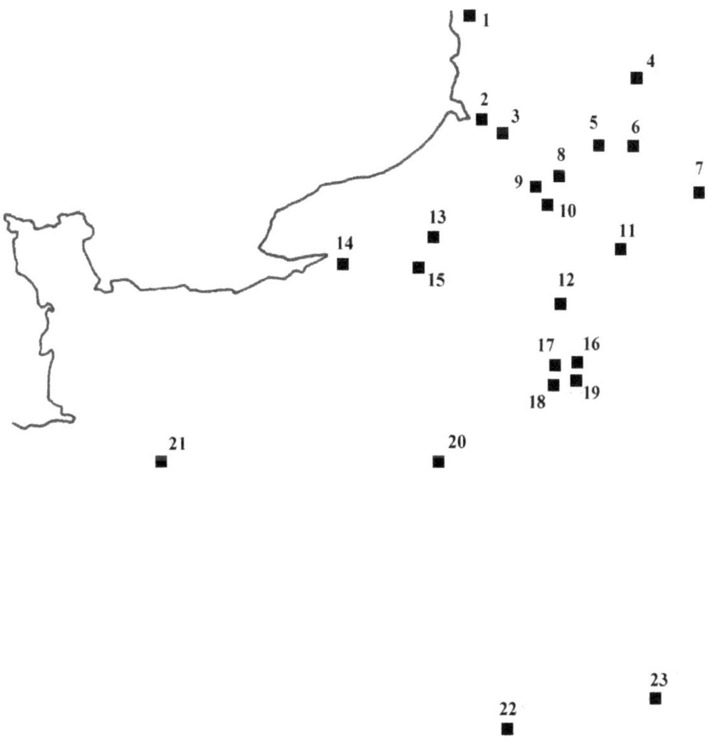

No	Site	Reference
1	Coquelles, Nord/Pais-de-Callais	Clavel 2001
2	L'Hôtel de ville, Abbeville, Picardie	Clavel 2001
3	Moreaucourt, Picardie	Clavel 2001
4	Douai, Nord	Vadet & Villete 1986
4	Parc St Julien, Nord	Leblay *et al.* 1997
5	Comble, Somme	Yvinec 1999
6	Savy, Picardie	Lepetz & Yvinec 2002
7	Noyon, Picardie	Clavel 2001
8	Amiens, Somme	Yvinec 1999
8	Saint-Germain, Amiens	Clavel 2001
8	ZAC Cathéderal, Amiens	Clavel 2001
9	Saleux, Somme	Lepetz & Yvinec 2002
10	Dury, Somme	Yvinec 1999
11	Compeigne, Picardie	Yvinec 1997
11	L'Hopital general, Compeigne, Picardie	Clavel 2001

No	Site	Reference
12	Villiers-le-sec, Picardie	Yvinec 1988
13	Grand Besele, Seine-Maritime	Lepiksaar 1966-78
14	Vatteville-la-Rue, Seine-Maritime	Collected personally
15	Rouen, Seine-Maritime	Lepetz & Yvinec 2002
16	Saint-Denis, Paris, Ile de France	Morel 1985
17	Roissy, Paris, Ile de France	Yvinec *pers. comm.*
17	Saint Michel, Paris, Ile de France	Clavel 2001
18	Rue de Collegiale, Paris, Ile de France	Clavel 2001
18	Rue de Lutece, Paris, Ile de France	Audoin-Rouzeu 1989
18	Cour Carrée, Paris, Ile de France	Clavel 2001
19	Vincennes, Paris, Ile de France	Clavel 2001
20	Chartres, Eure et Loire	Lepetz & Yvinec 2002
21	Château de Mayenne,	Powell n.d.
22	Ecuelle-Ravanne, Indre	Yvinec 1993
23	Charité-sur-Loire, Nievre	Audoin 1984

Figure 3 and Table 3: Northern French sites considered in this study

1.4iii: Other Continental Sites

The possibility that changes to post-Conquest England may have stemmed from other areas of the Norman Empire, rather than Normandy itself, demanded some consideration of the situation in these areas. Unfortunately, zooarchaeological data for many regions, especially Medieval Italy and Sicily, is scarcer than that for France. Work by Bossard (1984a and 1984b), Baker and Clark (1993) Baker (1993) and Audoin-Rouseau, (1993) has, however, permitted limited investigation of these regions.

In order to highlight trends that may have been unrelated to Norman influence, data from non-Norman areas of Europe have also been examined. All of the sites from areas outside England and Northern France are detailed below (Figure 4 and Table 4).

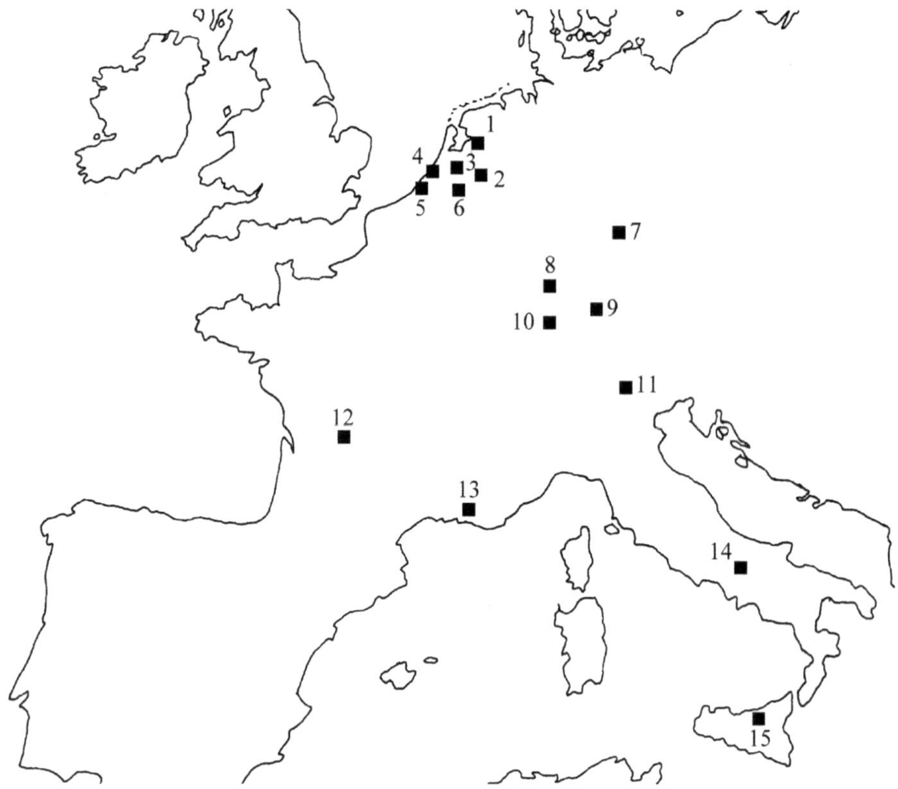

No	Site	Reference
1	Huis te Merwede, Netherlands	Clason 1967
2	Valkenburg, Netherlands	Prummel 1975
3	Ename Abbey, Belgium	Ervynck *et al.* 1994
4	Laarne, Belgium	Van Damme & Ervynck 1988
5	Dune Abbey, Belgium	Gautier 1984b
6	Tourinnes-Saint-Lambert, Belgium	Ervynck *et al.* 1999
7	Unterregenbach, Germany	Kuhnhold 1971
8	Munsterberg and Hochstetten, Germany	Schmidt-Pauly 1980
9	Heuneburg, Germany	McEneaney-Schneider 1979
10	Augusta Raurica, Switzerland	Schibler 1988
11	Schiedburg, Switzerland	Küpper 1972
12	Saint Avit-Seignieur, France	Gautier 1972
13	Marseille-La Bourse, France	Jourdan 1977
14	San Potito-Ovindoli, Italy	Bökönyi 1986
15	Brucato, Sicily	Bossard-Beck 1984a and 1984b

Figure 4 and Table 4: Continental sites considered in the text

1.5: Methods

The assemblages considered in this book were examined for five basic groups of zooarchaeological information: taxa ratios, age profiles, body part patterns, metrical data and butchery marks. Some research questions required the adoption of additional methods and these are detailed in the relevant chapters.

1.5i: Taxa Ratios

The primary aim of zooarchaeology has always been to obtain the relative frequencies of the different taxa within an assemblage. Numerous quantification techniques have been developed but after 30 years of debate there is still no consensus regarding standardisation (Casteel 1977; Chaplin 1971; Gautier 1984a; Grayson 1984; Klein and Cruz-Uribe 1984). In practice, two techniques have come to the fore: a basic fragment count, or NISP (Number of Identified Specimens), and the more interpretative MNI (Minimum Number of Individuals). Neither technique is without problem but MNI, which is heavily susceptible to inter-worker variation, was deemed unsuitable for this study, and NISP data was chosen for synthesis. Once collated, several different methods were used to calculate relative frequencies.

For basic studies of the main domesticates, NISP information was utilised only where the combined total for the three taxa exceeded 100 specimens. The relative percentages for each taxon were then calculated for each site. The percentages for each taxon, site type and period were summed and the mean percentage ascertained by dividing the summed figure by the number of sites that had contributed to it. By example, there are five assemblages from mid-eleventh to mid-twelfth century religious houses, for which pigs have relative frequencies of 36%, 18%, 18%, 19% and 29%. Summed, these percentages give a total figure of 120% which, when divided by five, gives an average of 24%.

For the lesser animal groups (domestic birds, wild birds and wild mammals) which on any single site are often represented by less than 100 specimens, the NISP was simply summed for each animal category, site type and period, and their relative frequency expressed as a percentage of the fragment count (excluding fish bones) for all the assemblages combined. It could be argued that, as a method for ascertaining the relative frequency of birds, this technique is problematic since bird and mammal remains seldom follow the same taphonomic pathway and are subject to different recovery biases (for instance Coy 1997b). However, the premise employed here is that the good sample sizes will mitigate the effects of any outliers. Thus, whilst the data presented do not reflect the actual dietary contribution made by wild birds they should provide a reliable, if somewhat basic, indication of inter-period and inter-site variation in their level of exploitation.

The decision to examine the bird and fish assemblages at a more detailed taxonomic level, introduced a number of logistical problems. Although it is recognised that rare components of an assemblage were probably the most significant socially, it was infeasible to consider every species recorded for Medieval Europe. Instead, I elected to study only the most commonly occurring taxa. In order to highlight inter-site and inter-period changes in the representation of individual wildfowl and fish species a modified presence/absence method was adopted. Wherever a species was represented within an assemblage it was noted simply as 'present' with an asterisk (see Appendix Idi). Its frequency of representation was then calculated as the percentage of assemblages (belonging to each site-type and phase) in which that particular species is represented: thus, if 'swan' or 'herring' are present in 10 of 20 mid-eleventh to mid-twelfth century elite assemblages their frequency for that period and site type would be 50%.

A similar presence/absence system was used as a proxy measure of the overall exploitation of 'fish', inter-taxa variations in recovery rendering the calculation of relative frequencies problematic. Inter-site and inter-period variations in fish consumption were calculated simply by adding up the number of 'present' asterisks for each date group or site type and then dividing that figure by the number that would have been achieved if all species were represented for that date group or site type. To explain, 20 fish taxa were considered in the study and, for the mid-fifth to mid-ninth centuries, four rural assemblages contained fish remains. If all taxa had been represented in each assemblage the number of 'present' asterisks would have been 80 (20 taxa x 4 assemblages). Looking at Appendix Idi it can be seen that three of these assemblages contained eel, with plaice, perch and pike being represented in one assemblage each: the total number of present asterisks is, therefore, 6. Fish 'representation' is then calculated by dividing the actual number of asterisks (6) by the potential number of asterisks (80). Whilst this quantification method does not reflect absolute quantities of bones or the frequency of fish relative to other animal groups, it does give an indication of the range and representation (here called a presence/diversity index) of fish taxa consumed during each date group or on site type, the theory being that levels of representation can be equated to levels of consumption. Results of the taxa representation studies are provided in Appendices I a-e.

1.5ii: Age profiles

Analysis of cull-patterns allows inferences to be made concerning a wide range of issues, including animal management and supply. Standardisation of ageing methods, by authors such as Silver (1969), Getty (1975), Payne (1973) and Grant (1975 and 1982), has meant that valid inter-site comparisons can now be undertaken. Inter-worker variation still exists, however, rendering bone fusion data unusable for a synthetic study of this kind. Dental ageing proved less problematic and it was possible to convert the majority of published data from the format in which it was presented. The system used for making the data comparable is shown in Table 5. Kill-off patterns were calculated for each period and site type as

Table 5: Correlation tables for a) cattle, b) caprine and c) pig dental ageing data

a) Cattle

Estimated Age	0-6 mths	6-12 mths	1-2 yrs	2-3 yrs	3-6 yrs	6-8 yrs	8-10 yrs
Halstead's Stage (1985)	A + B	C	D	E + F	G	H	I
Grant's MWS (1982)	1-7	8-16	17-29	30-40	41-43	44-47	48+
Definition based on	Dp4 <h	Dp4 f-j	Dp4 j-k	Dp4 m-n	P4 b-f	P4 f-h	P4 h
Grant's(1982)	M1 <a	M1 a-e	M1 f-g	M1 g-k	M1 j-l	M1 >k	M1 >m
wear stages		M2 <a	M2 a-f	M2 f-g	M2 j-k	M2 >j	M2 >l
			M3 <a	M3 b-g	M3 f-g	M3 >g	M3 >k

b) Caprine

Estimated Age	0-6 mths	6-12 mths	1-2 yrs	2-3 yrs	3-4 yrs	4-6 yrs	6-8 yrs	8-10 yrs
Payne's Stage (1973)	A + B	C	D	E	F	G	H	I
Grant's MWS (1982)	1-6	7-18	19-27	28-32	33-36	37-43	44-48	48+
Definition based	Dp4 <f	Dp4 f-k	Dp4 g-m	Dp4 h-m	P4 a-h	P4 g-l	P4 >k	P4 >l
on Grant's(1982)	M1 <a	M1 a-f	M1 f-g	M1 g-h	M1 g-l	M1 j-m	M1m-n	M1 h-m
wear stages		M2 <a	M2 a-f	M2 d-g	M2 f-g	M2 g-j	M2 j-m	M2 h-m
			M3 <a	M3 a-d	M3 c-f	M3 f-g	M3 h-j	M3 >k

c) Pig

Estimate Age	0-6 mths	6-12 mths	1-2 yrs	2-3 yrs	3 yrs +
O'Connor's Stage (1988)	Juvenile	Immature	Subadult	Adult	Elderly
Grant's MWS (1982)	1-7	8-16	17-26	27-40	41+
Definition based	Dp4 <f	Dp4 b-k	P4 b-d	P4 <g	P4 >g
on Grant's(1982)	M1 <a	M1 b-f	M1 c-j	M1 g-n	M1 m-o
wear stages		M2 <a	M2 a-d	M2 c-h	M2 h-m
			M3 <a	M3 a-d	M3 d-g

if all the mandibles derived from a single assemblage: the number of mandibles within each age category being summed and their relative frequency calculated. The raw data are provided in Appendix II.

1.5iii: Body Part Representations

Skeletal representations can provide useful information concerning provisioning systems and craft activities. Unfortunately, researchers seldom display data in the same way: they may be provided as raw NISP or MNE (Minimum Number of Elements) counts (Grant 1975, 1976, 1977); in a graphic percentage MNI format with no raw data supplied (Smith 1994); or as skeletal groups, for example 'head' elements, 'hind limbs' and 'forelimbs' (Noddle 1976; O'Connor 1984, 1988, 1989a). Such information is difficult to compare and, therefore, only those data presented, by element, as raw NISP or MNE counts have been utilised in this study. Because most of the data were derived from NISP counts, the degree of fragmentation-based variation was unknown. Although it was not possible to compensate for this, the inter-taxa variation in element frequency could be standardised by dividing the count for each element by the number of times it occurred in the body; this provided a basic MNE. In order to synthesise the various data, the MNE counts for each element have been summed. The element with the highest total MNE was then used as the denominator from which the percentage frequencies for all other elements were calculated (see Appendix III).

1.5iv: Metrical Data

Metrical analysis was used in numerous ways: to check fallow deer and rabbit identifications but also to make inferences concerning animal management regimes. Much of the data for cattle and caprines were derived from archaeological reports and from measurements accumulated by the Animal Bone Metrical Archive Project, (ads.ahds.ac.uk/catalogue/SpecColl/abmap).

Additional measurements were collected personally following the standards of von den Driesch (1976). For reasons of archaeological abundance and ease of measurement, the cattle and caprine studies were based on metapodia. Greatest length (GL) and smallest diameter (SD) measurements were selected as the primary variables on the basis of data comparability. Cattle wither heights were calculated using Foch's factors (cited in Driesch and Boessneck 1974).

To assist the detection of different sheep types, techniques were developed using skull and horn-core measurements from 134 modern individuals from twelve rare breeds (details are presented in Sykes 2001 and in prep.). Lack of measurable (mature) pig bones necessitated the collection of a new dataset. Archaeologically, pigs tend to be best represented by their teeth, thus mandibular deciduous fourth premolars (Dp^4), first and second permanent molars (M^1 and M^2) were targeted for metrical analysis. Measurements were taken following the standards set by Payne and Bull (1988) but the width of constriction (WC) and cervical length (CL) were suggested by Sylvia Warman (*pers. comm.*), the latter on the basis that it is unaffected by inter-tooth attrition. Three multi-period assemblages (Faccombe Netherton, the Cheddar Palaces and Portchester Castle) were re-analysed with the specific purpose of recording of these measurements.

Pig tooth measurements are provided in Appendix IVa with selected rabbit bone metrics available in Appendix IVb. Other metrical data can be found in Sykes 2001. The deer data will be published on-line at a later date.

1.5v: Butchery Patterns

For most zooarchaeologists, the recording and analysis of butchery evidence is not a prime objective. Whilst some researchers have considered butchery patterns (see for example Maltby 1979a; Dobney *et al* 1995; Luff 1994 and Audoin-Rouzeau 1987) or even synthesised personally-collected data to allow wider observations to be made (Peck 1986; Grant 1987; Maltby 1989) there exists no standard method for recording butchery marks. For this study butchery marks were recorded using Lauwerier's (1988) coding system, with additional marks being appended to the end of his list (see Sykes 2001). Multi-period assemblages from sites thought to have witnessed a period of Norman occupation – for instance Pevensey Castle, the Cheddar Palaces and Portchester Castle – were re-examined to obtain butchery data. In addition to these sites evidence was collected from the Guildford Castle and Steyning assemblages; the material from Vatteville castle allowing some cross-channel comparisons to be made. Summary tables are provided in Appendix V.

1.5vi: Historical Evidence

The events of 1066 inspired many writers to analyse the Conquest and the people involved. Whilst this work is founded in zooarchaeology, the wealth of documentary evidence for the eleventh and twelfth centuries must be taken into account.

Because most documents of the late eleventh century relate to the Conquest itself, they are less informative about quotidian life. The Anglo-Saxon Chronicle, Domesday Book and the Bayeux Tapestry provide some useful information pertaining to rural conditions but in general, pre-Conquest documents yield more evidence on this subject. For example, the Cotton Tiberius Calendar, held at the British Library, depicts numerous scenes of seasonal activities. Similarly, the *Rectitudines Singularum Personarum* (Rights and Ranks of People) also details agrarian conditions. Perhaps the most useful text concerning pre-Conquest life is Aelfric's Colloquy, an eleventh-century Latin schoolbook, which provides an insight to activities from hunting and fishing to trade and cooking. It is, therefore, evidence from these texts (albeit derived mostly from printed and secondary sources) upon which I have relied most heavily.

1.6: Chapter Outline

From the sections above, it can be seen that this book covers a wide spatial and temporal range, and deals with issues that are discrete but at the same time inter-linking. To avoid repetition it is structured so that Chapters 2 to 8 explore individual topics in their own right, whilst at the same time building a wider zooarchaeological perspective of the Conquest's impact.

In Chapter 2, the French dataset is examined in terms of the agricultural economy, provisioning system and wild resource exploitation. This overview forms the benchmark against which the English data are compared. In Chapter 3, for example, the pre- and post-Conquest economy of England is investigated to determine whether the Normans influenced systems of animal management or provisioning. Issues of Norman-instigated commercialisation are also examined by studying the development of animal-based craft activities. The possibility that the Conquest impacted on native animal stock is considered in Chapter 4.

Analysis of the animal economy is continued in Chapter 5, where the development of the fishing trade is viewed against wider patterns of wild resource exploitation. In this section, contextual analysis of wild animal representation is used to investigate whether the Norman introduction of Forest Law altered perceptions of either the hunting landscape or the social significance of game. Findings from this chapter are developed further in Chapter 6, where it is argued that shifts in deer skeletal representation reflect the introduction of new Norman hunting traditions.

The possibility that changes in wild mammal exploitation were somehow related to the introduction of fallow deer is examined in Chapter 7. In this section the archaeological evidence for this deer species is reconsidered to determine whether the Normans should be attributed with its introduction. This chapter also considers whether the Normans are correctly labelled as the importers of rabbits.

Chapter 8 combines all the evidence from the preceding chapters to determine whether the Normans impacted upon pre-Conquest cuisine. This is investigated from three perspectives: through the study of food ingredients and processing; by considering the development of dietary hierarchy, and by examining whether expressions of piety became more overt in the post-Conquest period.

The findings from all of these chapters are finally drawn together in Chapter 9 to re-examine the validity of the widely held belief, that the Norman Conquest is archaeologically invisible.

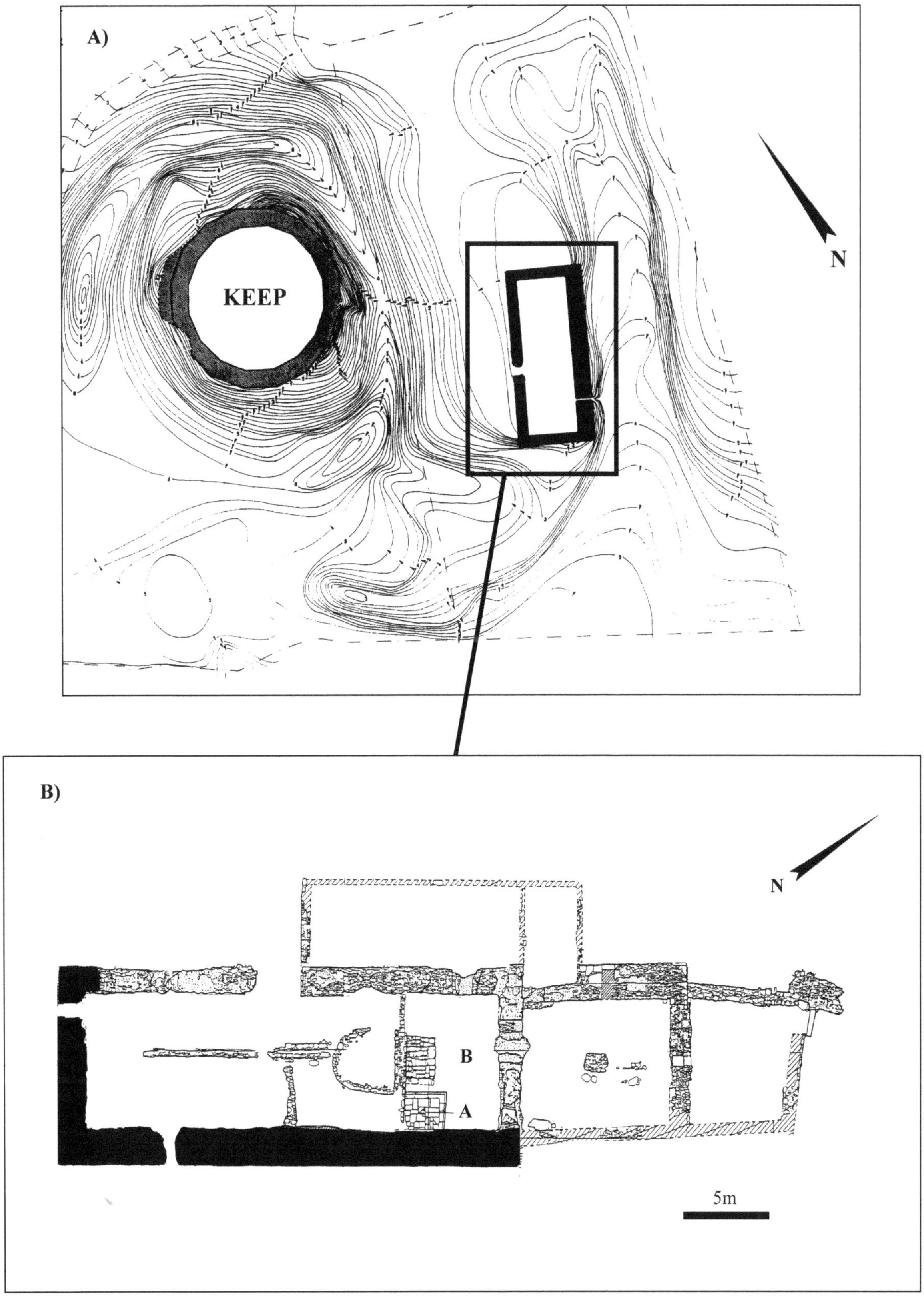

Figure 5: Plan of Vatteville-la-Rue's A) topography and B) excavations (after Flambarde Héricher 1996 and 1998)

It has long been recognised that the impact of invasions cannot be understood fully without examining evidence from the area whence influence was supposedly derived (Coy 1982: 107, Brown 1984, xvi). In this chapter zooarchaeological data from northern France are examined with the aim of gaining an insight into Norman systems of animal management and distribution. Particular attention is given to detecting socio-cultural practices and preferences, although these will become obvious only through comparison with the evidence from pre- and post-Conquest England. This chapter provides, therefore, an outline of the French situation, the implications of which will be developed in the succeeding chapters.

Central to this study is the animal bone evidence from the site of Vatteville-la-Rue (Figure 5a), a Norman castle constructed between 1059 and 1090 by Robert of Beaumont, who received the region from Duke William of Normandy (William the Conqueror). The site was excavated between 1994 and 1996 under the direction of Flambard Hériche (1996a, 1996b, 1998a, 1998 b): Figure 5b shows the structure that was the focus of the investigation. Situated within the bailey, the building had originally been interpreted as a chapel but excavation revealed it to be a residential complex containing a substantial kitchen. A large quantity of well-preserved animal bone was recovered by hand (the site was not sieved) and I examined the material from the better-dated contexts. It was possible to assign about one third to date-groups (eleventh/twelfth, thirteenth and fourteenth century) but the remainder was classified as twelfth to sixteenth century (Table 6). Sample sizes for the phased material were, for some research questions, large enough to allow valid inter-period comparisons. It proved necessary, however, to combine the phased data for the construction of cull and body part patterns.

Analysis demonstrated that the assemblage was composed almost entirely of food waste, thus providing a wealth of information concerning the diet of the Norman elite. But there is a considerable step from castle menu to the regional farming economy. To obtain evidence pertaining to the overall socio-economic set-up of medieval Normandy, it was necessary to view the Vatteville-la-Rue material in its wider regional setting. Unfortunately, only one other medieval assemblage from Normandy, that from the tenth to twelfth-century seignuerial site of Grand Besle (Lepiksaar 1966-68) was available for comparison. In order to increase sample sizes and obtain evidence for all major site-types, data from other regions of northern France have been synthesised and are also considered here.

2.1: Management of cattle, caprines and pigs

Most historians stress the role of pigs in the medieval economy, especially that of France where herds of animals could be maintained, at comparatively little cost, within the heavily wooded landscape (Charbonnier 2002, 52; Devailly 1973, 559; Forquin 1975, 305; Scully 2002, 54;). Zooarchaeological evidence would appear to support this contention, with pigs being the best represented species on most eighth- to twelfth-century sites (Figure 6 and see Audoin-Rouzeau 1997, 74).

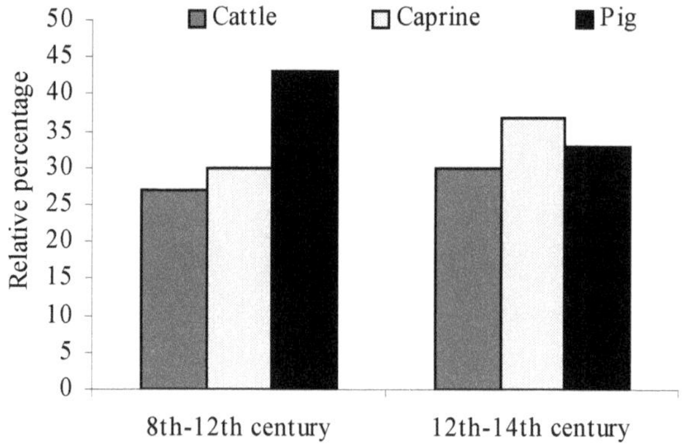

Figure 6: Relative frequency of the main domesticates in French assemblages

Certainly, pigs are the dominant taxon in the eleventh/twelfth-century assemblage from Vatteville-la-Rue (Table 6), where their remains account for 56% of the main domesticate total. The data for Vatteville-la-Rue suggest, however, that their relative frequency declined through time; that by the fourteenth century both cattle and caprines were more abundant than pigs. A temporal shift away from pig husbandry is also indicated by the synthesised data provided in Figure 6 which suggest that pig abundance in twelfth- to fourteenth-century assemblages was 10% lower than in the preceding period; a fall compensated by an increase in cattle and especially caprines. These changes in taxa ratio can be seen as a direct reflection of France's shifting farming economy, in particular the temporal increase in arable production (Comet 1997, 11). According to Bloch (1966, 5-8) the eleventh to thirteenth centuries saw large swathes of woodland put to the plough, a move with multiple connotations for animal husbandry. In the first instance, reduced availability of pannage curbed abilities to raise large herds of free-range pigs: indeed, by the fifteenth century most individuals were sty-reared on a smaller domestic scale (Comet 1997, 21). With new fields being laid out, cattle were required in increasing numbers for use in plough teams, although by the twelfth century the horse was also being used for traction (Postan 1966, 154). For the arable farmer, the greatest need soon became the maintenance of field fertility, a role for which sheep and goats are eminently suited, being capable of grazing (and thus manuring) poor quality stubble and fallow fields. Whilst caprines became an important source of fertilizer, it must surely be the growth in the wool industry that led to the increase in sheep frequencies. This theory is supported by the work of Yvinec (1998, 125) who noted that the average age-at-death for sheep increased during the course of the medieval period, suggesting that farmers were attempting to gain the maximum number of wool clips before slaughtering their animals. This importance of

	Date Group				
	11th/12th	13th	14th	12th-16th	Total
MAMMALS					
Domestic Pig *Sus* f.	218	28	16	824	**1086**
Domestic Cattle *Bos* f.	101	52	41	476	**670**
Sheep/Goat Caprine	71	19	17	270	**377**
Domestic Goat *Capra* f.				1	**1**
Domestic Horse *Equus* f.	2		2	9	**13**
Domestic Dog *Canis* f.	6	1		10	**17**
Domestic Cat *Felis* f.	1	2		33	**36**
Red Deer *Cervus elaphus*	57	15	40	311	**423**
Roe Deer *Capreolus capreolus*	21		2	40	**63**
Fallow Deer *Dama dama*				17	**17**
Brown Hare *Lepus europaeus*	2			30	**32**
Rabbit *Oryctolagus cuniculus*	1			7	**8**
Badger *Meles meles*				3	**3**
Sheep-Size	349	65	33	1137	**1584**
Cattle-Size	127	49	47	581	**804**
BIRDS					
Domestic fowl *Gallus* f.	164	18	23	589	**794**
Goose *Anser* spp	7		4	36	**47**
Duck *Anas/Aythya* spp		1		3	**4**
Teal *Anas crecca*	1			3	**4**
Swan *Cygnus* spp	3			4	**7**
Heron *Ardea* spp	1			1	**2**
Woodcock *Scolopax rusticola*	2	1		18	**21**
Pheasant *Phasianus colchicus*			4	9	**13**
Grey Partridge *Perdix perdix*	1			6	**7**
Pigeon Columbidae	1			5	**6**
Grouse Tetraonidae				3	**3**
Peacock *Pavo cristatus*				1	**1**
Lapwing *Vanellus vanellus*				2	**2**
Golden Plover *Pluviualis apricaria*				2	**2**
Curlew *Numenius arquata*				1	**1**
Godwit *Limosa limosa*				4	**4**
Peregrine falcon *Falco peregrinus*				1	**1**
Tawny Owl *Strix aluco*				12	**12**
'Falcon' *Falco* sp				2	**2**
'Gull' *Larus* sp				1	**1**
'Corvid'	1	1		8	**10**
'Small bird'				9	**9**
Unidentified	1155	284	185	5773	**7397**
Total	**2292**	**536**	**414**	**10242**	**13484**

Table 6: Composition of the assemblage from Vatteville-la-Rue (NISP)

	Total NISP	% Gnawed	% Burnt	% Butchered
Pig	1076	14	0.09	10
Cattle	670	15	0.1	28
Caprines	377	10	0.5	8
Birds	1310	0.07	1	1
Deer	488	8	0	22

Table 7: Taphonomic traits of the Vatteville-la-Rue assemblage

secondary products is also indicated by Figure 7, which shows cattle, caprine and pig cull patterns calculated from data from ninth to twelfth-century sites of all type. It indicates that almost 50% of sheep/goats were kept beyond two years of age with a significant number (30%) maintained beyond four years. That milk, as well as wool, was targeted is suggested by the percentage of animals slaughtered within six months of life, their death freeing their mothers' milk for human consumption. These juveniles would, of course, also have been a source of good meat, and the number of animals culled before two years demonstrates that the production of lamb and prime mutton was an economic concern. Thus, in medieval France, caprines were multi-purpose animals, being managed equally for their meat, milk and wool, with manure most likely being a very important by-product.

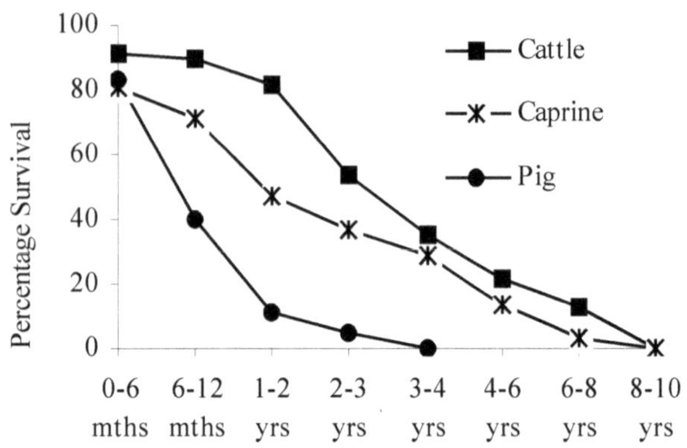

Figure 7: Cull patterns for the main domesticates in ninth to twelfth-century assemblages from northern France

In many respects, the ageing data for cattle (Figure 7) are similar to those for caprines, indicating that secondary products were as important as meat, if not more so. Use of cattle as dairy animals is hinted by the presence of juveniles but, in general, cattle were kept to adulthood with the majority of animals being slaughtered after three years of age. Whilst many of these older individuals would have been used both for breeding and dairying, it seems likely that their main role was as plough animals.

Pig cull-patterns are diametrically opposed to those for cattle and caprines, with 90% of animals culled before the end of their second year. This slaughter strategy is unsurprising given that pigs provide no secondary products to justify their maintenance beyond the point of their optimum meat-weight. Pigs were, without doubt, viewed and managed as providers of meat.

2.2: Production and supply: beef, mutton, pork

Whilst general patterns in taxa frequency and ageing data provide an insight into broad shifts in French medieval animal husbandry, they mask those inter-site variations that have the potential to inform on systems of production and redistribution. It is only when the evidence is considered by site type that these mechanisms can be explored and understood.

In a simple provisioning model, 'producer' sites are distinguished by the presence of young and old animals, representing infant mortalities and breeding stock, respectively. By contrast 'consumer' assemblages tend to contain a higher frequency of prime meat animals (Wapnish and Hesse 1988, 84). The cull patterns shown in Table 8, especially those for caprines, do conform to Wapnish and Hesse's model. Figure 8 shows that the majority (65%) of caprines on urban sites were between one and three years of age, with few young and old animals being present. Such a slaughter strategy would be insupportable on a self-sufficient site and the absence of neonatal animals indicates that individuals were not being raised in towns. Instead, it would seem that livestock were reared on rural settlements specifically for supplying consumer populations with prime meat animals; an idea supported by the lack of prime-aged animals in village assemblages. The selectivity of the cull-patterns for urban sites is suggestive of consumer demand, with the townsfolk requesting animals of a certain age. If prime meat was valued by the urban community, their desires appear to have had little sway over the quality of the beef they received: Table 8 shows that towns were regularly sent the oldest available cattle, whose meat would have been tough by modern standards. It may be for this reason that cattle figure small in urban assemblages, reflecting the town inhabitant's preference for prime pork and lamb.

Taxa representations for urban, elite and ecclesiastical assemblages (Figure 9) show little similarity to those from rural settlements, where cattle are the dominant domesticate, with pigs being present only in low numbers. This lack of pigs on rural sites is surprising when it is remembered that the availability of pannage would have enabled pork to be produced at little cost to the peasants. However, the composition of village assemblages suggests that, rather than consuming animals raised specifically for meat, the rural population obtained

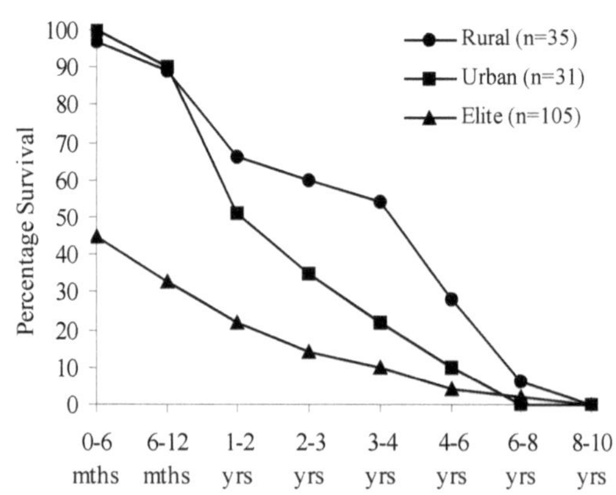

Figure 8: Inter-site variation in cull patterns for caprines in ninth to twelfth century assemblages from northern France

flesh-foods from animals that had spent several years providing secondary products. The diet of the village population appears, therefore, to have been restricted by the demands of the consumer populations and the wider agricultural economy. Whilst there is an obvious divide between producer and consumer assemblages, an element of status-based variation is also apparent. For example, although pigs are the most abundant taxon on all but rural sites, they are substantially better represented in assemblages from elite and ecclesiastical settlements. Furthermore, both the cattle and caprine cull-patterns (Figure 8 and Table 8) demonstrate that high-status sites were receiving larger numbers of animals aged under 6 months: in the case of sheep just 22% of individuals were over two years of age at the point of slaughter. Such a preponderance of juveniles could be a reflection of the elite's ability to demand the tender meat from animals slaughtered at an uneconomic point in their development. Alternatively, factors other than commercial trade may have been responsible for this provisioning pattern.

Historical evidence suggests that the elite received a significant percentage of their supplies through food rents systems, whereby the rural population paid animal tributes to seigneurial and religious authorities in return for land use or pannage rights (Le Blay 1993). In a situation where 'food' was the main concern, an emphasis on meat-animals is perhaps unsurprising. Pigs would have been the most suitable food rent animal because, as has been mentioned above, they could be raised cheaply and provide no secondary products so the peasants would have lost no income through their slaughter. Production costs may also explain why the elite were provided with large numbers of infant animals. Although at face value it appears that these animals were slaughtered at an 'economically irrational' point, O'Connor (1992, 104) has argued that surplus animals may have been culled because the potential cost of

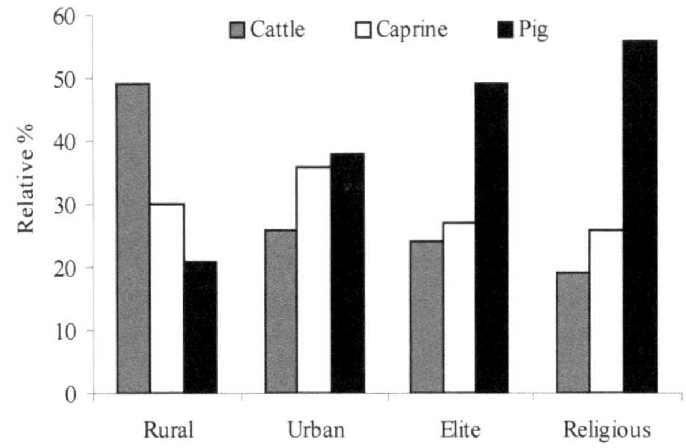

Figure 9: Inter-site variation in the relative frequency of the main domesticates, according to NISP

raising them to adulthood was too high. As such, individuals slaughtered as by-products of the dairy process would, like pigs, represent ideal food-rent animals. There exists, however, the possibility that the tender meat from these young animals would have fetched a good price if sold at market. That juveniles arrived at high-status, rather than urban, sites may indicate that the French aristocracy had some control, either legal or monetary, over their supplies.

Skeletal representation data support the idea that the elite were able to requisition the best quality foodstuffs. Few French reports provide anatomical representation data but Figures 10a-c show the body part patterns for Vatteville's eleventh/twelfth-century domesticates. Cattle are well represented by meat bearing bones, in particular those from the hind limb, with head and foot elements being scarce. A similar lack of extremities is demonstrated by the caprine assemblage, which is composed almost exclusively of scapulae, humeri, radii and tibiae. Since the Vatteville-la-Rue assemblage is derived predominantly

a) Cattle	**0-6 mths**	**6-12 mths**	**1-2 yrs**	**2-3 yrs**	**3-6 yrs**	**6-8 yrs**	**8-10 yrs**
Rural (n=9)			11	45	11	33	
Urban (n=30)		3	3	31	10	23	30
Elite (11)	27		9	9	37	9	9

b) Caprine	**0-6 mths**	**6-12 mths**	**1-2 yrs**	**2-3 yrs**	**3-4 yrs**	**4-6 yrs**	**6-8 yrs**	**8-10 yrs**
Rural (n=35)	3	8	23	6	6	26	23	6
Urban (n=31)		10	39	16	13	13	10	
Elite (N=105)	55	12	11	8	4	6	2	2

c) Pig	**0-6 mths**	**6-12 mths**	**1-2 yrs**	**2-3 yrs**	**3 yrs +**
Rural (n=8)		38	50		12
Urban (n=213)	35.5	28	33	3	0.5
Elite (n=227)	12	35	28	16	9
Religious (n=26)	19	77	4		

Table 8: Ageing data, shown as percentage of individuals killed at each stage, for a) cattle, b) caprines and c) pigs from mid-ninth to mid-twelfth-centuryFrench assemblages

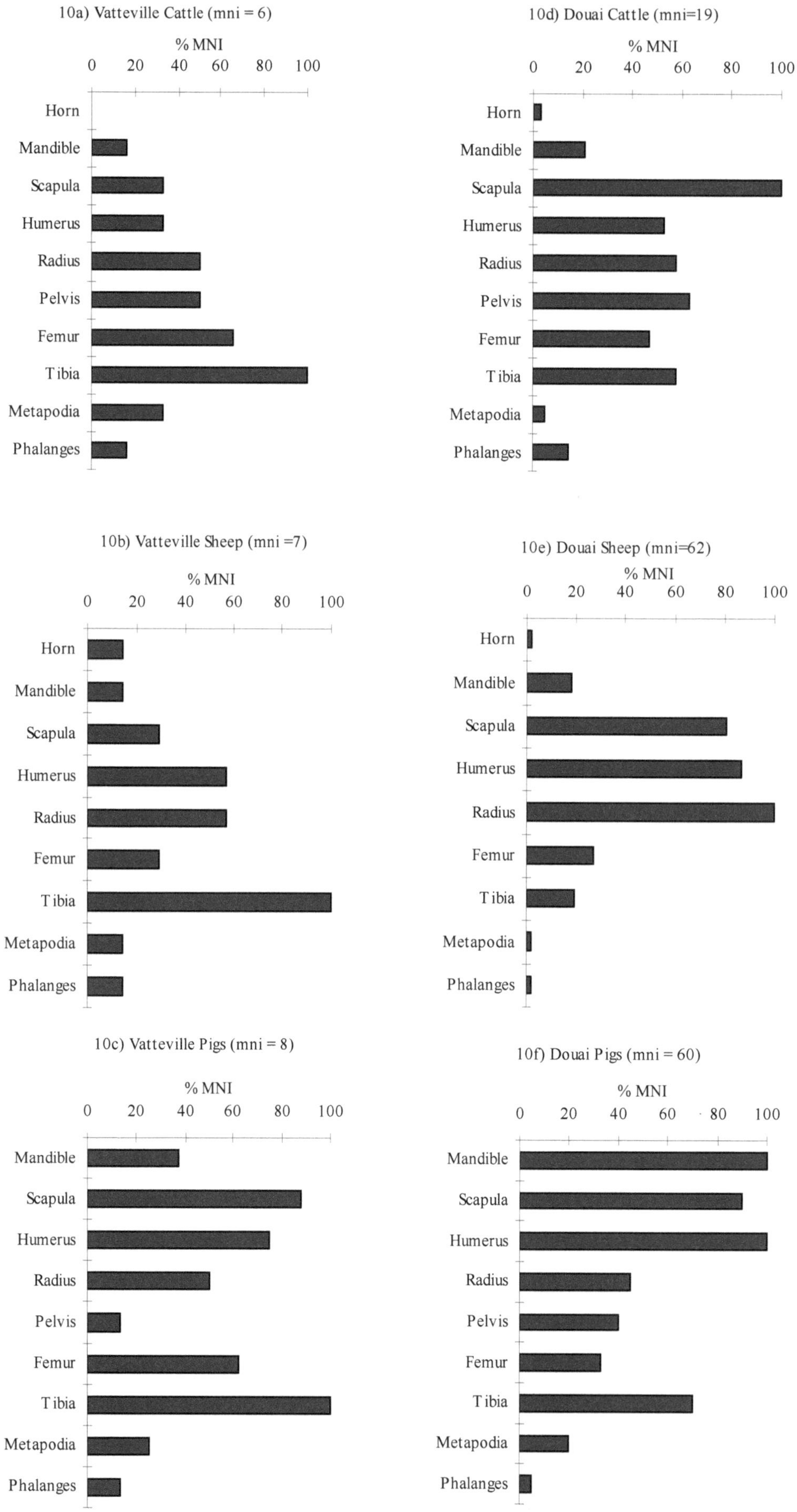

Figure 10: Anatomical representation patterns for the main domesticates from a-c) Vatteville and d-f) Douai

from food waste, the absence of head and foot bones is accounted for by the fact these elements are usually removed during primary butchery. The dearth of caprine pelves and femora is more surprising as it might be expected that the meat-bearing elements of the hind limb would be well represented. It seems possible that the observed patterns reflect inter-element variation in survival rates, perhaps due to carnivore activity: a high proportion (10%) of the caprine remains had been gnawed. It has been demonstrated repeatedly (e.g. Brain 1967; Lyman 1994) that dogs will consume bones, such as the femur, in preference to those of higher density: tibia and humerus. Whilst non-cultural processes may explain the anatomical frequencies at Vatteville, the similarity of their patterns to those from other French high-status sites suggests human involvement. At the castle site of Douai (Vadet and Villette 1986), for example, the assemblage was also characterised by a high frequency of sheep and cattle shoulder elements (Figure 10d-f). Again, neither animal was represented by head or foot bones. Further body-part data, from all site types, would be required before definite conclusions can be drawn but the current evidence suggests that cattle and sheep were sent to high-status sites in a pre-butchered form, with specific meat joints (the shoulders or haunches) being preferentially selected. Such an idea does not conflict with the theory of a food-rent system as meat portions, rather than living animals or complete carcasses, have historically been component parts of food rents (for instance Lasota-Mokalewska 1984).

The body-part patterns for Vatteville-la-Rue's pigs are different to those for cattle and caprines, with most parts of the skeleton being represented. Foot bones remain scarce but their absence is almost certainly due to issues of recovery. It seems probable that pigs were sent to high-status sites either as complete carcasses or on-the-hoof; although finds of neonatal remains at Vatteville, Douai (Vadet and Villette 1984) and in the high-status contexts from Place des Hallettes (Yvinec 1997, 199) indicates that some pigs were also raised on-site.

Butchery data, recorded during analysis of the Vatteville assemblage, reinforces the theory that pigs were obtained through a different mechanism to cattle and sheep. Sample sizes are insufficient to allow the reconstruction of species-specific butchery traditions but it is clear that patterns for pigs are unlike those for the other taxa. Most obvious is the variation in disarticulation method (Tables A and B, Appendix V). Whereas cattle and sheep bones display a high frequency of chop marks, most probably delivered by a meat cleaver, pigs were butchered more frequently with the use of a knife. This variation in tool use cannot be related to animal size, since both the largest (cattle) and the smallest (caprines) taxa were butchered primarily with a meat cleaver. Instead, the disparity in techniques may represent the difference between professional and non-professional butchery: pigs being butchered inside the castle itself whilst specialists were responsible for sheep and cattle disarticulation. That cattle carcasses were butchered by professionals is indicated by the number (28%) of vertebrae that had been saggitally split: the practice of dividing carcasses into equal sides is thought to have emerged under the influence of specialist butchers (Grant 1987, 56).

From the butchery, anatomical representation and ageing evidence it would seem that French high-status sites received meat supplies in two ways. Prime beef, veal, mutton and lamb were obtained as professionally-butchered joints; whether this was a monetary arrangement or one grounded in a food-rent system is uncertain. Pigs on the other hand, were perhaps supplied by the rural population, although some may also have been raised on-site. Arriving on-the-hoof, they may have been butchered 'in house' by non-specialists who used knives and cleaves haphazardly to break up the carcass.

Taken as a whole the evidence suggests that the French provisioning system was more complicated than simple market-based exchange. Nevertheless, the fact that high-status assemblages contained large numbers of meat-bearing bones from the most tender animals, indicates that the elite were in control of their supplies.

2.3: Domestic Birds - Chicken, Goose and Duck

Scully (2002, 56) has suggested that domestic birds made the greatest contribution to the meat diet of the medieval French aristocracy, and certainly their remains are well represented on elite sites (Figure 11). Religious houses in particular are characterised by an abundance of domestic birds, their high frequency surely reflecting monastic observance of the Benedictine Rule, which required abstinence from the meat of quadrupeds but permitted the consumption of birds.

Domestic fowl were the dominant taxon within Vatteville-la-Rue's avian assemblage, comprising 94% of the bird remains, and Figure 12 suggests that this pattern is typical of all elite French sites. The reason why domestic

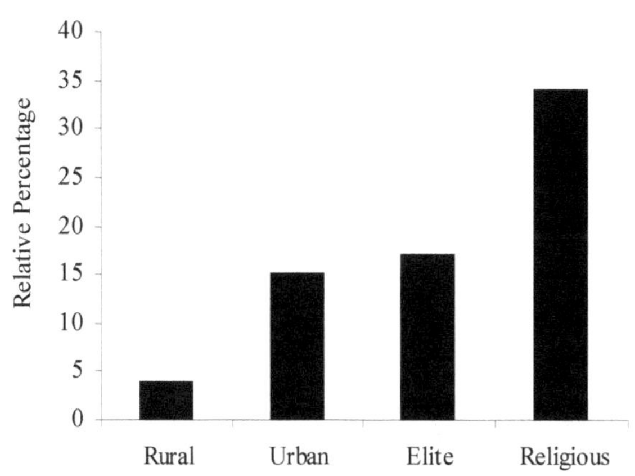

Figure 11: Inter-site variation in the representation of domestic birds (fowl, goose and duck) in assemblages from northern France

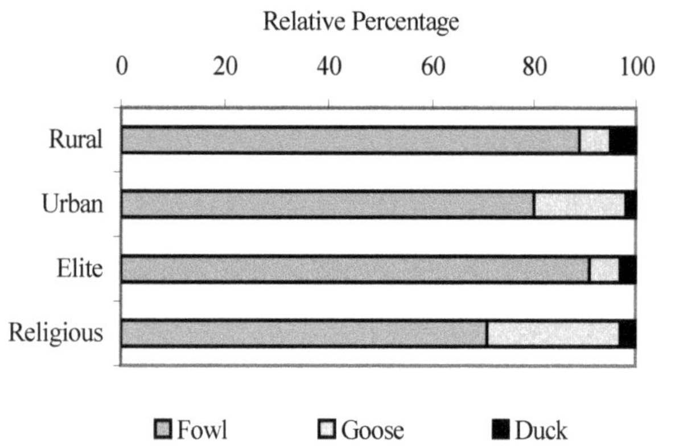

Figure 12: Inter-site variation in the relative frequencies (NISP) of domestic fowl, goose and duck in ninth to twelfth-century assemblages from northern France

fowl were reared in great number must be related to their unique value as egg producers. This idea is supported by the evidence from Vatteville-la-Rue: just twelve of the sixty-three tarsometatarsi recovered exhibited spurs, indicating that the majority (81%) of individuals were probably female. Although Sadler (1991) has warned against using tarsometatarsi as indicators of sex, the conclusion is corroborated by the large number of femora found to contain medullary bone, indicating that many of the birds (hens) were slaughtered whilst 'in lay'.

Figure 12 shows that villages exhibit domestic fowl:goose:duck ratios similar to those from high-status sites but the inter-site disparity in their overall frequency (Figure 11), suggests that domestic fowl formed but a small part of the peasant's diet whereas they were consumed regularly by the elite.

Town and monastic assemblages show much higher goose frequencies than the other sites. Geese may have been favoured within urban environments because of their suitability to being kept, and indeed bred, within confined spaces (O'Connor 1989b). Since religious houses would presumably not have been subject to space restrictions, the presence of goose may reflect other factors; perhaps goose flesh was considered a suitable food for men of the cloth. Domestic ducks are infrequent on all site-types, although marginally better represented in rural contexts, perhaps indicating that their exploitation and consumption was little valued.

Butchery marks were present on just 1% of Vatteville-la-Rue's bird bones (Table 7), suggesting that carcasses were cooked whole, after which the meat would have come away from the bone without requiring a knife. Roasting would have been a suitable method of cooking, and the fact that a higher percentage of birds than mammals (although still only 1%) exhibited burning indicates that this practice may occasionally have been employed.

2.4: Wild Mammals

Wild mammals are well represented in the Vatteville-la-Rue assemblage, conforming to wider patterns of wild animal distribution: Figure 13 shows that their remains tend to be far more numerous in elite assemblages than those from other site types. Whilst the dearth of game on lower-status settlements may reflect the lack of time the farming and urban population could devote to hunting, there is historical evidence to suggest that other restrictions may also have been in place. According to William of Jumièges (*Gesta Normannorum Ducum* cited in Hilton 1992, 161), Norman peasants revolted in 996 because their woodland and fishing rights were being appropriated by the elite. This revival of the Carolingian Forest system is interesting because Wickham (1994, 161) has argued that Europe's history of hunting closely mirrors that of its socio-political change. That the re-imposition of the Forest system occurred at approximately the same point that Normandy's social hierarchy was re-organised – with Frankish families, such as the Tosnys and the Bellêmes being incorporated into the aristocracy (Bates, 1982, 34) – suggests that these elite groups were deploying hunting rights as a mechanism through which they could assert their new power and authority.

The social impact of hunting would, of course, have depended upon the quarry – the symbolism and meaning of wild boar hunting being entirely different to that surrounding the capture of hares. Lepetz *et al.* (1995) have made the case that wild boar (*Sus scrofa*) were targeted primarily by the seignuerial class, who hunted these most dangerous of beasts as a demonstration of martial valour. It can be difficult to separate the remains of wild and domestic pigs but metrical analysis of the suid teeth from Vatteville-la-Rue indicated the presence of at least six individuals of a size comparable with wild boar (Figure 14). Other high-status sites, such as that excavated at 'la place des Hallettes' (Yvinec 1997), have yielded wild boar in greater abundance but, in general, elite assemblages contain cervid (red and roe deer) and hare remains in equal or higher frequencies (Figure 15).

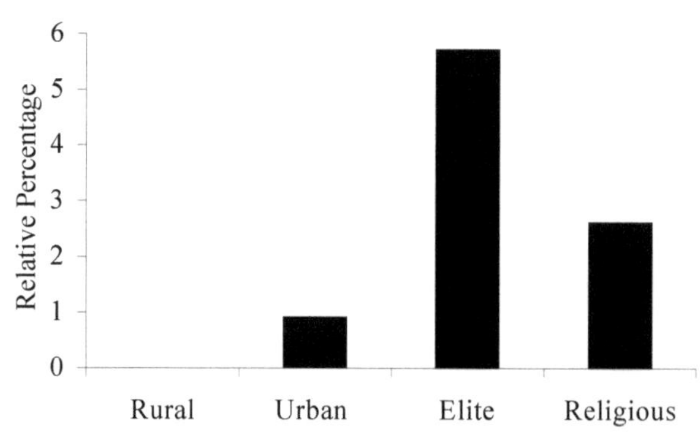

Figure 13: Inter-site variation in the representation of the main game mammals in ninth to twelfth-century assemblages from northern France.

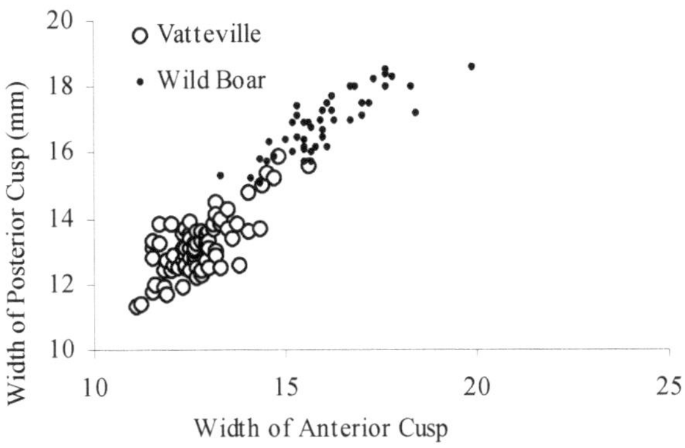

Figure 14: Mandibular second molars measurements of pigs from Vatteville-la-Rue shown against measurements for north European Mesolithic wild boar (taken from Albarella et al. *forthcoming)*

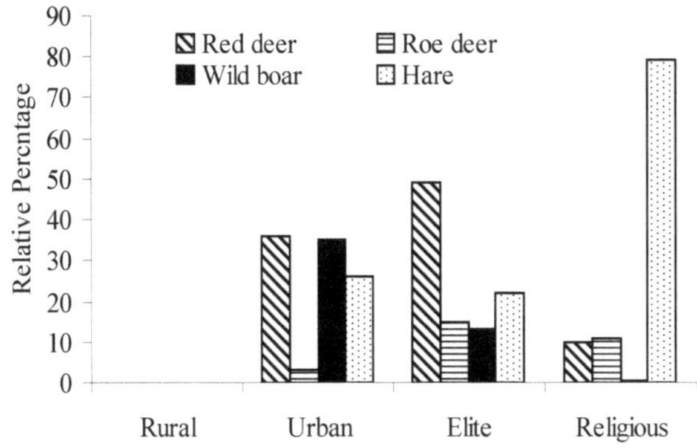

Figure 15: Inter-site variation in the representation of the main game mammals in ninth to twelfth-century assemblages from northern France

Wild boar appear to make a greater contribution in urban wild assemblages, where it is often as well represented as red deer. In terms of presence/absence, however, hare is the most consistently represented game species in urban contexts, being present, albeit in low numbers, in 75% of assemblages (Appendix Ia). Hares are particularly abundant in assemblages from religious houses: Larioux (1988) has linked this to the fact that their consumption was authorized for ecclesiastics, on medical grounds, in a number of medieval penitentials. By contrast, monastic assemblages show a dearth of wild boar, which Yvinec (1993, 498) suggests is related to the violent methods by which they were procured; aggression not being actively promoted within the religious community.

Examination of Vatteville-la-Rue's deer butchery and body part data (synthesised for all periods) revealed some interesting patterns. In terms of butchery, it was clear that deer carcasses were not processed in a similar way to cattle and caprines. Rather than a meat-cleaver, they were butchered using a knife: cuts accounted for 63% of the disarticulation marks. From this it might be suggested that, as with pigs, butchery was not undertaken by professionals but rather by either the hunters themselves or kitchen workers: the 'dalle' ('A' on Figure 5b), has been interpreted as a butchery area. At a more detailed level, the butchery patterns demonstrate some marks that were considerably better represented on deer bones than those of other taxa. One mark in particular (detailed in Figure 16) occurred consistently on deer tibia. It is, according to Lauwerier (1988), a trace indicative of de-boning, and was found on 50% of butchered deer tibiae compared to only 17% of cattle and 13% of those belonging to caprines.

Figure 16: Butchery mark(shaving of tibia tuberosity) found commonly on deer tibiae from Vatteville-la-Rue

Whilst the cervid remains show butchery patterns different to those of the main domesticates, their skeletal representation data are similar to those for cattle, showing an over-representation of hindlimb elements compared to those of the forelimb (Figure 17). Since the metacarpals and metatarsals (elements with the same bone density) are not represented in equal numbers, the possibility that this pattern is caused by variation in bone preservation can be ruled out. Cultural taphonomy would seem to be responsible and it is feasible that shoulders of venison were exported whilst the haunches were retained on site. Interestingly, a similar pattern is evident for the red deer assemblage from Douai (Vadet and Villette 1986, 102), although the sample size is too small to conclude that this was a widespread phenomenon.

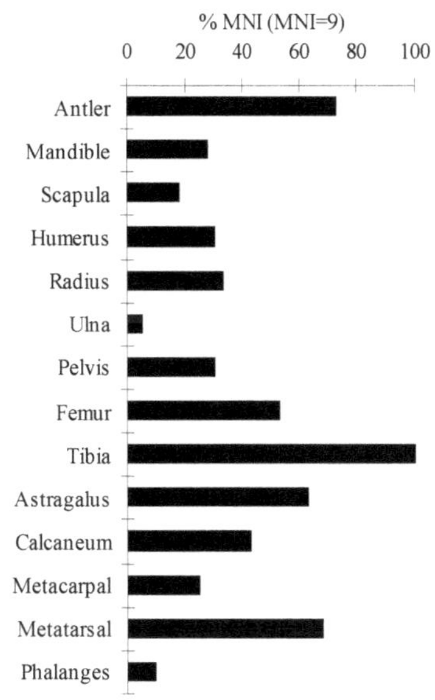

Figure 17: Anatomical representation pattern for deer (red, roe and fallow) from Vatteville-la-Rue

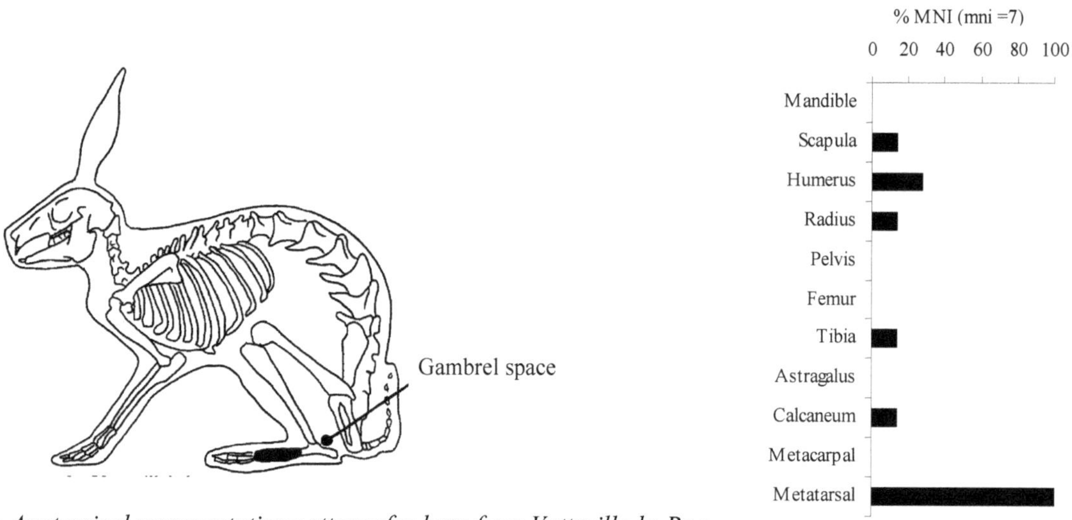

Figure 18: Anatomical representation patterns for hare from Vatteville-la-Rue

Unusual body part patterns were also noted for the Vatteville hares, their skeletal representation is shown in Figure 18. It is immediately apparent that the hind foot bones greatly outnumber all other skeletal elements. Spatial analysis of the hare bone distribution revealed that 15 of the 20 hare foot bones (MNI 5, all from the right foot) came from the kitchen area ('B' on Figure 5b). No other hare specimens were recovered from this location. This situation has resonance with the eleventh/twelfth- century kitchen deposit from the Cluniac monastery of Charité sur Loire (Audoin-Rouzeau 1986) where a large number of hare hind feet, in this case all from the left side, were recovered. Yvinec (*pers. comm.*) has suggested that these body part patterns may reflect the custom of hanging game, in order to improve its flavour. Traditionally, hare are suspended by the ankle, often by insertion of a hook into the gambrel space (see Figure 18b). It seems possible that such side-specific accumulations of foot bones could occur if the same person was responsible for hanging and cutting-down all individuals, the severed paw falling to the ground and remaining *in situ*.

2.5: Wild Birds

Overall, wildfowl comprised 2.3% of Vatteville-la-Rue's identifiable assemblage, a proportion higher than the average for elite sites (Figure 19). Status-based patterning in the representation of wild birds is less obvious than for wild mammals, with religious houses showing the highest frequency, although this may be an artifice of sample size: the large wildfowl assemblage from Charité sur Loire seems to have skewed the result (see Appendix Ia).

That wild birds were incorporated into expressions of status is better indicated by the taxa representations for each site type (Figures 20-22). Unfortunately information pertaining to rural sites is scarce. Excavation of the eleventh/twelfth-century farm settlement at Argentan (Orne) produced some partridge remains (Carpentier 2002, 88) but Lepetz *et al.* (1995, 179) have shown that rural assemblages seldom contain taxa other than corvids and field birds, indicating that opportunistic trapping, rather than deliberate bird-catching forays, was the main method of acquisition.

Urban assemblages exhibit a wider range of species, partridge and pheasant being the best represented (Figure 20). This abundance of pheasant is surprising because neither Audoin-Rouzeau's (1993) nor Clavel's (2001) syntheses of zooarchaeological data showed pheasants to be well represented in Medieval France. Re-examination of the data in Appendix Ie, demonstrates that all the purported pheasant specimens came from the site of Saint-Denis, where they were said to be present in great number (Morel 1985). This species is notoriously difficult to separate from other galliforms and it seems possible that the 'pheasants' could actually be misidentified domestic fowl. If, however, their classification is correct, the fact that pheasant was absent on all other sites (with the exception of a single specimen from Charité sur Loire) may suggest that the inhabitants of Saint Denis had access to rare foodstuffs – perhaps local demand for these birds encouraged their importation from other regions of Europe. In general, however, most of the wild birds brought to urban centres would have come from the surrounding countryside, as is attested by the presence of pigeon (*Columba* spp.), and woodcock (*Scolopax rusticola*). Wetland birds – duck

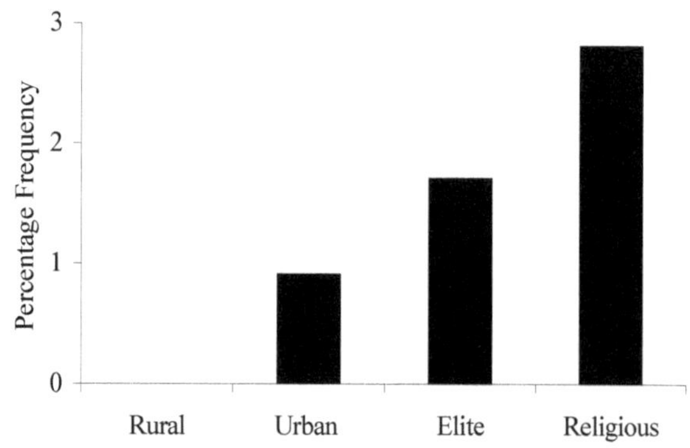

Figure 19: Inter-period variation in the representation of wildfowl in assemblages from northern France

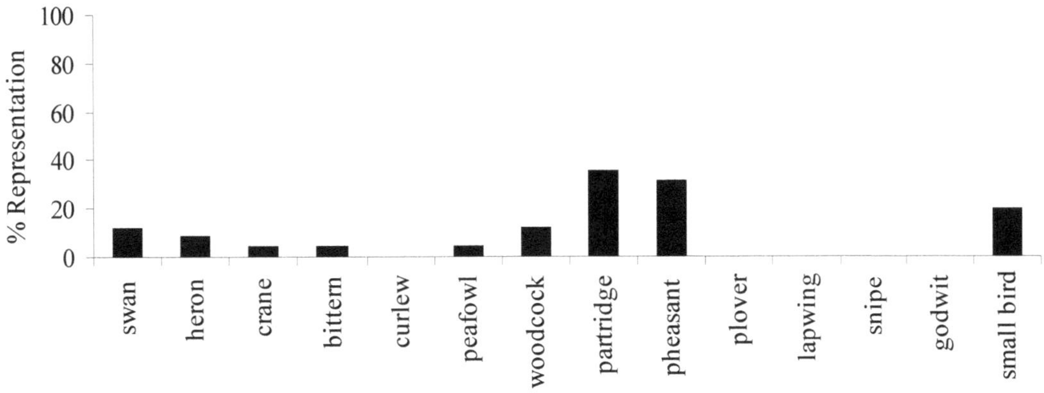

Figure 20: Wildfowl representation in urban assemblages from northern France

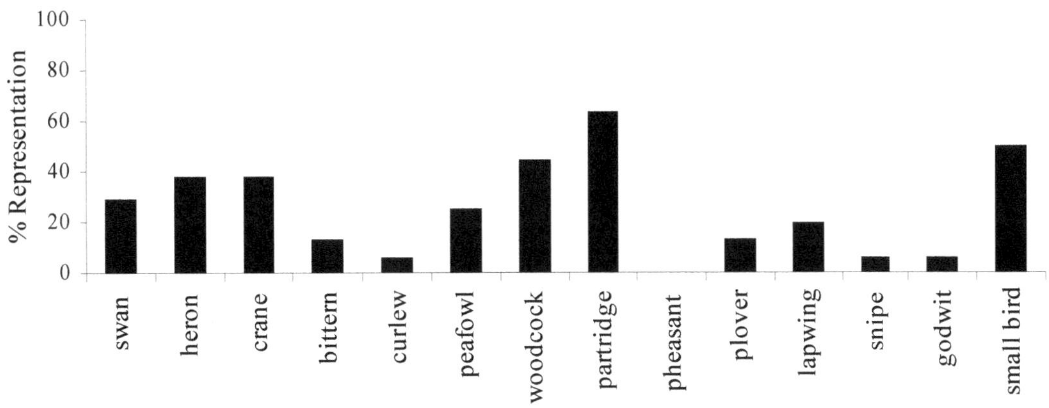

Figure 21: Wildfowl representation in assemblages from elite sites in northern France

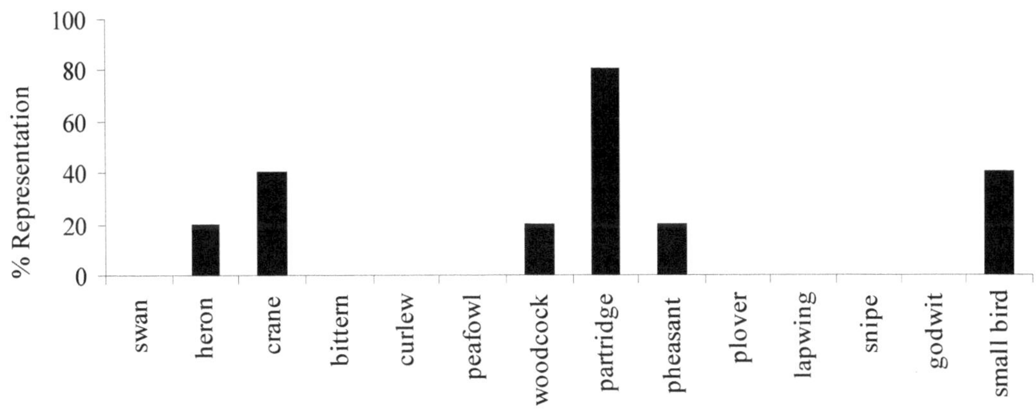

Figure 22: Wildfowl representation in assemblages from religious houses in northern France

(*Anas* spp.), swan (*Cygnus* spp), heron (*Ardea* spp.) and crane (*Grus* spp.) – may have been caught on urban waterways.

Swan, heron and crane, together with bittern, also feature in high-status assemblages (Figure 21). Another species notable by its presence in elite assemblages is peafowl (*Pavo cristatus*). To date, their remains have, with the exception of a single specimen from an urban context in Compiègne (Yvinec 1997, 178), been recovered only from seigniorial sites: in addition to those mentioned in Appendix Ie peafowl remains were also found at Andone in south-west France (Giles 1991). The high-status associations attached to peafowl are supported by historical evidence: on the Bayeux Tapestry two peacocks are depicted only on the roof of William I's palace (Thorpe 1973) and even in the time of Charlemagne these birds are said to have been kept on royal estates (Harvey 1982).

Most frequently represented on elite sites are partridge and woodcock, although a wide range of game birds – grouse (*Lagopus lagopus*), plover (*Pluvialis* sp.), lapwing (*Vanellus vanellus*), snipe (*Gallingo gallingo*), godwit (*Limosa* sp.) and curlew (*Numenius arquarta*) – are also present. There are many techniques by which these species could have been caught but there is some evidence to suggest that the elite regularly practiced falconry. William the Conqueror is depicted carrying a hawk on the Bayeux Tapestry and William of Poitiers mentions that he regularly 'diverted himself with the flight of falcons and more often with that of hawks' (*Gesta Guillelmi ducis Normannorum et Regis Anglorum*, 16-19; cited in Brown 1984, 21). Furthermore, the remains of raptorial birds have been found at several elite sites: three falcon bones, one positively identified as peregrine falcon (*Falco pereginus*), were recorded at Vatteville-la-Rue (Table 6) and two sparrow hawk (*Accipiter nisus*) bones were recovered from a thirteenth-century deposit at Cour Carrée (Clavel 2001, 12). For the later period Clavel (2001, 117) refers to the early fourteenth-century site of 'la rue Fromenteau' where the remains of numerous hunting birds were recovered from a deposit associated with the royal mews.

Hunting birds are less common on ecclesiastical sites, although the kestrel remains from 'l'Abbaye de Saint-avit Senieur' could feasibly represent an individual used in hawking (Gautier 1972, 372). Whilst Figure 19 suggests that wildfowl are abundant in assemblages from religious houses, Figure 22 demonstrates that the range of species represented is actually quite restricted, including little other than partridge, a pattern that contrasts with that seen for religious houses in England (Chapter 5, page 64)

2.6: Fish

Strategies of fish procurement, distribution and consumption in medieval France are considered in detail by Clavel (2001, 131-86) and here it is sufficient to present just an overview of the situation.

Temporal developments in the fishing industry can be gauged from the taxa frequency data (Appendix Idiii); the post-thirteenth-century increase in the representation of species, in particular marine fish – mackerel (*Scomber scombrus*), whiting (*Merlangius merlangius*) and haddock (*Melanogrammus aeglefinus*) – reflecting evolution in the organisation and technology of off-shore fishing. Variations between site-types are also apparent. Figure 23 provides a tentative indication of inter-site variation in fish presence/diversity: the problems associated with this kind of analysis have been outlined in Chapter 1, section 1.5i. Fish are best represented on urban sites, not only do they have the highest presence/diversity index but they have also been recovered from 100% of the assemblages considered in this study (Appendix Idiii). Their presence must, to some extent, reflect the nature of urban deposition and the ease with which town pits can be sampled (see Chapter 4.1). But fish were also more readily available within the urban environment; in part because towns were generally located on large rivers or estuaries, thus giving townsfolk direct access to aquatic resources, but also because catches of fish, both freshwater and marine, would have been brought into towns for processing and redistribution. Historical evidence suggests that herring-salting was taking place in Dieppe by 1030 (Cutting 1955) and men from Rouen are said to have been exporting fish to London from at least the tenth century (Robertson, 1939). At Compiègne, significant numbers of marine fish were recovered, despite the site being some distance inland.

If the high frequency of fish in urban assemblages is linked to proximity to supply, it might be expected that abundance of fish would drop-off with distance from the source. This does appear to be the case as, by contrast to towns, rural assemblages show fewer fish remains, especially those dating from the ninth to twelfth centuries. By the thirteenth century rural-urban exchange

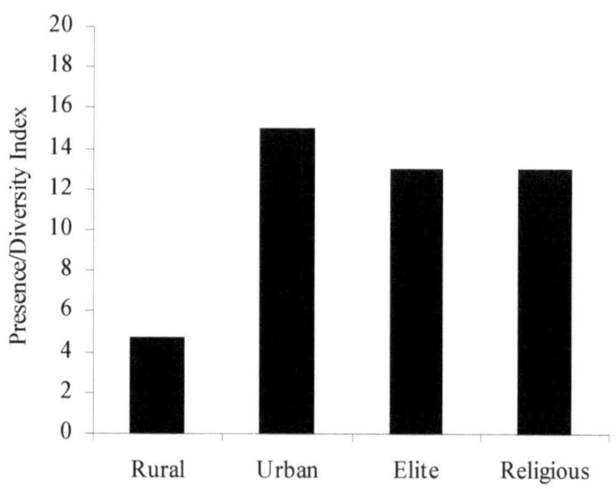

Figure 23: Inter-site variation in the presence/diversity of fish in assemblages from ninth to twelfth-century France

was taking place, as is indicated by the presence of marine species on village sites such as Dury and Saleux (Yvinec 1999, 250) and Coqulles (Clavel 2001, 55). Interestingly, there is an absence of freshwater taxa on rural sites, suggesting either that the peasant population was not exploiting local rivers and streams, or that they were passing catches, for profit or from obligation, to local consumer groups.

The eleventh-century elite's seizure of fishing rights betrays the status-associations that must, at this time, have been linked to freshwater species. This is also clear from taxa representation data for elite assemblages, in which cyprinids and, in particular, sturgeon (*Acipenser sturio*) are far better represented than in those from any other site-type (Figure 26). By the thirteenth century sturgeon become less numerous, their place being taken by carp (*Cyprinus carpio*), an introduced species whose presence attests to the growing popularity of artificial fishponds (Appendix Idiii). The spread of both the fishpond tradition and carp has been linked to monastics, keen to establish supplies of fish to cater for fast-day meals. This possibility is supported by the zooarcheological data, for although sample sizes are small, assemblages from religious houses are skewed towards the representation of freshwater taxa (Figure 27 and Appendix Idiii).

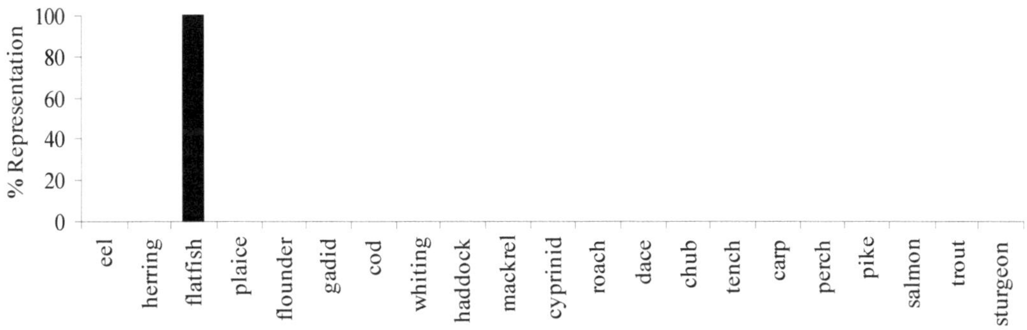

Figure 24: Representation of fish taxa in ninth to twelfth-century assemblages from rural sites in northern France

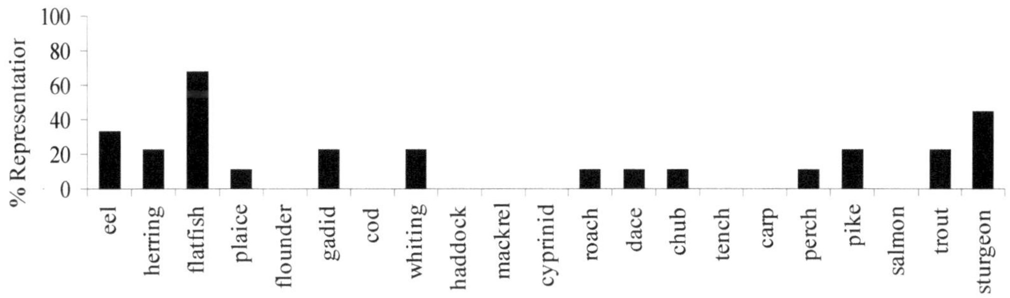

Figure 25: Representation of fish taxa in ninth to twelfth-century assemblages from urban sites in northern France

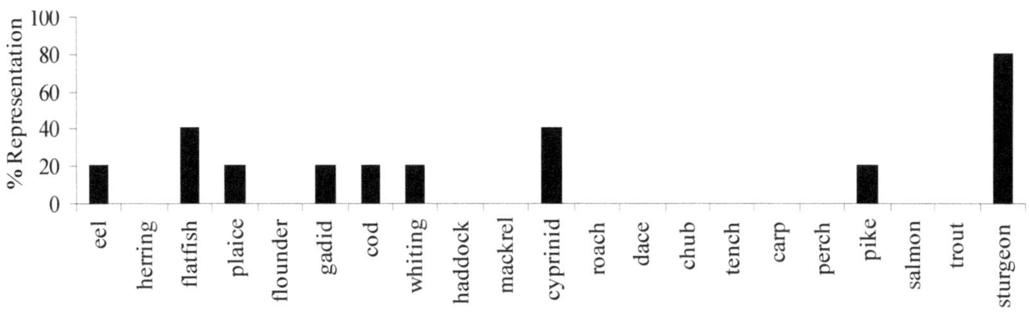

Figure 26: Representation of fish taxa in ninth to twelfth-century assemblages from elite sites in northern France

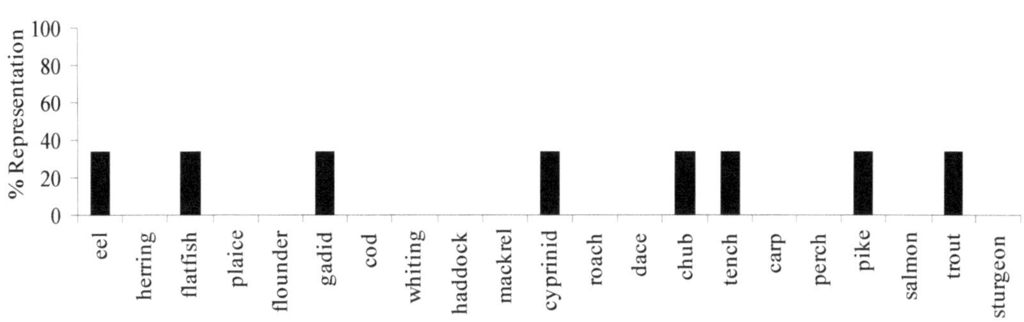

Figure 27: Representation of fish taxa in ninth to twelfth-century assemblages from religious houses in northern France

2.7: Conclusion

Through consideration of the French dataset it has been possible to gain an insight into the role of animals in French medieval society. Human-animal relationships clearly varied by social milieu and differences in assemblage composition can be viewed as a reflection of economic and status-based forces.

Rural assemblages are the most restricted, being dominated by the remains of mature cattle and caprines and containing few domestic birds or wild animals. The almost complete absence of game species suggests that the rural community devoted little energy to hunting, trapping and fishing, due to either lack of time or legal prohibition. Alternatively, it may be that wild animal resources were exploited but that peasant hunters sold their bag in nearby towns. The rural set-up does appear to have been geared towards meeting consumer demands for food: whilst peasant farmers clearly derived profit from grain and wool, they did not concentrate on these products to the detriment of 'quality' meat, both urban and elite populations being provisioned with quantities of prime-aged animals.

That by the later medieval period supplies were also moving in the opposite direction – from town to country – is indicated by the presence of marine fish within land-locked rural contexts. Fish was perhaps the one foodstuff in which townsfolk were self-sufficient. Although some domestic birds may have been maintained within the urban environment, the bulk of food supplies must have arrived in towns from elsewhere. Consumer profiles are even clearer for elite assemblages, which generally contain high frequencies of domestic animals aged under six months. Potentially an uneconomic point of slaughter, these cull-patterns hint at an aristocracy capable of demanding meat that, by modern standards, was of the best quality. An emphasis on fine meats would appear to characterise elite assemblages, dominated as they are by the remains of animals whose role, from an economic perspective, was simply to provide flesh foods. Pigs, the premier meat animal, are particularly abundant in elite assemblages. The presence of foetal and neonatal remains indicates that some animals were raised on-site. More frequently, however, pigs appear to have been supplied – perhaps as a food rent – on the hoof or as partially dressed carcasses. Similarly, the data from Vatteville-la-Rue and Douai suggest that cattle and caprines were also supplied as pre-butchered joints. However, unlike pigs which seem to have been processed on site, the butchery data indicate a level of professionalism in the processing of cattle and caprine carcasses and it is tempting to suggest that elite households obtained cuts of beef and mutton from urban butchers.

Elite residences were not entirely reliant on supplies from external sources, the range and frequency of wild taxa suggesting that hunting and hawking were popular activities of the French aristocracy. Wild boar is often the best represented wild animal on seigneurial sites but at Vatteville-la-Rue red deer were the favoured quarry and venison must have made a significant contribution to the diet of the castle's occupants. At this site the absence of cervid forelimbs may even indicate that venison was re-distributed, although there are insufficient data from other sites to suggest that this was a regular practice. What is clear from the composition of high-status assemblages is that the elite had a meat diet that was more varied, both in terms of its content and methods of procurement, than that consumed by other sections of society.

Having gained an over-view of the situation in medieval France, it is now possible to examine the English dataset. In particular it will be used to ascertain whether the Norman Conquest of England was accompanied by the imposition of continental traditions of animal management, resource exploitation and distribution, or the introduction of French dietary preferences.

CHAPTER 3: THE ANIMAL ECONOMY – CONTINUITY OR CHANGE?

The English medieval economy has received considerable attention from both historians and archaeologists (for example, Astill and Grant 1988; Astill and Langdon 1997; Britnell 1993; Hinton 1990; Miller and Hatcher 1995). Since the 1980s zooarchaeological evidence has played an increasing role in economic investigations: animal bone analysis has been central to the theories surrounding *Wic* provisioning (Bourdillon 1994; O'Connor 1991, 1994, 2001); early town development (O'Connor 1992, 1994) and agricultural systems (Biddick 1984a). Vince's (1994) study of urban economies emphasised the evidence from animal remains, and Astill (1991, 111) concluded that 'faunal remains may provide a key to [understanding] urban development'.

That zooarchaeology has provided a significant amount of economic information is perhaps predictable. For many sections of society, not only were animal products their main source of income, they were also accepted units of currency. Although the Anglo-Saxon economy was money-using from the late seventh century, Britnell (1993, 29) has estimated that even by the eleventh and twelfth centuries the total currency circulating in England did not exceed £120,000 (less than 1s per person). As a result, coins would have been too valuable for day-to-day transactions and Britnell (1993, 30) suggests that over half the population obtained their provisions independent of the market system, in particular through the receipt and redistribution of food rents.

Considering the fundamental importance of livestock in the medieval economy, and the potential information that can be obtained from the study of their remains, it is surprising that little attempt has been made to examine the Saxo-Norman situation from a zooarchaeological perspective. It is, perhaps, for this reason that most scholars believe the Norman Conquest to have caused little change to the Saxon economy (Hooke 1998; Reynolds 1999; Rowley 1999). Furthermore, where change has been identified, it has often been classified as reflecting long-term trends rather than resulting from the Conquest. This opinion may well be accurate; certainly Late Saxon England witnessed dramatic socio-economic shifts. For example, the creation of *burhs* in the late ninth century fundamentally changed urban development (Astill 1991; Vince 1994). At the same time, rural settlements were re-organised, with multiple estates being broken up in favour of smaller manors. This move towards nucleated villages was invariably accompanied by the adoption of the open-field system, which took more land into cultivation (Rowley 1981; Lewis *et al.* 1997). There is still no consensus of opinion concerning the impetus for these changes but whatever their cause, it is clear that they began prior to the Conquest and were not complete until at least the twelfth century (Hooke 1998, 115).

Against this background it is, understandably, difficult to detect changes related specifically to the Normans but this should not be equated with absence of Conquest-related change. Miller and Hatcher (1995, 38) have proposed that the consequences of 1066 were somewhat contradictory. On the one hand, the initial devastation caused by William's army would have upset the local economy and arrested international trade, with William I's imposed taxation placing further demands on the system. On the other, the net effect of the Conquest was to strengthen continental links, the Norman institution of markets, towns and cash rents (instead of food rents) accelerating the commercialisation of the English economy.

In order to determine whether the events surrounding the Conquest made any zooarchaeologically-perceptible impact on the Saxon period economy, it is necessary to examine the pre- and post-Conquest data at a number of levels. In this chapter attention is first given to systems of animal management: whether taxa ratios and age profiles display any inter-period variation. The possibility that provisioning systems were re-arranged as a result of Norman influence is then explored at a site level, with particular attention being given to butchery and body part data. Finally, evidence for butchery, tanning and bone working are examined to inform on the development of these specialist industries. Cattle, caprines and pigs are central to these studies but domestic birds are also considered. It is recognised that other domesticates – horses, dogs and goats – had an economic function but the irregularity by which these animals are recovered and detailed in zooarchaeological reports prompted the decision to omit them from the study. The economic importance of wild resources is examined in Chapter 5.

3.1: Trends in domestic animal management

By synthesising animal bone data from a large number of sites it has become clear just how dynamic the animal economy was during the medieval period, with husbandry regimes shifting considerably through time and space. Outlined below are some of the more significant trends.

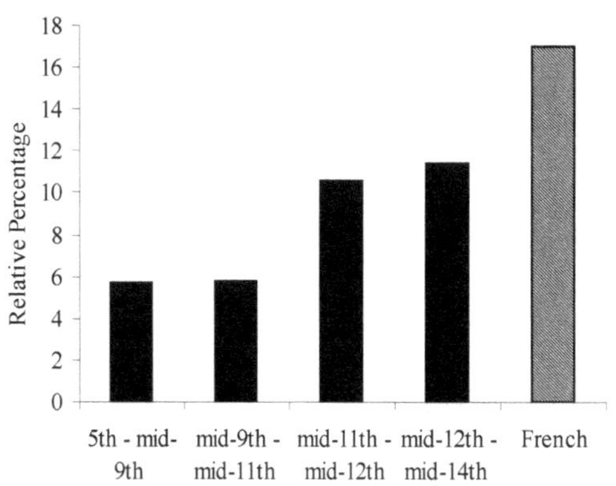

Figure 28: Inter-period variation in the representation of domestic birds (chicken, goose, duck) as a percentage (NISP) of the whole assemblage

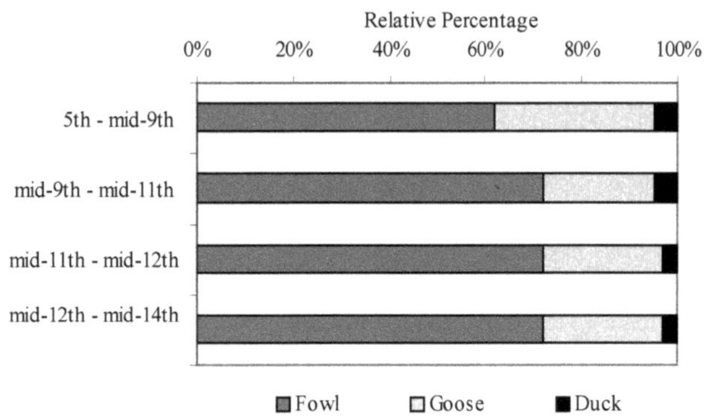

Figure 29: Inter-period variation in the relative frequencies of domestic fowl, goose and duck

3.1i: Domestic fowl, geese and ducks

Having begun with the assumption that Conquest-related change would be difficult to detect, the shifts observable for domestic bird representation come as a pleasant surprise. As Figure 28 indicates, the mid-eleventh century heralded a distinct break in patterns of domestic bird exploitation, their frequency increasing twofold from the pre-Conquest situation. An unprecedented move towards domestic birds is interesting in itself but when it is remembered that domestic bird levels were also high in medieval France, it is tempting to ascribe the shift to Norman influence. This was the conclusion drawn by McCormick (1991, 49), who noted a similar rise in domestic bird frequencies following the Anglo-Norman Conquest of Ireland. He attributed this to the presence of the army, documentary sources suggesting that troops were supplied with large quantities of fowl: a similar explanation could account for the situation in post-Conquest England.

Pre- to post-Conquest changes in the ratios of domestic bird taxa are also perceptible (Figure 29). In Saxon period assemblages, domestic ducks are moderately well represented. They are occasionally mentioned in English place-names, some of which – e.g. *Ductune* – even suggest specialised duck farms (Smith 1964, 111). After the Conquest, however, ducks become noticeably less abundant, their frequency in eleventh/twelfth-century deposits is half that of the preceding period. In the previous chapter it was suggested that the flesh of domestic ducks was eschewed in medieval France and it seems possible that this avoidance was transported to post-Conquest England. Definitely by the later medieval period ducks were deemed to be bad eating, as is indicated by the lyrics of a Middle English drinking song which states '*nor bryng us in no dokes flesche, for thei slober in the mer*' (Hagen 2002, 12).

There are several other observable shifts in taxa ratio but none of these coincide with the Conquest and must reflect different socio-economic factors. Figure 29 shows, for instance, an Early/Mid- to Late Saxon period decrease in the frequency of geese. The reason for the decline is unclear and may simply reflect the relative increase in domestic fowl frequencies. Egg production seems the most likely reason for the move towards domestic fowl exploitation. Finds of eggshell are common on mid-ninth to mid-eleventh century sites: large quantities were recovered from Coppergate (Bond and O'Connor 1999, 355), St Martin-at Palace Plain (Murphy 1987, 113) and the kitchen floor deposits from Eynsham Abbey (Ayres *et al.* 2003). Studies of chicken remains confirm their importance as egg producers. Ageing and sexing evidence indicates that most archaeological specimens derive from adult females, and the presence of femoral medullary bone demonstrates that most of these hens were killed whilst in lay (Ayres *et al.* 2003).

3.1ii: Cattle, caprines and pigs

Inter-period variation in the relative frequencies of the three main domesticates are shown in Figure 30. They support the conclusions of earlier studies (see for example Noddle 1975a; Clutton-Brock 1976; Grant 1988a), suggesting that throughout much of the medieval period cattle were generally the dominant taxon, followed closely by caprines, with pigs being the least abundant animals. Whilst these relative frequencies are well recognised zooarchaeologically, they contrast with the documentary evidence. In medieval texts cattle are mentioned rarely comparative to the other domesticates; in the Domesday Book they appear less frequently than horses and, in some cases, even donkeys. By contrast, pigs are frequently referred to in Saxon texts, a fact observed by Clutton-Brock (1976, 378), and sheep are recorded as the most numerous animal in post-Conquest documents, they outnumber all other animals in the Domesday Book (Lennard 1957; Darby 1973).

Other authors have noted the disparity between the historical and the zooarchaeological record, Albarella (1999b) suggesting that this may be due to the preferential preservation and recovery of cattle remains compared to those of caprines. Quantification techniques can also bias zooarchaeological interpretation. NISP counts, for instance, tend to inflate artificially the frequency of cattle relative to pigs and caprines, due to inter-taxa variation in bone fragmentation. By contrast pigs and caprines are often suggested as the dominant taxa when minimum numbers are calculated. Since this study has utilised NISP data it could feasibly account for the disagreement between the two sources of evidence. It seems likely, however, that historical inaccuracies are also partly responsible for the discrepancy. Rackham (1997, 75) argued convincingly that, rather than reflecting actual animal frequencies, the Domesday figures represent 'imaginary' animal units, used to indicate land area. Consequentially, the Domesday Book is likely to be of limited value as an indicator of livestock frequencies and should certainly not be used to retrodict early medieval farming regions (Lennard 1959; Darby 1977).

Distribution maps showing the zooarchaeological frequencies of the main domesticates (Figures 31-34) may provide a more reliable insight into regional husbandry regimes and their changes through time. For

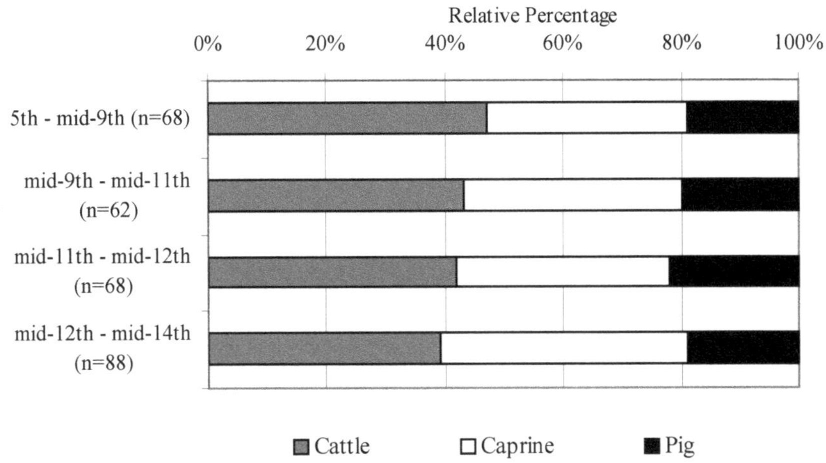

Figure 30: Inter-period variation in the relative frequencies (according to NISP) of cattle, caprines and pigs. N.B. n= number of assemblage on which frequencies are based.

the post-Conquest period several caprine-dominated assemblages have been noted in East Anglia (Figure 33b), perhaps supporting the widely-held belief that the fenlands in this region were 'dedicated' to sheep farming (for instance Round 1903; Lennard 1959; Darby 1977; Hooke 1998; Rowley 1997). Prior to the Conquest, however, the situation was entirely different. Fifth to mid-ninth-century assemblages from across the country typically contain cattle in high frequencies, West Stow being the only East Anglian assemblage containing caprines in abundance (Figure 31a and b). A few assemblages in the central midlands and along the South Downs also demonstrate high sheep/goat frequencies but, elsewhere, cattle predominate, especially in areas of fenland.

Maps for the mid-ninth to mid-eleventh centuries (Figures 32a-c) hint at a regional divide: cattle remains dominating assemblages from the east and north of the country, whilst caprines and pigs are better represented in central and southern areas. Environmental variation is the most obvious reason for this regional difference: cattle farming being suited to areas of lower relief and marshland, caprines being raised preferentially on wold and downland, with pig husbandry occurring principally in densely wooded areas. That landscape influenced husbandry regimes is undeniable and it is clear that similar patterns of animal husbandry date back to at least the Iron Age (Hambleton 1999). It is, however, deterministic to suggest that the environment dictated the medieval economy; if this were the case taxa ratios and regional distributions would have remained entirely static, which Figures 31-34 demonstrate they did not. It is necessary to recognise that the patterns observable in the zooarchaeological record reflect *human choices*. With this in mind, it is interesting to note that the regional divide apparent for the Late Saxon period equates approximately to the areas of Danelaw and English settlement.

The idea that cattle were culturally important to the Vikings is not new. Richards (1991, 71) noted that even when faced with unsuitable conditions in Iceland, the Vikings stubbornly continued to keep cattle. An association between Vikings and cattle has also been suggested by O'Connor (1989a) who recognised that Scandinavian-influenced assemblages from sites such as York, Lincoln and Dublin demonstrated very high frequencies (generally between 60-80%) of cattle. Because these assemblages were from urban centres – site-types that generally contain large numbers of cattle – O'Connor was unable to conclude absolutely that the taxa ratios resulted from Viking influence. On average, however, north/east village sites display 5% more cattle than their south/west counter-parts, indicating that the regional variation is not a product of site-type. Nevertheless, it cannot be argued that the Vikings introduced a cattle-based economy to the areas they settled, Figure 31a indicates that this was already in place when they arrived. Instead, the north/east of the country was simply slower to follow the move towards caprine-based husbandry, which began in the central midlands towards the end of the ninth century and gradually spread across the country (Figures 33b and 34b).

In the last few decades it has been recognised that the late ninth and tenth centuries witnessed fundamental changes in social structure, settlement patterns as well as rural and urban economies (Astill and Langdon 1997; Hinton 1990; Hooke 1998; Lewis *et al.* 1997; Reynolds 1999). One aspect of this transformation was an increase in arable production: pollen analysis for Hockham Mere in Norfolk indicates a rise in cereal cultivation between AD 650-850 (Bennett 1983). This seems to have been accompanied by a shift from spelt to rivet wheat (Astill 1997, 199). Any move toward intensive arable production brings with it a reduction in animal pasture but, more particularly, an increased need to maintain soil fertility. In some areas of Late Saxon England – the inner Midlands and Cotswolds especially (Roberts and Wrathmell 2002, 145) – new regimes of field management were adopted to combat these problems: the so-called 'Midland' or 'open-field' system. Under this regime, arable land was divided into two or three strip-fields, one being left fallow each year to act as rough pasture and regain fertility. It has been suggested that open-field agriculture was associated with village nucleation, with communities coming together to farm in cooperation. Plough sharing is one possible reason why the relative frequency of cattle fell in the core midland zone, although a relative increase in caprines could equally account for the observed shift.

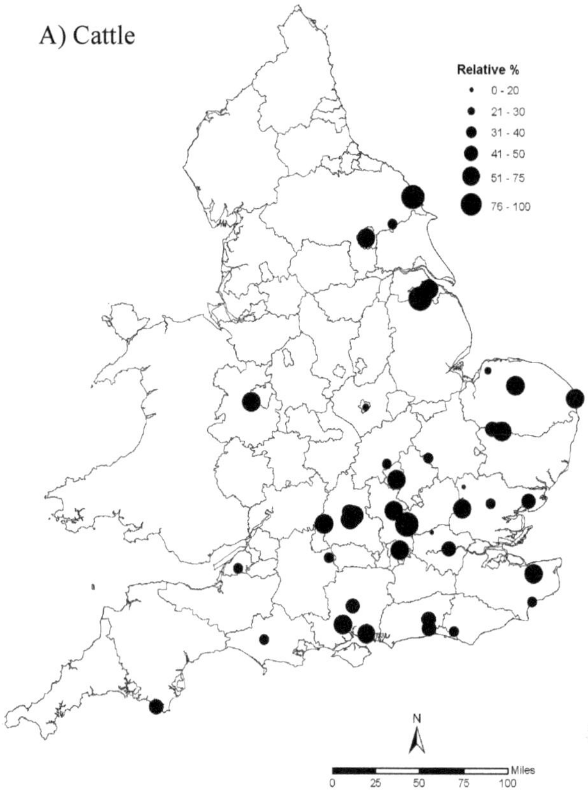

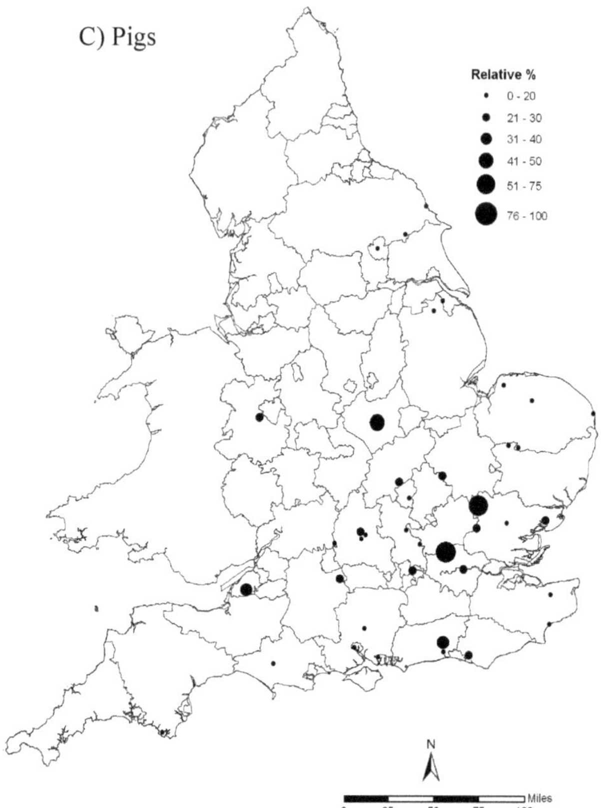

Figure 31: GIS maps showing the distribution of fifth to mid-ninth-century assemblages and their relative percentages of A) cattle, B) caprines and C) pigs. Maps generated by R. Jones.

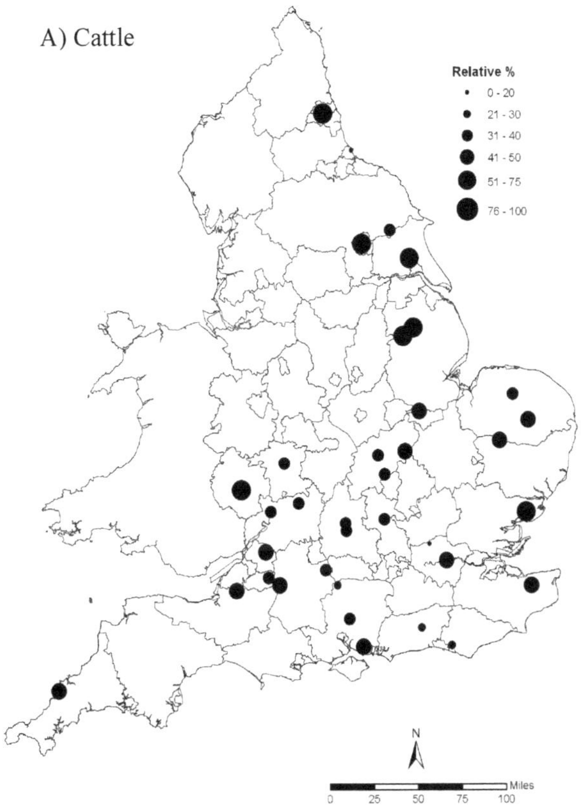

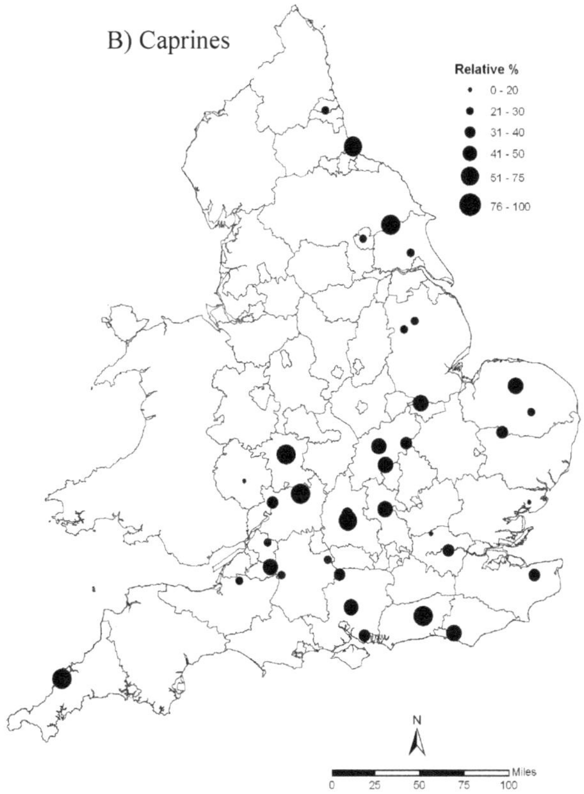

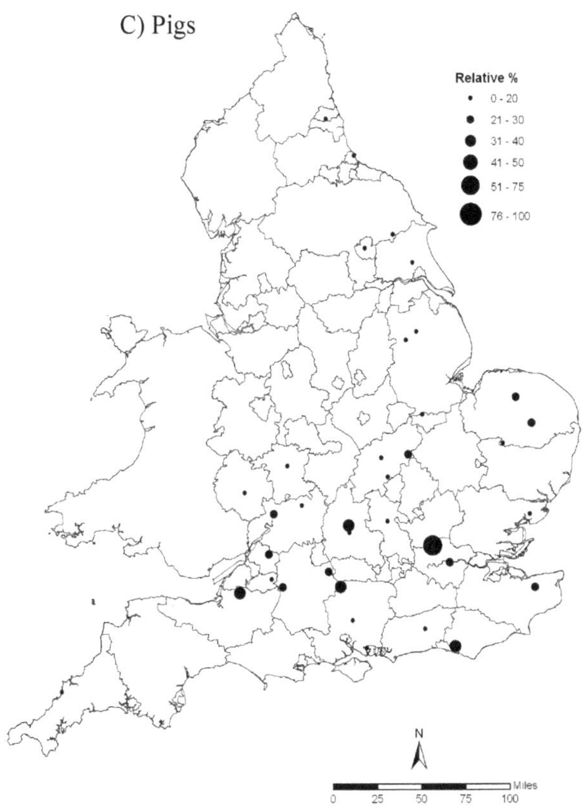

Figure 32: GIS maps showing the distribution of mid-ninth to mid-eleventh-century assemblages and their relative percentages of A) cattle, B) caprines and C) pigs. Maps generated by R. Jones.

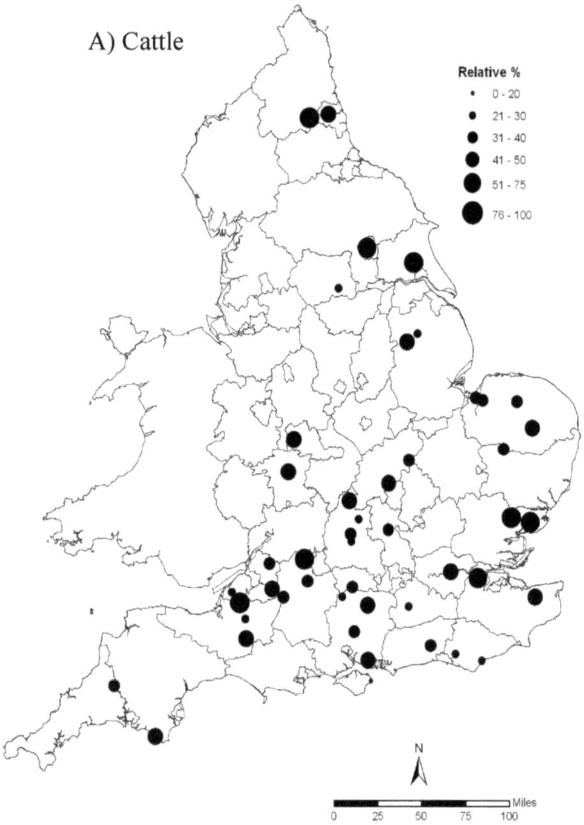

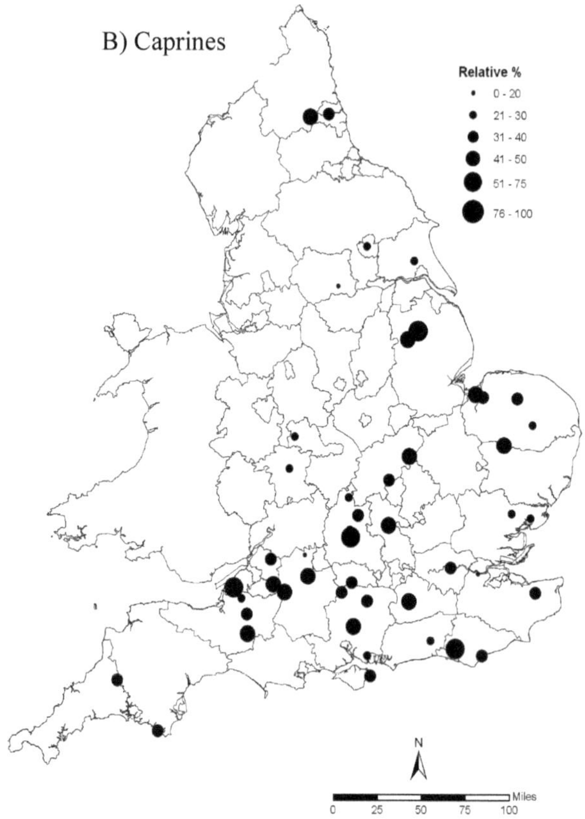

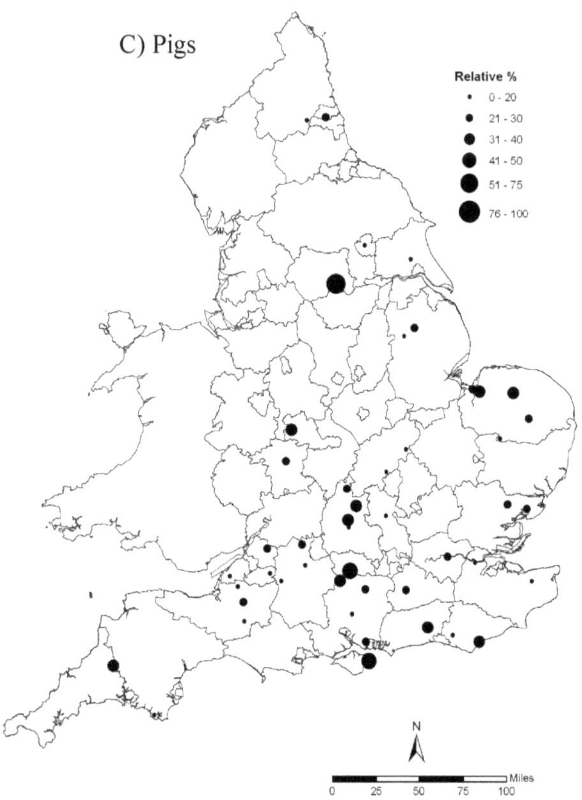

Figure 33: GIS maps showing the distribution of mid-eleventh to mid-twelfth-century assemblages and their relative percentages of A) cattle, B) caprines and C) pigs. Maps generated by R. Jones.

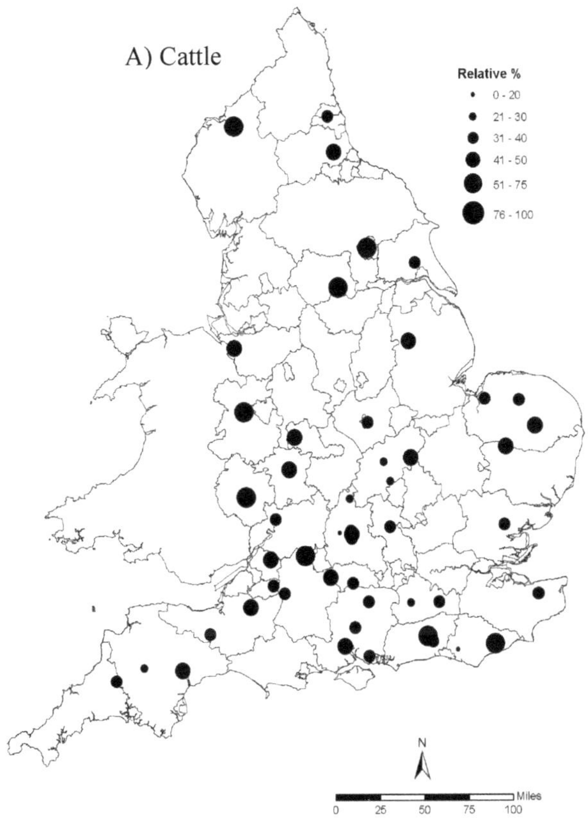

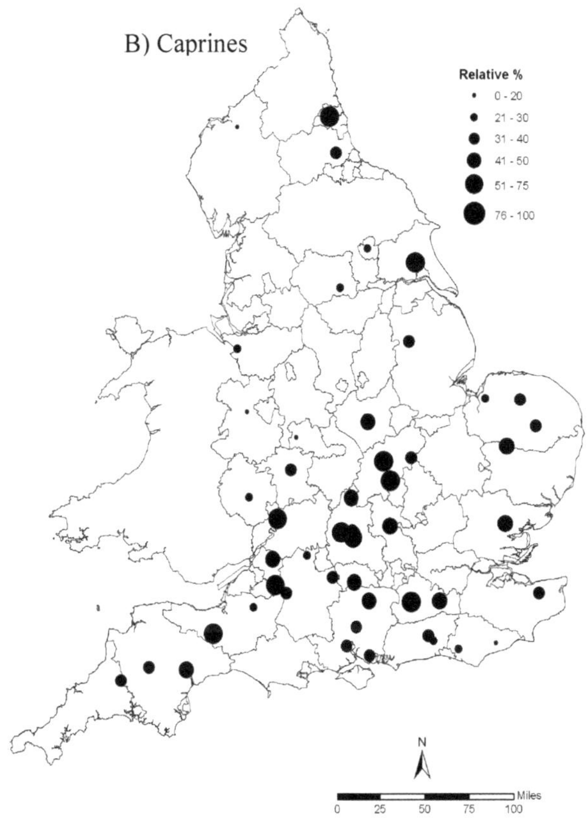

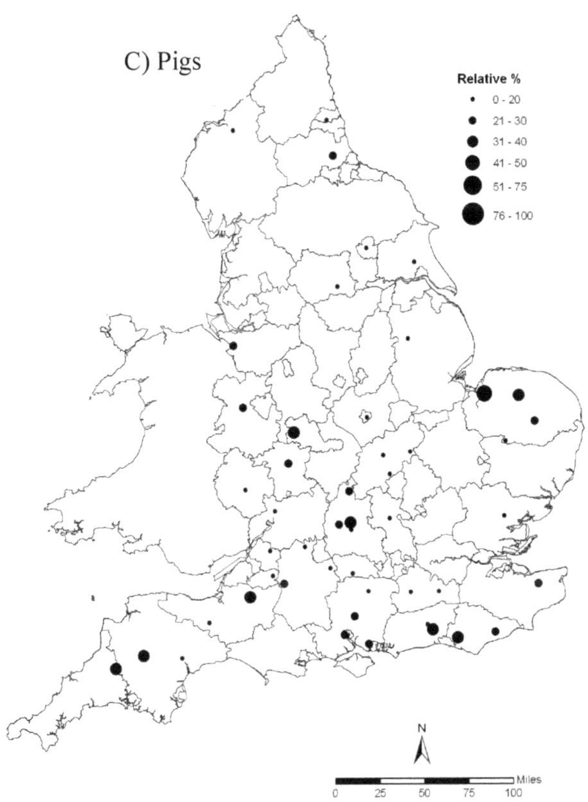

Figure 34: GIS maps showing the distribution of mid-twelfth to mid-fourteenth-century assemblages and their relative percentages of A) cattle, B) caprines and C) pigs. Maps generated by R. Jones.

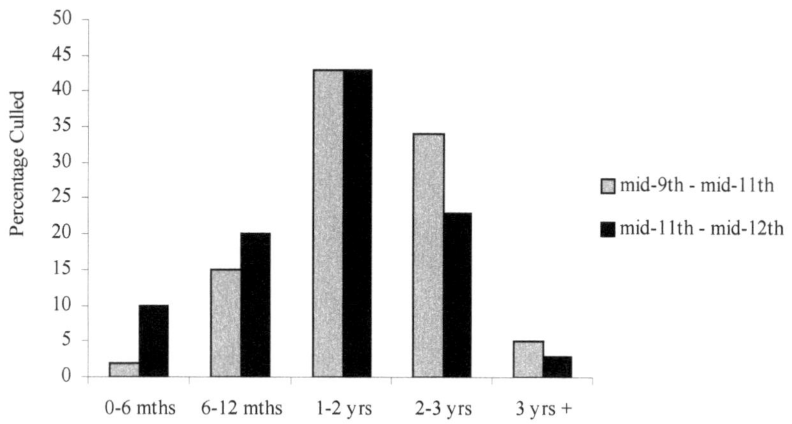

Figure 35: Percentage of pigs culled in each of five age classes, mid-ninth to mid-eleventh and mid-eleventh to mid-twelfth centuries

Several researchers have noted that expansions in arable farming are often accompanied by a rise in sheep rearing (McCormick 1991, 46; Cunliffe 1978, 184). This is seemingly related to the fertilising properties of sheep manure, which, according to White (1970, 127), contain higher frequencies of nitrates and phosphates than the dung of other animals. It seems likely, therefore, that the Late Saxon period changes in animal distribution reflect the adoption of open-field agriculture. This situation, although interesting, is beyond the scope of this book. Suffice to say that these mechanisms pre-date 1066, as has been demonstrated by occasional finds of ridge and furrow marks beneath Norman earthworks – for instance Hen Domen in Montgomeryshire (Barker and Lawson 1971) and Sandal Castle in Yorkshire (Mayes and Butler 1983).

There is no evidence to suggest that the Norman systems of animal management were imposed wholesale on post-Conquest England: the taxa ratios demonstrated either side of the Channel remain entirely different. One aspect of change, which could feasibly be related to Norman influence, is the slight post-Conquest increase in pig abundance. Figures 30 and 33c indicate that, relative to other taxa, pigs attained their highest frequency during the Norman period. Although their representation increases only by 2%, the large sample sizes involved in this study suggests the shift to be significant. This is interesting because, as was seen in Chapter 2 (Figure 9), pigs were the dominant animals in medieval France, comprising more than 70% of some assemblages (see Appendix Ib). It seems possible that the new Norman lords of post-Conquest England encouraged pig husbandry in an attempt to satisfy their appetite for pork. This idea is supported by evidence from post-Conquest Ireland: McCormick (1991, 48) noted that whilst pigs were poorly represented in the Early Christian period, they became abundant after the Conquest. Twelfth-century documents, in particular Pipe Rolls, record numerous instances where large quantities of pigs were sent from England to supply the Anglo-Norman armies in Ireland. Whilst cultural dietary preferences may have led to increased pork production, the amount of woodland in the Irish and English landscape would not have permitted pig husbandry on the continental scale. This may explain why pig frequencies, although rising by a few percent in post-Conquest England, never attained the level seen in medieval France.

In addition to showing a rise in pig frequencies, cull patterns for this taxon also suggest that the Norman period witnessed a shift in slaughter strategy: Table 9 and Figure 35 demonstrate that pigs were being killed at a younger age than in the preceding period. A logical explanation would be that as pigs became more plentiful, the need to retain breeding stock, and thus the average age of slaughter, was reduced. An alternative, and more interesting, possibility is that the shift reflects a move from free-range to sty maintenance.

There is ample evidence that Saxon swine husbandry relied on systems of 'denbera' (known as 'pannage' in the post-Conquest period) whereby animals were driven to woodland where they might forage for acorns, beechnuts and other treats. In Anglo-Saxon England the denbera season lasted four months, traditionally from August 29th to 31st December, giving animals plenty of time to fatten up either for slaughter or in preparation for a winter of abstinence (Wiseman 2000, 32-33). Maintained in this semi-wild state, sows would have farrowed away from settlements rendering any infant mortalities archaeologically invisible. Juveniles could also be underrepresented due to the difficulties of catching them, their camouflaged coats making them difficult to spot (Noddle 1980, 400). After 1066, however, the Norman's pre-occupation with woodland management, saw the pannage season reduced; in some counties it was restricted to just six to eight weeks (Wiseman 2000, 33; Albarella 2006, 79). The need to find alternative sources of fodder may have encourage pig-owners to keep their animals closer to home, feeding them waste from crop processing together with kitchen scraps. If domestic husbandry became more common in the post-Conquest period this would account for the rise in juvenile animals: infant fatalities would be present on sites and young animals would have been accessible for slaughter. Sty management would also have brought pigs under human control and it is conceivable that the Norman period marks the point at which selective breeding of pigs occurred. This possibility will be examined in the next chapter.

9a) Cattle	0-6 mths	6-12 mths	1-2 yrs	2-3 yrs	3-6 yrs	6-8 yrs	8-10 yrs
5th - mid-9th century (average %)	**6**	**15**	**21**	**17**	**21**	**13**	**6**
Rural (n = 239)	12	20	24	18	10	11	5
Urban (n = 303)	4	13	26	24	22	8	3
High Status (n = 318)	3	11	14	10	31	21	10
mid-9th - mid-11th century (average %)	**8**	**12**	**18**	**21**	**12**	**19**	**10**
Rural (n = 28)	18	32	14	7	4	14	11
Urban (n = 433)	3	4	20	36	16	18	3
High Status (n = 123)	3	1	21	20	15	24	16
mid-11th - mid-12th total (average %)	**0.3**	**3.0**	**20**	**24**	**20**	**22**	**11**
Rural (n = 12)			33	34	8	17	8
Urban (n = 94)		8	17	23	29	22	1
High Status (n = 204)	1	2	10	15	23	25	24
mid-12th - mid-14th total (average %)	**10**	**11**	**10**	**18**	**19**	**26**	**6**
Rural (n = 78)	6	24	8	13	13	33	3
Urban (n = 168)	11	7	16	20	23	20	3
High Status (n = 178)	14	2	7	22	21	24	10

9b) Caprine	0-6 mths	6-12 mths	1-2 yrs	2-3 yrs	3-4 yrs	4-6 yrs	6-8 yrs	8-10 yrs
5th-M9th century (average %)	**4**	**18**	**18**	**14**	**13**	**22**	**7**	**3**
Rural (n = 548)	8	26	22	11	14	12	5	2
Urban (n =874)	2	14	20	22	14	21	6	1
High Status (n = 313)	2	15	12	9	10	34	11	7
M9th-M 11th century (average %)	**7**	**12**	**17**	**21**	**22**	**17**	**3**	**1**
Rural (n = 84)	11	12	15	23	24	13	1	1
Urban (n = 663)	7	9	23	22	20	15	3	1
High Status (n = 198)	4	15	14	17	21	22	5	2
M11th-M12th (average %)	**5**	**7**	**21**	**21**	**16**	**16**	**8**	**5**
Rural (n = 34)	3	9	18	29	17	12	6	6
Urban (n = 638)	5	5	26	17	15	21	10	1
High Status (n = 239)	8	8	20	18	17	14	8	8
M12th-M14th (average %)	**6**	**8**	**18**	**20**	**17**	**21**	**8**	**2**
Rural (n = 136)	1	13	27	20	13	15	10	1
Urban (n = 821)	5	6	18	19	23	21	7	2
High Status (n = 182)	12	6	8	20	16	26	8	3

9c) Pig	0-6 mths	6-12 mths	1-2 yrs	2-3 yrs	3 yrs +
5th - mid-9th century (average %)	**10**	**20**	**28**	**36**	**6**
Rural (n = 162)	13	17	28	37	5
Urban (n = 305)	6	22	29	36	7
High Status (n = 378)	12	21	27	34	6
mid-9th - mid-11th century (average %)	**2**	**15**	**43**	**34**	**5**
Rural (n = 11)		10	45	45	
Urban (n= 320)	3	20	42	29	6
High Status (n = 239)	4	15	44	27	10
mid-11th - mid-12th total (average %)	**10**	**20**	**43**	**23**	**3**
Rural (n = 38)	8	29	42	21	
Urban (n = 217)	10	16	44	22	8
High Status (n = 206)	13	15	43	27	2
mid-12th - mid-14th total (average %)	**11**	**21**	**34**	**25**	**9**
Rural (n = 20)	20	25	20	20	15
Urban (n = 105)	6	21	43	25	5
High Status (n = 169)	7	18	38	30	7

Table 9: Inter-period and inter-site variation in the ages of a) cattle, b) caprines and c) pigs, shown as the relative percentage of mandibles in each age group.

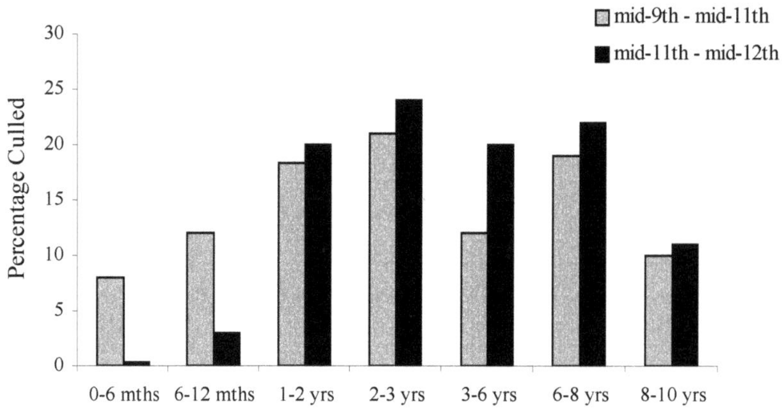

Figure 36: Percentage of cattle culled in each of seven age classes, mid-ninth to mid-eleventh and mid-eleventh to mid-twelfth centuries

Pre- to post-Conquest changes in animal management are also indicated by the cattle and caprine cull-patterns (Figures 36 and 37). Table 9 shows that Saxon period assemblages contain sizable numbers of animals aged under six months which, if not natural fatalities, hint at dairy production. According to the *Rectitudines Singularum Personarum* (Lemanski 2005, 36-7) cattle, sheep and goats were all milked. Aelfric's Shepherd states that 'I lead them [the flock] back to their fold and milk them twice a day, I make cheese and butter' (Swanton 1975, 109). Dairying is also referred to in the Leechdoms (Cockayne 1851) and the Exeter Riddle Book (Crossley-Holland 1993). French and Frisian documents suggest that cheese was one of Saxon England's main exports, together with grain and cloth (Loyn 1970, 100).

These three commodities – dairy, wool and grain – are economically compatible within a mixed farming regime. Adult cattle may be used for traction alongside milking and breeding: one of the responsibilities of Aelfric's Oxherd was to take his cattle to the ploughman (Swanton 1975, 109). Similarly wool and/or milk may be cropped from the same sheep responsible for manuring arable fields. The zooarchaeological data for mid-ninth to mid-eleventh-century England are consistent with this scenario. It was clearly a successful regime and, with the introduction of the novel treddle loom (Hinton 1990, 123), Late Saxon England came to be renowned for its woollen goods (Loyn 1970).

Whilst the production of agricultural surplus became central during the Late Saxon period, by the Norman period sheep/goats and cattle were seemingly managed for little other than wool, manure and traction. Cull-patterns for post-Conquest cattle and sheep/goats show that many more animals were maintained beyond six years, with fewer than 5% of caprines and 1% of cattle culled before six months. By raising the majority of juveniles to adulthood, milk yields would have been reduced, allowing little dairy production beyond the domestic scale. This may explain the scarcity of *vaccaria* mentioned in the Domesday Book (Lennard 1959, 265) and the fact that, by contrast to the Late Saxon period, dairy products – especially blue varieties of cheese – were actually imported to Norman England (Lennard 1959, 265; Hagen 1998, 30).

Demand for wool must have been heightened by the arrival of numerous Norman ecclesiastics, whose need for cowls and habits prompted many religious houses to maintain large flocks of sheep (Postles 1984). Norman-established markets would have assisted redistribution, with new towns acting as a focus for production: the Pipe Roll of 1130 records Weavers Guilds in London, Oxford and Lincoln (Bartlett 2000, 339). Interestingly both Oxford and Lincoln show post-Conquest increases in caprine frequency, as do assemblages from Beverley, a town for which weavers and fullers are recorded by 1189.

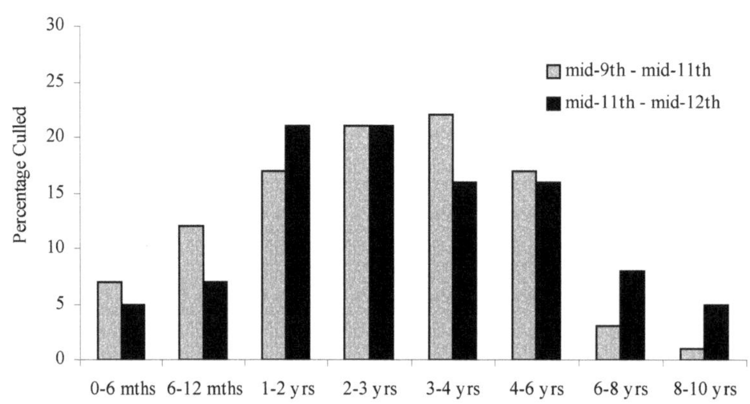

Figure 37: Percentage of caprines culled in each of seven age classes, mid-ninth to mid-eleventh and mid-eleventh to mid-twelfth centuries

These post-Conquest changes in animal management seem to represent a second wave of agricultural intensification but the degree to which the Normans can be held responsible for them is debatable. Hooke (1998, 131) suggests that the 'devastation' caused by William's army may have facilitated agricultural change, with settlements being rebuilt in a nucleated layout conducive to the uptake of the open-field system. Arable production on marginal lands may equally have been assisted by the post-Conquest introduction of the heavier French plough, although Langdon (1986; 1988, 88) does not consider that its introduction can be dated to the Norman period with certainty. Archaeobotanical studies have identified a new type of free-threshing tetraploid wheat (*Triticum turgidum/durum*) that appears in England shortly after the Conquest and could potentially have been imported by the Normans (Moffett 1989). Aside from these changes, there is little to suggest that the Normans were responsible for any great economic innovation, the zooarchaeological evidence suggesting that the post-Conquest regime evolved simply from the Saxon set-up. However, it is commonly cited (for instance Rowley 1984, 21) that the Norman's success came from their ability to maximise the profitable aspects of societies they conquered; in a period when wool and grain prices were high it seems possible that new Norman lords encouraged tenants to concentrate on these commodities. According to McCormick (1991, 46), this was the tactic used in post-Conquest Ireland where the native economy was re-organised so that wool and grain became the dominant products.

On the basis of the zooarchaeological evidence the most substantial pre- to post-Conquest shift seems to have come in the form of increased production, be it in terms of pig husbandry, wool yields or arable output. Examination of synthesised national data has provided a general impression of these changes but if we are to understand how these demands for produce were made and met by different sections of society it is necessary to consider the data at the level of site-type.

3.3: Provisioning Systems

That provisioning systems did change from the pre- to post-Conquest period can be inferred from historical and archaeological evidence. For example, although towns were well developed by the Late Saxon period, the success of some Norman-founded towns and markets caused the decline and abandonment of older trading settlements: the Norman foundation of Arundel in Sussex, for instance, led to the demise of the nearby Anglo-Saxon *burh* of Burpham (Sutermeister 1976). Linguistic evidence – that much of the modern terminology of commerce is derived from French vocabulary – also indicates that the Normans influenced the mechanics of trade (Platt 1976, 21). But market exchange was just one aspect of commerce. The rural population of Anglo-Saxon England was largely self-supplied, with goods being obtained through services, renders and barter (Miller and Hatcher 1995, 17). Landlords received food-rents from their tenant farmers, and all estates were required to contribute food and drink for the upkeep of the royal court (Stafford 1980; Hagen 2002). However, from the time of Edward the Confessor, and particularly under Norman rule, food renders were gradually abolished in favour of cash payments (Britnell 1993, 41; Lennard 1959, 136; Stafford 1980). This, combined with William I's taxation demands may have encouraged the move towards a coin-based economy. National taxes, such as the Danegeld, were known before the Conquest but were suspended between 1051 and 1066 (Britnell 1993). In 1067 William I reinstated the geld, the Anglo-Saxon Chronicle recording it as having been 'very severe' (Garmonsway 1967, 200). By 1068 John of Worcester described the tax as 'insupportable' and there is some evidence that non-payment even resulted in the confiscation of land (Williams 1997, 13-19).

It may be hypothesised that since the availability of currency was greatest in towns, markets would have become the intermediary for the distribution of animals and their products. The addition of an extra stage in the

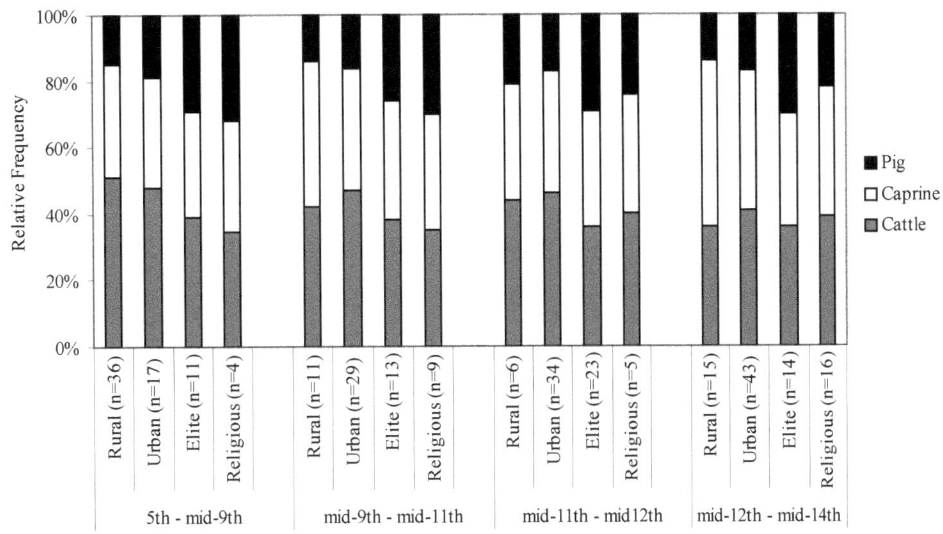

Figure 38: Inter-period and inter-site variations in the relative frequencies of the main domesticates, according to NISP

provisioning process may have altered fundamentally producer-consumer relationships, with implications for the wider fabric of society. Should such changes have occurred, it seems possible that they could be detectable in the zooarchaeological record. In this section, taxa ratios, cull patterns and anatomical representation data are examined, by site-type, to determine if Conquest-related shifts in the provisioning systems are discernible.

3.3i: Rural sites: farmsteads, hamlets and villages

Zooarchaeological evidence is least abundant for rural sites, rendering studies problematic. Assemblages from these settlement types do, however, display some traits typical of 'producer' sites (Figure 38). For instance, caprine and cattle remains dominate rural assemblages from both the Late Saxon and Norman periods, reflecting the wool and grain economy that characterised medieval England. By comparison to other site types, these rural assemblages also contain higher frequencies of caprine remains; presumably preferential export of cattle to urban sites (see below) left the rural population with comparatively more sheep/goats to eat up. There is pre- to post-Conquest continuity in these patterns, with no dramatic shift in taxa representation occurring at the point of 1066. The only notable change is the 5% rise in the frequency of pigs; this is an increase greater than is seen for the other site types. Small sample size may be responsible for the apparent anomaly, especially since rural pig frequencies return to their former level by the mid-twelfth to mid-fourteenth century, but it is tempting to view the shift as a short-term consequence of Norman rule. When pig frequency data are examined with the cull-patterns for rural sites (Table 9), which show a substantial post-Conquest increase in juvenile animals, it seems conceivable that the patterns reflect an intensive breeding programme intended to supply greater quantities of pork to Norman settlers. But this is perhaps stretching the evidence too far given that all site types demonstrate an increased representation of juvenile pigs. Instead, the shift in age profiles may simply reflect the change from free-range to sty husbandry proposed in the previous section. The post-Conquest reduction in pannage period would have seen more pigs live and die on-site, an increased presence of natural fatalities could itself account for the post-Conquest rise in rural pig frequencies.

It is difficult to isolate changes pertaining to provisioning from those relating to wider economic shifts. Cull-patterns for rural cattle and caprines (Table 9), for instance, reflect the general move away from dairying and towards the exploitation of other secondary products (wool and traction). Nevertheless, close examination of the cull patterns reveals some subtle inter-period variations potentially indicative of Norman-instigated changes to the provisioning system. Slaughter profiles for the Late Saxon period cattle demonstrate that rural sites contain by far the highest percentage of animals under 12 months, with urban sites having fewer individuals of this age, and elite sites the least. To some extent these inter-site differences are unsurprising; the presence of juveniles (infant mortalities) on 'producer' sites is to be expected. However, it seems unlikely that natural deaths can account for all the juveniles on rural sites. Another possibility is that the farming population were retaining these tender young animals for their own consumption whilst sending the older stock, with their tougher meat, to consumer populations.

Patterns for the Norman period are entirely different. Most of the variation can be explained by the agricultural intensification discussed above but it is notable that the representation of juveniles is reversed from the preceding period: urban and elite assemblages contain more cattle aged under 12 months than rural sites. Similar pre- to post-Conquest trends are demonstrated by the caprine data suggesting that, although sample size is small, the pattern may accurately reflect the development of market forces; consumer populations gaining the power to demand and obtain tender meat from younger animals.

Consumer demand for meat may equally explain the decline in the relative frequency of domestic birds on rural sites, the farmers of Norman England perhaps supplying markets with ducks, and geese in particular (Figures 39 and 40). It is difficult to prove this conclusively because, as Coy (1989b, 31) has pointed out, birds were probably traded live and thus anatomical patterns are ineffective as indicators of trade and exchange. Skeletal representation is studied more profitably for the larger domesticates, where mechanisms of animal movement – whether on-the-hoof or in a pre-butchered form – can be determined.

Figures 41a-f show skeletal data for Late Saxon and Norman cattle, caprines and pigs from rural sites. Anatomical representation is not equal in either period, with the body part patterns, especially those for mid-ninth to mid-eleventh-century cattle, showing an over-representation of foot and head elements. Potentially, this may reflect a situation where villagers butchered their own animals but sent the meatier portions to the consumer

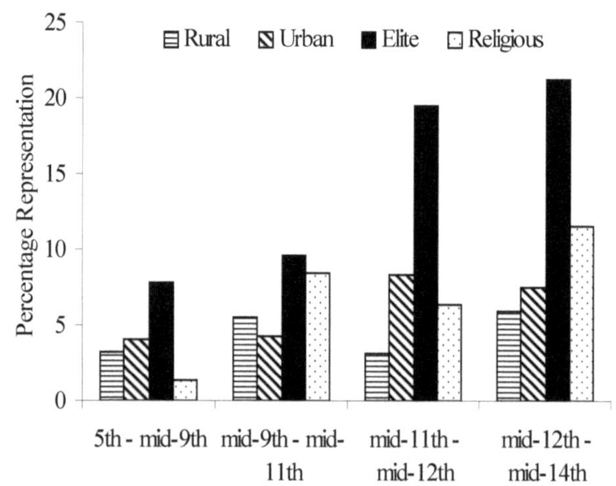

Figure 39: Inter-period and inter-site variation in the representation of domestic birds (chicken, goose and duck), shown as a percentage (NISP) of the total assemblage.

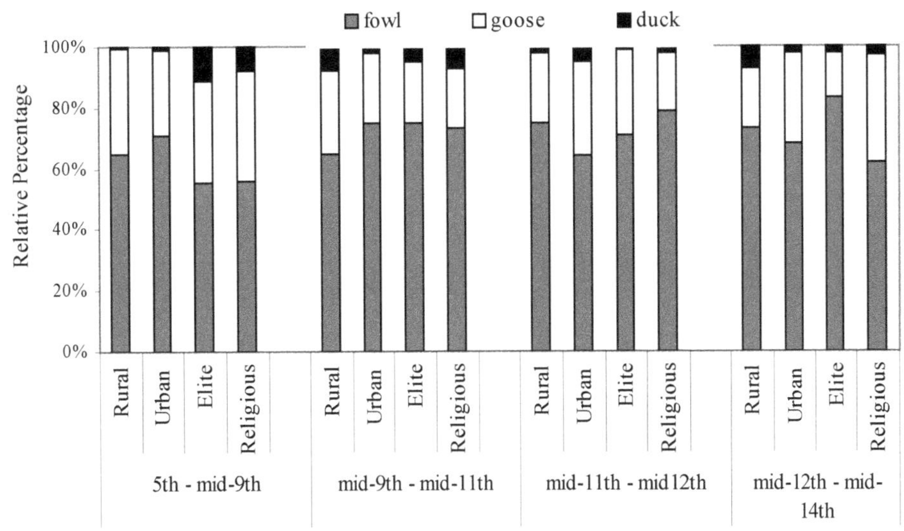

Figure 40: Inter-period and inter-site variation in the relative frequencies of domestic fowl, goose and duck

population. This was the conclusion drawn by Chaplin (1971, 127) in his study of the Late Saxon Whitehall Farm site. Alternatively, the abundance of extremities could feasibly be explained by factors of rubbish disposal. In the medieval period, refuse from rural sites was often accumulated in middens, so that it might later be spread on the fields as fertiliser (Astill and Grant 1988, 2). Because refuse was left exposed, scavengers would have had access to the material. Brain (1967) has shown that dogs will preferentially consume the lower density, meat-bearing bones, leaving behind the less palatable elements such as mandibles and metapodia. Carnivore activity could, therefore, have masked anthropogenic influences. Whatever the case, the anatomical patterns provide little to suggest that carcass processing changed from the pre- to post-Conquest period.

3.3ii: Urban: wics, trading settlements and towns

By contrast to village refuse, medieval urban waste was usually placed in pits, which would have provided good preservation conditions and perhaps also protected the material from scavengers (Astill and Grant 1988, 4). This may explain why cattle and caprine body-part graphs show equal representation of skeletal elements (Figure 42a-f). There is no over-representation of meat-bearing elements to suggest that either pre- or post-Conquest towns were supplied with ready-butchered joints. Instead it would seem that animals were either raised in the urban environment (as is indicated by the ninth-century animal byres found in Gloucester - Heighway *et al.* 1979) or brought from the agricultural hinterland on-the-hoof. Regardless of how they arrived, they were apparently slaughtered and butchered within the town: an idea supported by the butchery data (Section 3.4i). The fact that cattle are easier to drive than sheep and pigs, and also carry a greater meat weight, may explain why they are relatively more abundant on towns than other sites (Figure 38). It is possible that cattle were supplied due to consumer demand but there is little other evidence to suggest that the urban population had any control over their supplies. This is clearly demonstrated by comparison with the French data which, in Section 2.1iii, I argued reflects a consumer-led provisioning system. In France, producer and consumer assemblages were markedly different, suggesting that agricultural production had little impact on urban supply. By contrast, English villages and towns show a great similarity in taxa range, and the fact that English towns show a significant pre- to post-Conquest increase in sheep frequencies indicates that urban supplies were influenced more strongly by economic factors (wool production) than the consumer's dietary preferences.

Since pre- to post-Conquest shifts in taxa frequencies conform to the wider patterns of economic change, it is perhaps surprising that the ageing data demonstrate somewhat contradictory trends. Although cull-patterns are typical for consumer sites, showing neither a large number of very young or very old animals, they do not demonstrate the same dramatic post-Conquest move towards delayed slaughter exhibited by other site-types. Indeed the slaughter profiles for sheep/goat actually suggest that Norman towns were receiving more 1-2 year olds than in the preceding period. This pattern has also been recognised for Norman period Flaxengate, Lincoln, and O'Connor (1982a, 49) suggested that this short-lived shift could represent the dietary preferences of the Norman soldiers who were stationed in the area.

Arrival of the Norman army must have had some effect on the Saxon provisioning system but, if the Lincoln pattern reflects Norman preferences, it might be expected that high-status sites, which saw the heaviest Norman influence, would also show a shift towards young animals. As will be seen below, this is not the case. Another explanation must, therefore, be sought. It seems possible that the decline in urban animal age somehow corresponds to the increase in age noted for village sites, perhaps reinforcing the idea that post-Conquest farmers sent a higher frequency of prime meat animals to market. Britnell (1993, 51) has argued that the peasant's main impetus for the sale of produce was to obtain cash to pay taxes.

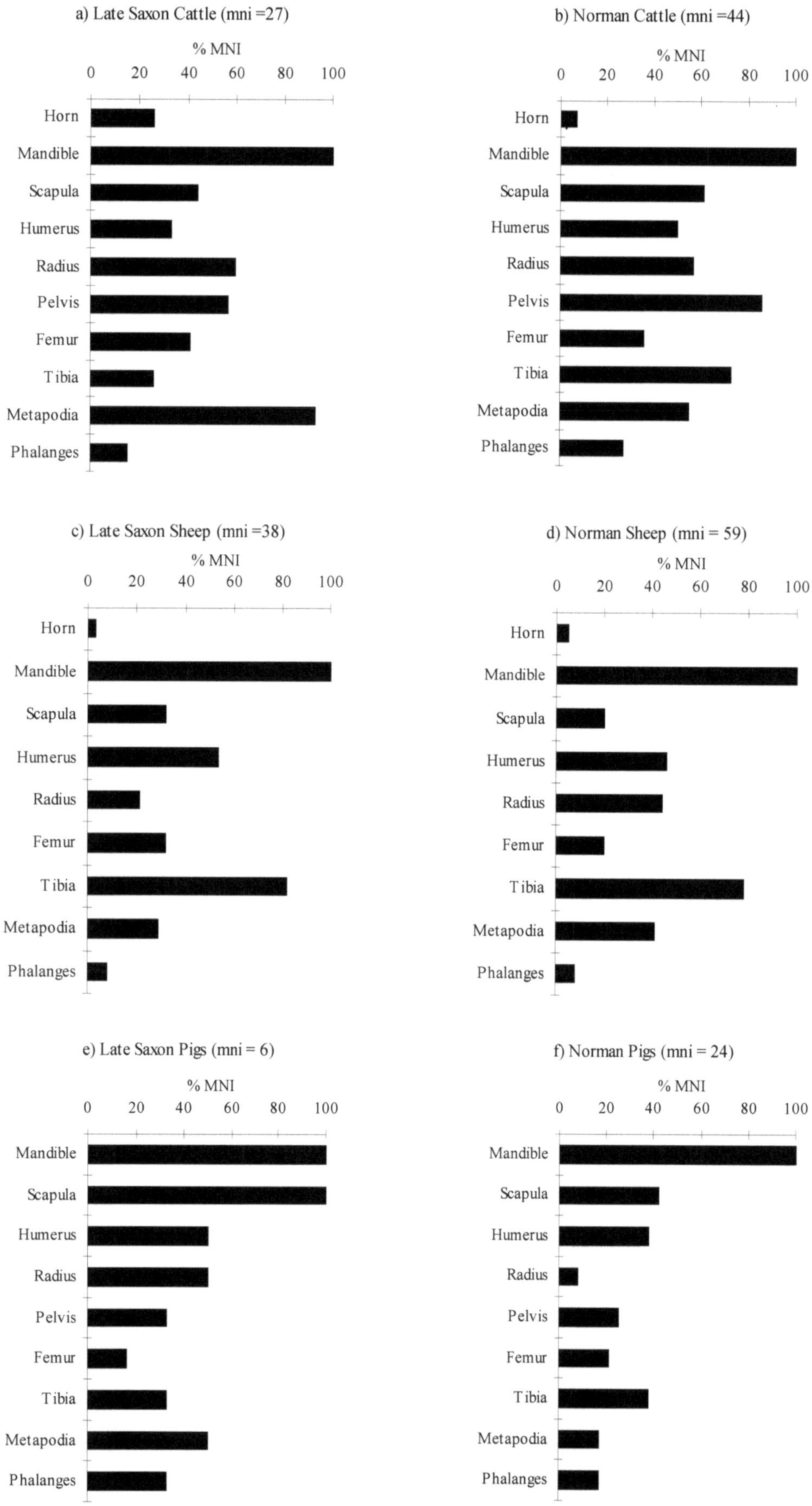

Figure 41: Anatomical representation patterns for the main domesticates from rural assemblages dating to the Late Saxon (left) and Norman period (right).

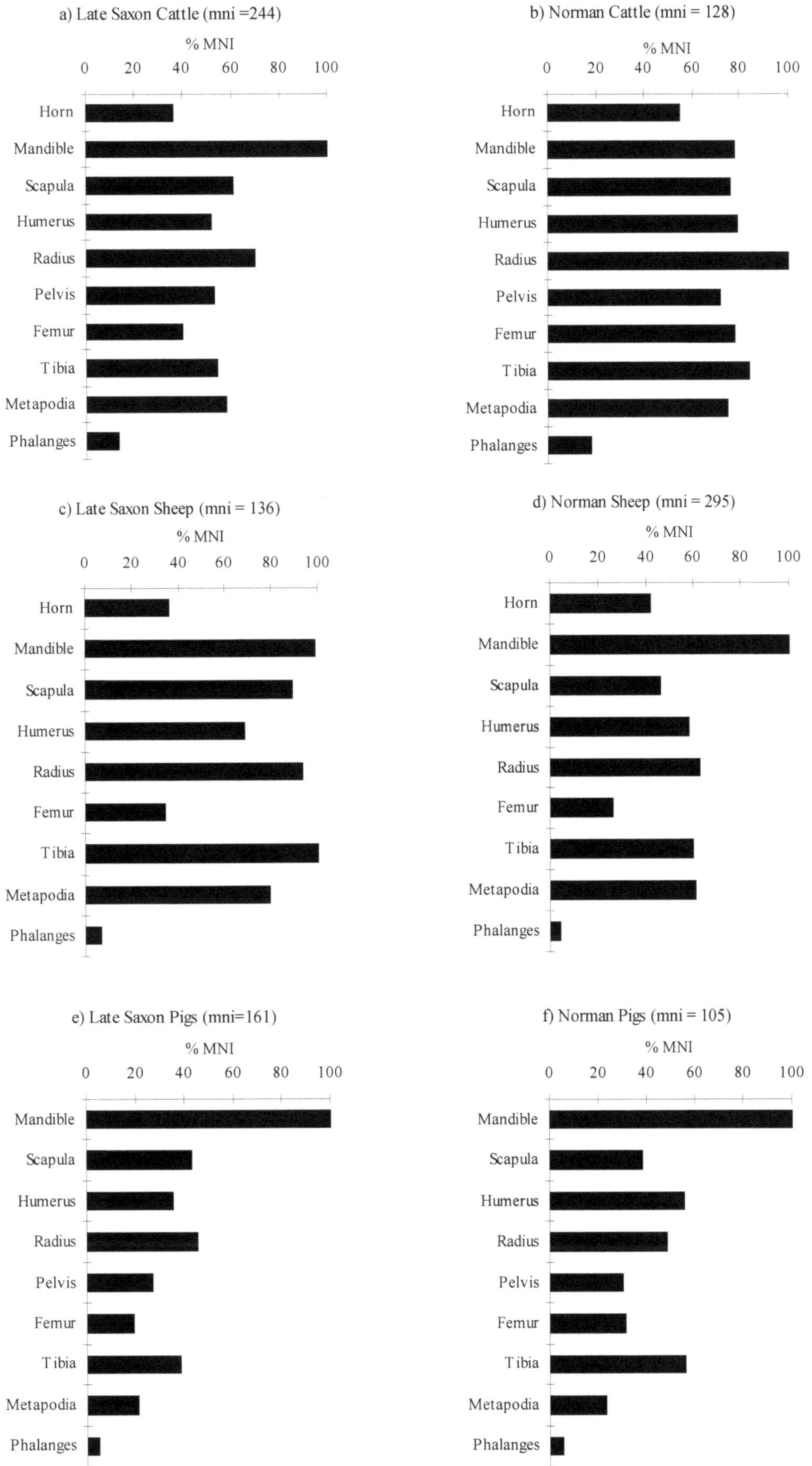

Figure 42: Anatomical representation patterns for the main domesticates from urban assemblages dating to the Late Saxon (left) and Norman period (right).

It has already been noted that the post-Conquest period saw tax burdens increase and the need to obtain money could certainly have encouraged the rural population to market many of their prime-aged animals. This could also explain the post-Conquest rise in domestic birds on urban sites, which occurred at the same point that bird frequencies on village sites declined (Figure 39). According to Carlin (1998, 28), the late eleventh and twelfth centuries saw the proliferation of professional cooks who produced ready-made food for the urban population. By the 1170s William Fitz Stephen was able to record that London had a vendor of cooked poultry (Douglas and Greenaway 1953, 958).

The data for pigs show perhaps the least evidence of inter-period change: there is continuity in their relative frequencies, anatomical patterning and their age profiles. The only slight difference is that, as for other site types, a higher percentage of juvenile individuals are represented in the post-Conquest assemblages, perhaps suggesting that domestic pork production became more common in the Norman period. This possibility is supported by recent dental microwear and bone isotope analyses which have indicated that pigs were raised increasingly within urban households, being fed on domestic rubbish rather than foraging free-range (Albarella 2006, 79).

3.3iii: Elite: castles, manors and religious houses

Whereas pigs generally constitute less than 20% of rural and urban assemblages, they are usually well represented on high-status sites, a fact first recognised by Grant (1988b). The association between pigs and elite settlement is demonstrated well by the assemblage from Goltho in Lincolnshire (Beresford 1987; Jones and Ruben 1987). During its earliest phase the site comprised a modest farmstead, and cattle and caprines dominate the bone material derived from it, with pigs contributing just 8%. By the late ninth century, Goltho had become a fortified *thegnly* residence, and this change in function is reflected by a dramatic increase (to 31%) in pig frequency (See Appendix Ib).

That pigs were obviously important to the elite may explain why this animal is over-emphasised in the documentary evidence, the authors of the texts preferentially mentioning those animals they favoured. Why this pig-status relationship should originally have developed is less clear. It is commonly cited that because pigs are a meat-animal, providing little by way of secondary products, only the rich could afford their upkeep (Biddick 1984b; Grant 1988a). This view is valid if it can be demonstrated that resources were expended for their maintenance; for example, the accounts for later medieval Peterborough Abbey indicate that considerable quantities of grain were purchased for pig fodder (Biddick 1984b). However, pigs might equally be viewed as the peasant's animal, since they could have been fed kitchen refuse or put out to pannage. Bennett (1971, 92) states 'no animal was so easily fed, and no animal so easily put on flesh and was so soon prepared for slaughter'. That pigs hold such a dichotomous position suggests that the association between pigs and high-status sites may relate to something other than wealth.

In Chapter 2 it was suggested that pigs were used as a food-rent animal in France, and the same may have been true in England. Documentary evidence indicates that pigs were often paid as a fee for land use: in 866 for example, land at Seckley in the west Midlands was leased in return for, amongst other things, fifty swine (Hagen 2002, 261). As was mentioned in Chapter 2, pigs would have been the most suitable rent animal because not only were they cheap to raise but they would also have provided the biggest range of meat products (ham, pork, bacon, sausages). The Domesday Book records numerous swine-rents, paid for pannage rights, so the food rent system was clearly operating efficiently in the decades immediately after 1066 (Lennard 1959, 259). Cull patterns for pigs from elite sites, however, suggest that the post-Conquest aristocracy were also practicing some pig husbandry in their own back yards, indicated by the increased representation of neonatal and juvenile animals (Table 9). The site of Faccombe Netherton provided clear evidence for on-site husbandry, with six compete neonatal skeletons being recovered from a Norman period cesspit (Sadler 1990b, 481).

The consumption of home-grown pork in addition to pigs supplied in rent may explain why pig frequencies are higher on elite sites dating to the mid-eleventh to mid-twelfth century than those of the preceding period. Why the post-Conquest aristocracy wanted to breed their own pigs may be related to Norman dietary preferences, a desire to maintain high levels of pork consumption, but it could equally reflect an awareness that provisioning system were changing: payments for pannage were moving away from swine-rents and towards cash fees.

If, as historical evidence suggests, food rents were gradually being commuted for cash payments, it might be expected that provisioning patterns would have changed in response, with animal products being channelled increasingly through urban markets. Ageing and body part patterns for elite sites do provide some evidence to support this contention. Anatomical representation data for the main domesticates are provided in Figures 43a-f. Those for the Late Saxon period, in particular those for caprines and pig, demonstrate a low frequency of foot elements, indicating that rather than arriving on-the-hoof animals were brought in a pre-butchered form. Some individual sites, such as the Abbeys of Eynsham and St Albans, seem to have received quantities of meat-joints (see Appendix IIIa-c) but the synthesised data suggest that, in general, elite assemblages are characterised by an abundance of cranial elements. This is particularly evident for pigs (Figure 43e) which are represented by little other than mandibles. Admittedly, pig assemblages usually demonstrate an over-representation of jaw bones

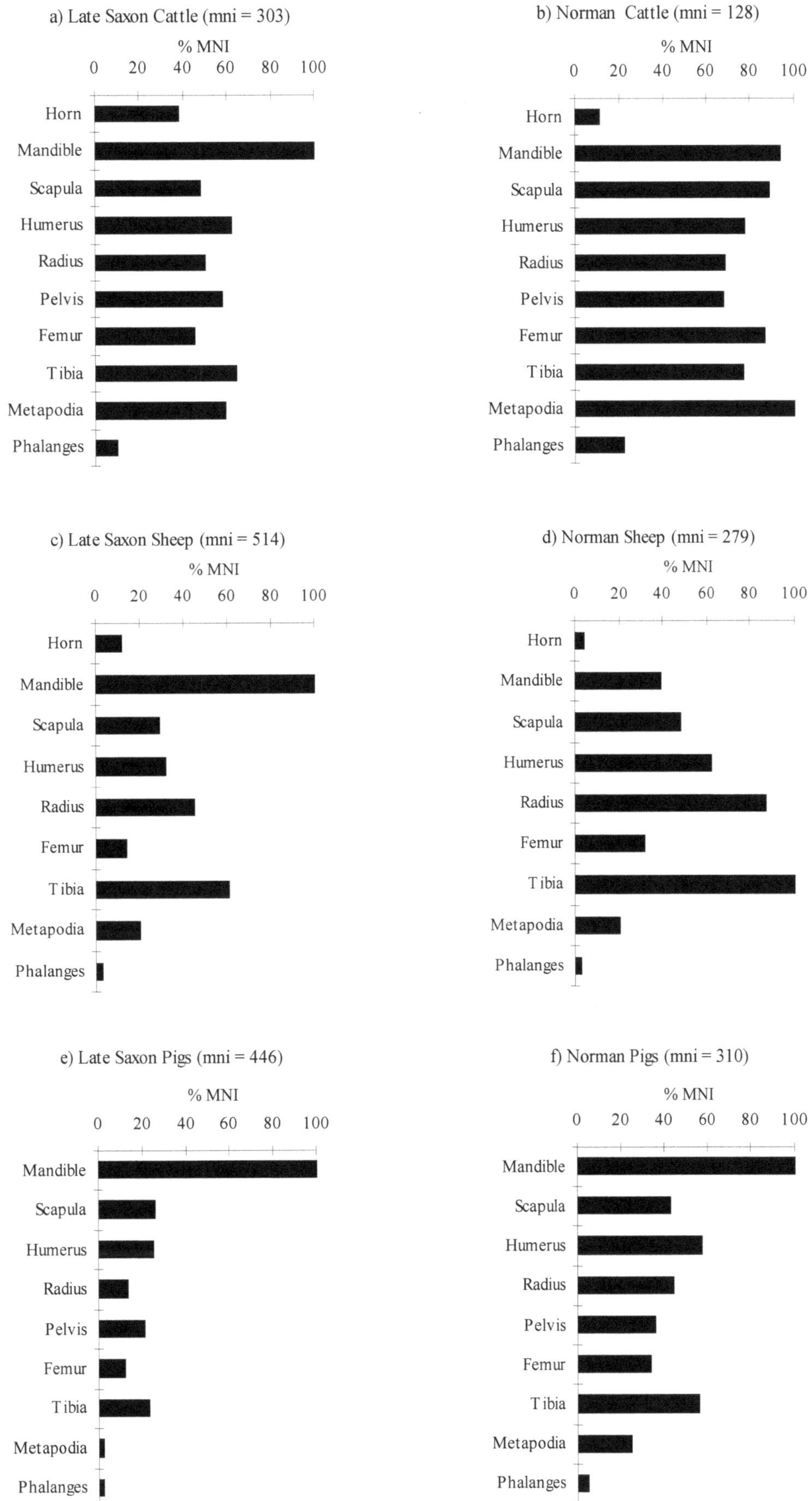

Figure 43: Anatomical representation patterns for the main domesticates from elite assemblages dating to the Late Saxon (left) and Norman period (right).

but, by comparison to rural and town assemblages the ratio of mandibles to post-cranial bones on elite sites is much greater. It seems possible that either heads were selectively imported to elite sites or that animals were butchered at these sites and their meat redistributed. What is clear is that the body part patterns for Norman animals, in particular sheep (Figure 43d), show a far higher frequency of meat-bearing bones compared to jaw bones. Could this suggest that greater quantities of pre-butchered joint were now being purchased from markets?

The inter-period variation is even more interesting when it is recognised that the post-Conquest patterns are very similar to those from high-status French sites (Figures 10a-f). It was seen in Chapter 2 that the assemblage from Vatteville-la-Rue and Douai consisted primarily of meat-bearing joints and I argued that the occupants of the castle were being provisioned with specific cuts of meat: shoulders of mutton and the shoulders and haunches of beef. Hams may also have been imported but there is evidence that pigs were supplied to the elite via a different mechanism to sheep and cattle.

The fact that anatomical representations change from the Saxon to Norman period but also come to resemble those for French sites, may suggest that the continental provisioning system was installed in post-Conquest England, whereby pre-butchered beef and mutton was obtained from specialist butchers, rather than direct from the rural population. Certainly this theory is supported by the butchery evidence for high-status sites, which demonstrates a post-Conquest shift towards more professional techniques (see below, Section 3.4i). According to McCormick (1991, 48), similar changes to the native provisioning system were made after the Anglo-Norman Conquest of Ireland. As in England, food rents were replaced and traditional methods of animal butchery and distribution were legislated against.

If the post-Conquest elite were demanding the types of animals that they were accustomed to in France, it might be expected that the cull-patterns for Norman period cattle and caprines would indicate a move towards the representation of juvenile animals: it was seen in Chapter 2 that 20-30% of animals consumed by the French elite were under six months of age. Table 9 demonstrate no overwhelming decline in average slaughter age, indeed, the data for Norman cattle indicate a significant rise in age. But, close examination of the slaughter profile for post-Conquest caprines reveals that the percentage of animals under six months of age is double that for the preceding period, and there are also many more animals aged between one and two years. Against an economic backdrop of increasing caprine age, this shift towards the consumption of younger animals seems significant, perhaps suggesting that the elite were exercising greater control over the supplies they received. This possibility will be discussed further in Chapter 8 but at present it is necessary only to recognise that evidence for a post-Conquest change in provisioning is apparent and that the Normans were feasibly responsible for its introduction.

3.3iv: Summary

In this section it has been seen that whilst some aspects of provisioning system were unaffected by the Conquest, others altered considerably. For example, while taxa ratios show little variation, beyond that associated with the wider economy, ageing and body-part data indicate a shift in the mechanisms of animals distribution.

Although the picture is far from clear, there is evidence to suggest that, in the post-Conquest period, peasants marketed greater quantities of their livestock, including juvenile cattle and sheep as well as domestic birds. This decision may have been prompted by Norman demands for money payments, in the form of taxation, but also as a result of the abolition of food-rents.

That the Norman elite were not provisioned in the same way as their Saxon counterparts can be inferred from body part data, in particular the rise in the frequency of meat-bearing bones. It seems possible that rather than obtaining supplies directly from the rural population, the Norman lords utilised their new source of money to buy specific meat cuts from towns.

3.4: Craft Specialisation

It has already been seen that the textile industry was well developed by 1066, and scholars such as Hall (1988), Hodges (1989), Wade (1988) and Wilson (1976) have demonstrated that Saxon period England supported numerous bone workers, butchers and tanners. It is clear that these specialists proliferated and became more centralised during the post-Conquest period but whether the Normans motivated this process is less apparent. Norman foundations of new markets and towns must have encouraged the process, especially since many were laid out with accommodation for potential craftsmen: in Nottingham, for example, thirteen houses were built in the new borough, with a further forty-eight merchants' houses being erected in the space between the new and old towns (Miller and Hatcher 1995, 41). Nevertheless, other factors such as international trade and population increase must also have been influential.

Smith (1988) has argued that the eleventh to thirteenth centuries witnessed a European wide expansion in population: he estimated that whilst there were 1.5-2.5 million people in England at 1086, this figure had risen to 3.8-7.2 million by the end of the thirteenth century. As more families had little or no land they were forced to seek income as service providers. Eleventh- and twelfth-century documents record folk movement away from rural areas to the newly-created urban settlements (Rowley 1984, 124). In larger towns, specialists of the same craft began to live in communities and by 1130 trade guilds, such as those for butchers, were well established (Sabine 1933; Pooley 1945; Britnell 1993).

3.4i: Butchery

Historical evidence demonstrates that specialist slaughterers, flayers (*hyldere*) and butchers (*flaesc tawere*), were known to the Late Saxon population (Hagen 1998). Tenth-century legal codes state that cattle had to be sold and slaughtered in the presence of two witnesses (Whitelock 1955) but no specifications regarding the location of these activities were laid down (Britnell 1993). Before the Conquest, specialist butchers may have been itinerant, undertaking slaughter and butchery on the consumer's premises: this is the indication provided by the synthesised body part data for mid-ninth to mid-eleventh-century towns (Figure 42). The patterns are not indicative of commercial butchery as they show neither a dearth nor over-abundance of head or foot elements, instead most parts of the skeleton are represented equally. Similar patterns are provided by several individual assemblages, such as those from Coppergate (O'Connor, 1989a) and Bedford (Grant 1974, in Hagen 1998) The Wigford site in Lincoln, however, did provide a more selective assemblage; the anatomical data for both cattle and caprines indicated an over-representation of metapodia and mandibles relative to other body parts (see Appendix IIIa and b). Dobney *et al.* (1995) suggested that this may be evidence for professional butchery.

If butchery was becoming more specialised or centralised in the Late Saxon period, it might be expected that Norman period assemblages would exhibit even greater evidence for the trade's development. But skeletal representation data for the post-Conquest period actually show patterns less, rather than more, indicative of systematic butchery. One explanation to account for the absence of distinct deposits could be that, in the early medieval period, there was no such thing as butchery 'waste'. Most parts of an animal's carcass would have been valued, be it nutritionally, functionally, or symbolically, and it seems unlikely that heads and feet would have been discarded un-utilised. There is ample evidence to suggest that brains were eaten (cleaved skulls are often found amongst food debris) and that metapodia were processed for their marrow (see Chapter 8.1). Such activities are perhaps more likely to have been undertaken in the kitchen rather than at the place of butchery. For these reasons the status of the butchery profession may be better deciphered from butchery marks than the body part patterns.

The butchery data obtained for Saxon and Norman period sites are provided in a summarised format in Appendix V. At the most basic level, inter-period variation in the ratio of chop to cut marks suggests a post-Conquest shift in butchery tools, with meat cleavers becoming the more commonly used. On its own, this information has little interpretative value but when viewed in conjunction with the butchery patterns the evidence becomes more informative. Butchery diagrams for the pre- and post-Conquest periods are shown in Figure 44. While variation in the location of butchery marks is minimal, with most being concentrated around the articular surfaces, methods of carcass reduction were significantly different in the Saxon and Norman periods.

Perhaps the most obvious change, and one that has been noted by many authors (for example, Grant 1987, Hagen, 1998, Jones, 1993 O'Connor 1982a, 1984), is the way in which carcasses were split into sides. In the earlier periods the vertebral column tended to be split along one or both sides of the centrum but by the later medieval period carcasses appear to have been divided into equal sides along the medial-sagittal line. Table F (Appendix V) demonstrates that the post-Conquest period saw a higher percentage of butchery marks indicative of sagittal splitting: 31% of cattle vertebra butchery in the Norman period was attributable to this method, compared to just 8% in the Saxon period.

Methods of separating the foreshank (the meat surrounding the radius) from the upper arm were also different in the pre- and post- Conquest periods. In the Saxon period the distal humerus was chopped through in a way which removed its angle of epicondylus, the dorso-medial part of the proximal radius and part of the ulna's olecranon (Figure 44a). By contrast, post-Conquest period butchers adopted a mirror opposite technique: chopping through the trochlear of the humerus, causing the volar-medial part of the proximal radius and proximal ulna to be removed (Figure 44b). Since these differences would not have produced significantly different cuts of meat, they seem to reflect style rather than function.

According to Sackett (1986) stylistic variation in butchery may be an indicator of cultural difference. However, the possibility that differences between the pre- and post-Conquest butchery patterns reflect the introduction of Norman techniques cannot be substantiated. Figure 45 demonstrates that humerus-radius butchery at Vatteville-la-Rue was actually more akin to the techniques observed for pre- than post-Conquest England. Neither can the Normans be attributed with the introduction of vertebra splitting; indeed, there is little to suggest that these techniques were commonly practised in France before the twelfth century. Although the Vatteville assemblage demonstrated a high frequency of split vertebrae, poor dating of the material reduces the applicability of the data to the eleventh/twelfth century. Charité-sur-Loire produced no evidence of cleaved cattle or pig vertebrae until the fourteenth century, although a small number of eleventh-century sheep vertebrae had been cleaved (Audoin-Rouzeau 1987). The animals from the tenth-twelfth century assemblage from Douai (Vadet and Villette 1986) were also split by cutting either side of the centrum, rather than through the middle.

Without further evidence from France the situation cannot be clarified but at present the Saxon to Norman changes in butchery style cannot be related to Norman influence. Instead, it is more probable that the shift from careful but random butchery, towards standardised but heavy techniques reflects the emergence of specialist butchers.

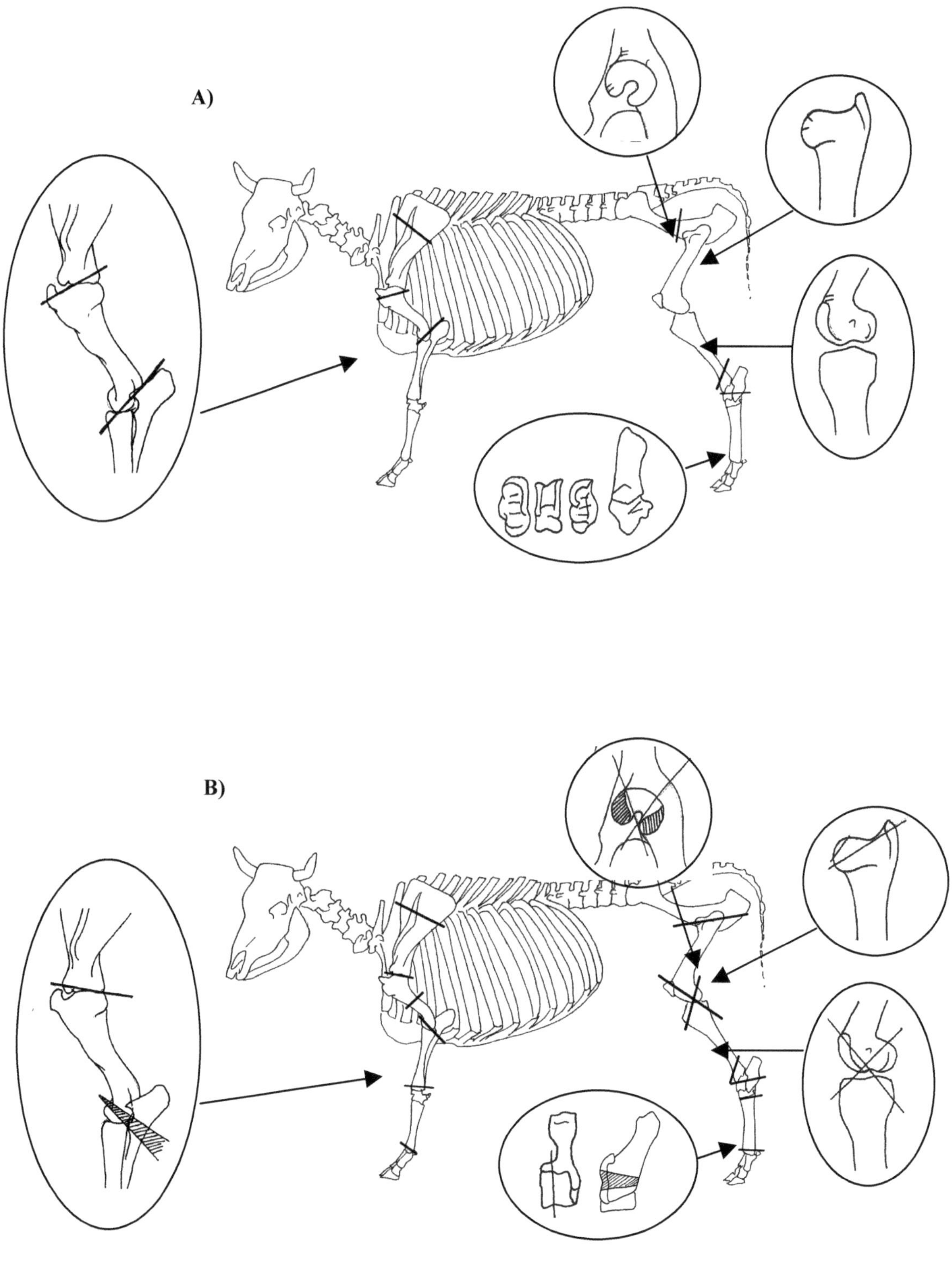

Figure 44: Differences in the location and style of butchery as evidence by A) mid-ninth to mid-eleventh and B) mid-eleventh to mid-twelfth century assemblages

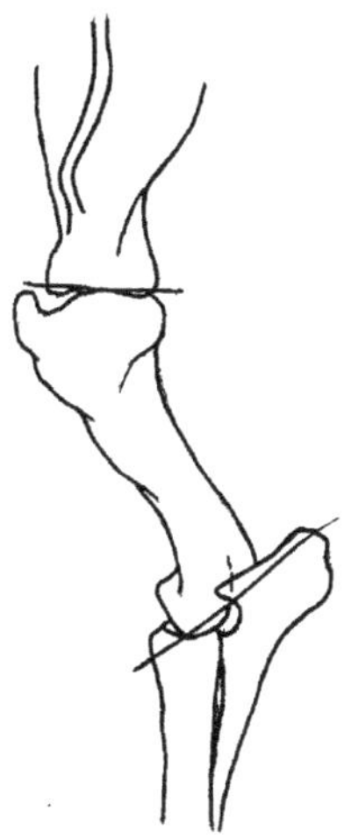

Figure 45: Pattern of fore-limb butchery displayed by the assemblage from Vatteville-la-Rue

There is some evidence from Steyning (Sussex) and Flaxengate in Lincoln (O'Connor, 1982a) that 'professional' techniques were practised in pre-Conquest towns but they do not appear to have been used on other site types of this date. Absence of urban-style butchery in elite assemblages reinforces the idea that the pre-Conquest aristocracy were supplied from their own estates or via the rural population, rather than from professional butchers. The fact that butchery patterns for post-Conquest elite sites conform more closely to those from town assemblages supports the idea (proposed in Section 3.3iii) that, following the Conquest, the elite began to obtain higher frequencies of meat from urban markets. Such a move would have increased the need for professional butchers and may have led to the establishment of meat stalls, such as those owned by William I's half-brother, the Count of Mortain (Clapham 1966, 60).

3.4ii: Tanning

Shaw (1993) has summarised the structural evidence for tanning in medieval England, and here the evolution of the tanning industry will be considered from a zooarchaeological perspective. In the absence of tanning pits and hide off-cuts, evidence for the leather industry can be difficult to recognise. Skins can be removed without influencing the underlying carcass and thus, the tanning process need leave no osteological trace. The tables in Appendix V show that skinning marks were recorded on both pre- and post-Conquest bones but this provides little evidence for the development of the tanning industry, since all animals, perhaps with the exception of pigs, would have been skinned before butchery.

In many instances tanning is better indicated by skeletal representation data. In the medieval period skins were sent to the tanners with certain elements (usually a combination of phalanges, metapodia, skulls and/or horns) still attached; Schmid (1973) suggests this was either because their weight stopped the hide from shrinking or so that the tanner could ascertain the age and identity of the animal. The tanner's first job was to strip these elements from the skin, resulting in deposits with distinct body part patterns. Unfortunately, the similarity of these deposits to those created by other craft industries has often complicated their interpretation: the assemblage from St Peters Street in Northampton has been interpreted as both butchery (Harman 1979b) and tanning waste (Serjeantson 1989).

It was seen in the previous section that there is little evidence for assemblages dominated by metapodia or phalanges. One explanation would be that Saxon tanners, like those of the modern day, worked skins that had no body-parts attached. Alternatively, the dearth of selective deposits may suggest that during the Saxon period craft activities were largely integrated with domestic life. Such an explanation has been proposed for ninth- to eleventh- century Winchester, which, according to Serjeantson (2000a) produced no bone groups indicative of specialised activity. Material dating to the mid-eleventh century, however, provided the first indication of professional tanning: the Henley's Garage cattle assemblage was dominated by crudely butchered metapodia (Serjeantson 2000a, 185). The tannery at Lower Bridge Street, Chester, is also dated to the mid-eleventh century (Mason 1985, 23), a date after which tanning deposits appear more regularly. It might be conjectured that the Conquest was somehow responsible for the intensification of the industry but the difficulties of dating archaeological deposits makes it uncertain whether these changes occurred shortly before or just after 1066.

3.4iii: Antler, Bone and Horn Working

There is ample evidence for bone and, in particular, antler working in the pre-Conquest period. Pits at Hamwic produced quantities of antler shavings and sawn bone (Bourdillon and Coy 1980; Driver 1984), and many Late Saxon towns, such as Lincoln (O'Connor 1982a) and York (O'Connor 1991) have yielded collections of partially worked shed antlers. Despite this, it is generally believed that production was on a domestic scale as antler, horn and bone working waste has seldom been found in quantities sufficient to suggest full-time workshops (MacGregor *et al.* 1999, 205). On the basis of regional similarities in artefact style, MacGregor (1991, 356) suggested that antler workers may have been itinerant, serving several communities. Again Chester provides an exception to this as the remains of an antler and bone working industry are reported to have been discovered at Abbey Green (McPeake *et al.* 1980, 31).

Whilst antler was the predominant raw material in the Saxon period, MacGregor (1991) and MacGregor *et al.* (1999) have demonstrated that by the Norman period it had been superseded by bone and horn. At the same time, the scale of artefact manufacture seems to have increased from domestic to commercial production. For example, a late eleventh/early twelfth-century pit at St Alband house in Wood Street, London, was found to contain numerous worked cattle metatarsals thought to have been associated with the manufacture of finger rings (Vince 1989, 175).

To some extent the change in raw materials and intensification of the industry can be accounted for by urban development. The growing urban population would not only have increased demand for artefacts but the remains of the animals they consumed would also have produced reliable supplies of bone and horn for working. In a situation where production rates were increasing, seasonal supplies of antler may have been insufficient, causing artisans to adopt the more reliable sources of bone. It may equally have been this abundance of raw materials in towns that encouraged itinerant craftsmen to become sedentary.

Post-Conquest town foundation could again be cited as a contributory factor in these changes but it is possible that the shift away from antler working was even more directly linked to Norman influence. MacGregor (1991) cites the twelfth-century Miracle of St Cuthbert, which tells of an artisan who, having discovered a stag's carcass, fantasised about the profits he could make from crafting the animal's antler. This story suggests that antler was still a valued and sought after commodity in the post-Conquest period, thus avoidance of its use may suggest that it was subject to restrictions. After the introduction of forest law (see Chapter 5) the penalties for poaching could feasibly have dissuaded artisans from collecting shed antler. If this hypothesis is correct it would suggest that the Normans were partly responsible for the uptake of bone and horn working.

Horn itself is known to have been a valuable commodity throughout the medieval period: the laws of Ine mention that 'the horn of an ox is valued at tenpence' and the 'horn of a cow twopence' (Whitlock 1955). Cattle body part patterns (Figure 41-43) provide a valuable insight into the development of the horn-working trade. The graphs for the Late Saxon period, particularly those of cattle, demonstrate that horn cores are represented on all site types in similar frequencies. Patterns for the post-Conquest period, however, indicate a very different situation, with horn-cores becoming poorly represented on village and high-status sites whilst their frequency in town assemblages increases considerably. This suggests that the inhabitants of rural settlements were collecting horn-cores and sending them to urban specialists. Such a situation has implications for the study of urban horn-core assemblages: it should not be assumed that deposits, such as those examined in Chapter 4 (see Figure 52), came from animals of the same flock or even the same geographical region.

At a site level, it is only from the mid-eleventh century that distinct deposits of horn-cores appear regularly. At York, for example, concentrations of goat horn-cores have been recovered only from sites post-dating the eleventh century (O'Connor 1989a). An eleventh-century site in Thetford did, however, produce an assemblage dominated by cattle horn-cores (Jones 1984). This deposit overlay a concentration of worked antler and bone, seemingly waste from comb manufacture. The difference between these two layers may reflect the transition from antler to horn working. Potentially, this could suggest that the shift away from antler working occurred just before, rather than as a result of, the Norman Conquest. Nevertheless, there are many more horn-dominated assemblages dating to the Norman period, such as those from Skeldergate (O'Connor 1984), and the post-Conquest rise in specific craft deposits indicates that, even if the Normans did not instigate centralised craft specialisation, they certainly accelerated it.

3.5: Conclusion

In this chapter the Saxo-Norman economy has been examined at a national, regional and, site-type level. The animals, their management, as well as the production and distribution of their primary and secondary products, have all been considered for evidence of Conquest-related change. In many respects continuity is apparent: rural sites are characterised by high frequencies of caprines, with towns dominated by cattle remains and high-status sites demonstrating an abundance of pigs. This surface continuity may be the reason why scholars have found little evidence for Norman impact upon the animal economy of the Saxon period. When the situation is studied in more detail, however, a slightly different picture emerges.

All of the categories of zooarchaeological data considered in this chapter have provided some evidence of pre- to post-Conquest change. Perhaps most obvious is the intensification of the agricultural economy. Cull-patterns demonstrate a significant shift suggesting that cattle and caprines were targeted increasingly for traction and wool respectively. When this shift is viewed against wider temporal trends it is clearly part of an ongoing process which began at least a century earlier. It was seen in Section 3.1 that the Mid- to Late Saxon transition was accompanied by significant economic change: domestic bird assemblages demonstrate that chickens began to be raised preferentially to geese and caprine frequencies increased relative to cattle, a move I proposed was linked to the adoption of the open-field system.

It was after the Conquest that these new husbandry regimes became particularly defined, with cattle and sheep/goats being kept to older ages, presumably to provide traction, wool and manure. Other evidence for an intensification of arable production is provided by the appearance of a new wheat variety and possibly a new plough type. Nevertheless, evidence for direct Norman involvement in this process has remained elusive. Since economic change is closely linked to social and political structures, the likelihood of extrapolating Norman influence from such a complex equation is low. However, it seems possible that intensification may have occurred as a result of forced re-organisation, similar to that which occurred in Anglo-Norman Ireland. Another possibility is that increased taxation under the Norman kings encouraged concentration on commodities (namely grain and wool) that could be sold for cash. The need for money may also explain the decrease in animal age in town assemblages: it seems possible that the farming population was selling, rather than consuming, any available meat-animals in order to raise the necessary funds.

Post-Conquest shifts in pig and domestic bird frequencies can perhaps be more closely tied to Norman influence, with their relative percentages rising towards levels found in France. The idea that this reflects dietary preferences of the Normans is reinforced further by the fact that similar shifts have been recognised for post-Conquest Ireland.

Conquest-related change may also be indicated by the anatomical representation data for cattle and caprines from elite sites. By contrast to the patterns for the Saxon period, those for the Norman period show an over-representation of meat-bearing bones, suggesting the receipt of pre-butchered beef and mutton. On the basis that the latter resemble anatomical representations from France, I have suggested that the Normans installed continental provisioning systems in post-Conquest England. This theory is supported by the butchery data. Although butchery styles in France and post-Conquest England were not identical, they were similar in the sense that methods were standardised, relying heavily on meat-cleavers. Saxon methods were seemingly more random and the knife was used more frequently. I propose that the post-Conquest shift in butchery practices is related to the change in provisioning systems and that this, in turn, may reflect the replacement of food rents, which occurred increasingly under Norman rule.

Abolition of food rents would have been possible only with an alternative source of provisions. It may be for this reason that the Norman lords founded numerous towns and markets, although the desire to centralise economic and political power must also have been a driving force. The creation of these markets, laid out with accommodation for traders, must have encouraged the movement of artisans into the urban context. Zooarchaeological evidence for the presence of sedentary craft specialists is provided by the post-Conquest appearance of horning, bone working and tanning deposits, all of which are largely absent before 1050. In the case of bone workers, the promise of a reliable source of raw materials may have seen them gravitate towards towns. This availability of bone and horn in the urban context could equally have caused antler to become obsolete, although there is historical evidence to suggest that it remained a sought after commodity. The possibility that the Normans restricted access to antler is interesting and deserves further examination.

In summary, the effect of the Conquest was not the introduction of the Norman agricultural and economic system; instead the Normans appear to have exploited and intensified the existing set-up. In this sense, the zooarchaeological data concur with historical opinion, that the seeds of change were already sown by the Late Saxon period. However, the idea that the Norman Conquest made no impact on the economy is difficult to uphold. There is much evidence for late eleventh-century change and, in most cases, change appears to have been in a common direction: towards commercialisation. It is difficult to know how the agricultural intensification, craft specialisation, urbanism, and provisioning systems developed in relation to one another. Without further research, this question may not be answerable but from the coincidence of the changes with the Conquest suggests that the Normans were involved to some degree.

Whether these shifts in the animal economy resulted from, or led to, the appearance of new forms of cattle, sheep or pigs is considered in the next chapter.

CHAPTER 4: NEW NORMAN BREEDS? STUDIES IN ANIMAL SIZE AND CONFORMATION

Figure 46: Cattle, sheep and pig corbels from twelfth-century Kilpeck Church, Herefordshire (Photos by R. Jones)

Domestic animals are essentially artefacts, their form and temperament modelled not simply by environmental conditions but by human choices. The human-animal relationship involved in the selection and breeding of stock is complex and, by necessity, close since years of improvement can be undone within a single generation. Selective breeding is always undertaken for a reason, be it economic or social, and husbandry choices have the potential to reveal much about the ideology of the humans responsible for them. We see this clearly from the proliferation of breeds in the nineteenth century, a period in which husbandry became as much an elite pastime as an economic necessity. And it is to the post-Medieval squires that the emergence of true breeds should be attributed; prior to the seventeenth century the Gloucester Spot pigs, Southdown sheep and Longhorn cattle were present in undeveloped forms only.

In the absence of distinct breeds, any attempt to assess the impact of the Normans on domestic stock may seem futile, especially when based on zooarchaeological evidence that lacks the details – coat colour, stance, the distribution of meat and fat – that are so often used to characterise breeds. For these reasons the term 'breed' is actually a misnomer when applied to archaeological material, the best we can hope to achieve is the detection of animal 'types'. For instance, studies of Iron Age and Roman animal bones from across Europe have highlighted gross increases in the size of cattle, sheep and horses. These were roughly coincident with areas and periods of Roman occupation and it has been suggested that the Romans either imported or raised-up new animal types (Teichert 1984). Unlike the Roman Conquest, most agricultural historians believe that the Norman Conquest had little impact on the animal breeds of Saxon England, Trow-Smith (1957, 46) concluding that there was no general importation of livestock following 1066. However, it is also widely accepted that the rural economy endured unchanged from the pre-to post-Conquest period and, as was argued in the previous chapter, this was not the case. There is clear evidence that animal husbandry became more intensive with cattle and caprines being raised increasingly for their secondary products whilst pigs were culled at a younger age. It seems feasible that these shifts in animal management were accompanied by attempts to either introduce or breed animals with superior qualities. Alternatively, the destruction and disruption brought by the Conquest may have interrupted husbandry regimes, causing the degeneration of breed lines.

There is some zooarchaeological evidence to support the idea that a post-Conquest shift in animal stock occurred: several multi-period assemblages such as Faccombe Netherton (Sadler 1991 and *pers. comm.*), North Elmham (Noddle 1980) and Southampton (Bourdillon 1988) exhibit a reduction in the size of Norman period animals. Researchers often link diminution to stock deterioration but this equation may be unfounded, especially if animals were being bred for qualities other than absolute size, for instance fine wool, high milk yields or rapid maturity. Rather than skeletal size, improvement may be detected in anatomical conformation, with particular elements demonstrating a change in shape. In this chapter both size and shape are examined to ascertain whether there is any evidence to suggest that the Normans influenced cattle, sheep and pig stock, for better or worse.

4.1: Cattle

Many livestock historians have sought to explain the evolution of cattle breeds but the developments that occurred during the medieval period remain little understood (see for instance Wilson 1909; Trow-Smith 1957). In his review of medieval stockbreeding, Bökönyi (1995, 43) argued that cattle were fairly uniform across Europe, the zooarchaeological and iconographic evidence both suggesting the widespread presence of primitive, slenderly built animals with short-horned *longifrons* skulls. Certainly metrical data from Britain and the continent are similar, particularly in terms of the broad temporal trends that they demonstrate: most studies for northern Europe have shown that cattle size dropped off gradually from the end of the Roman period with animals recovering their size only in the post-medieval period (Audoin-Rouzeau 1991; Lepetz and Yvinec 1998; Clavel 2001). Whilst this decline is frequently attributed to the withdrawal of the Romans and their breeding know-how, other factors must have played a role because some of the most dramatic diminution occurred in areas without Roman occupation (Bökönyi 1995, 42). Climatic change and its impact on vegetation seems a likely contributor, with the large

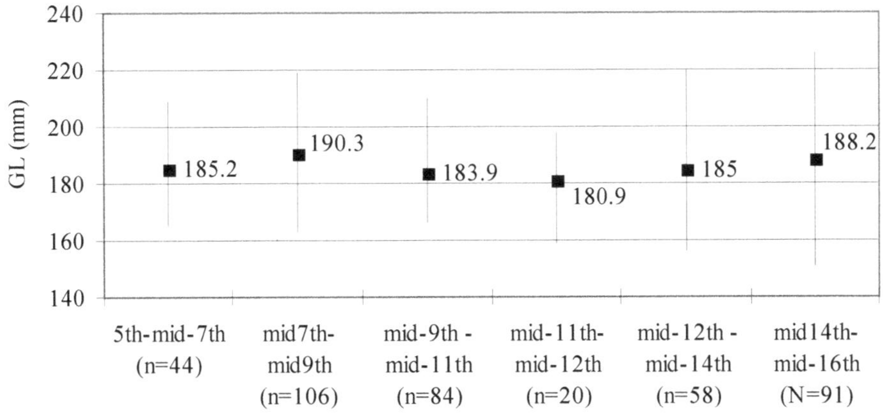

Figure 47: Inter-period variation in the greatest length (GL) – range and mean – of cattle metacarpals

Roman cattle resulting from the warm conditions of the third and forth centuries and the dwarfing of stock mirroring the climatic deterioration that occurred between the fifth and tenth centuries (Lamb 1981; Dark 2000, 27). However, local conditions – whether environmental or socio-economic – must also have exerted some influence because data from different countries demonstrate slight variations within the overall trend.

The situation in medieval England is exemplified by Figure 47 which shows the inter-period variations in the length of cattle metacarpals. The range and mean plots conform to the European trend towards diminution, although shifts in animal size were not equal between all periods. For instance, data for the mid-seventh/mid-ninth century actually suggest a short-lived increase in cattle size. Most of the specimens dating to this period derive from emporia or *wic* sites, supporting Bourdillon's (1994, 123) idea that these trading settlements were able to tap good resources from the countryside. For the period under consideration, size decline between the mid-seventh/mid-ninth and mid-ninth/mid-eleventh century is perhaps the most dramatic. Interestingly, it is coincident the introduction of the open-field system, which reduced the amount of quality land available for pasturing stock (Astill 1997, 205). It seems feasible that as premium grazing diminished, so too did the size of the cattle due to nutrient deficiency. Similar conclusions have been drawn by Audoin-Rouzeau (1991) for France, and Berke (1997) for Germany.

In the previous chapter it was seen that the frequency of cattle relative to caprines also decreased at this Mid- to Late Saxon junction, possibly because of a move towards plough sharing. If the open-field system reduced the number of plough oxen required, this may offer an additional explanation for the inter-period decline in cattle size: that fewer large males are represented within assemblages. To some extent, this possibility can be checked. Studies by Higham (1969) and Thomas (1988; 1989) have shown that cattle metapodia, in particular metacarpals, are sexually dimorphic: those of females are short and slender, with males being short and wide, whereas the metapodials of castrates are long and slender. Figures 48a and 48b show shape indices plots for the Mid- and Late Saxon cattle metacarpals. The plot for the Mid-Saxon cattle appears to fall into two discrete clusters, which might reasonably be interpreted as a male-female division. By contrast, the plot for the Late Saxon cattle is less clear-cut, with no obvious clusters. Albarella (1997b) has demonstrated that breed differences may mask sexual variation. With this in mind it is tempting to see the ill-defined plot for the mid-ninth to mid-eleventh century as reflecting the presence of numerous cattle types, a departure from the generic cattle forms of the mid-seventh to mid-ninth centuries. If this is the case it is significant in terms of breed history but it does not assist the recognition of sexual composition.

In order to highlight inter-period shifts in herd structure without the impact of breed-based variation, Figures 49a

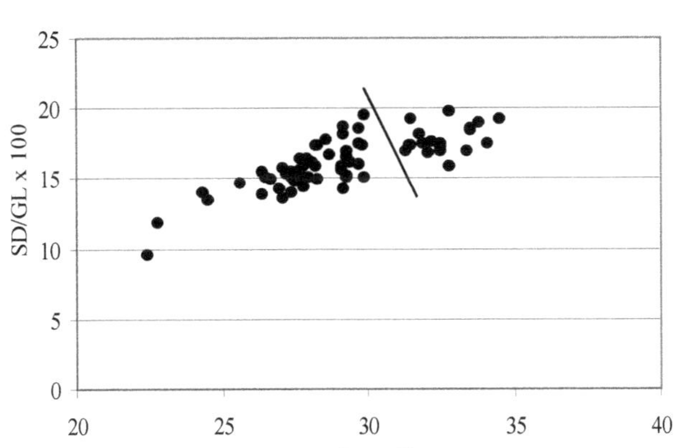

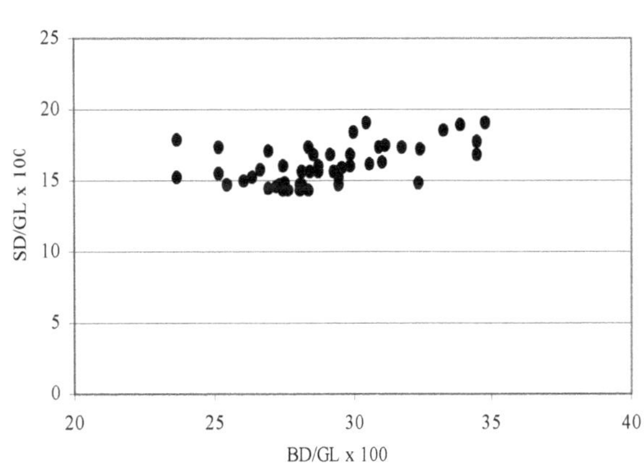

Figure 48: Plots of shape indices (smallest diameter (SD) divided by greatest length (GL) x 100 against the distal breadth (BD) divided by greatest length (GL) x 100 for mid-seventh to mid-ninth century (left) and mid-ninth to mid-eleventh century cattle metacarpals (right)

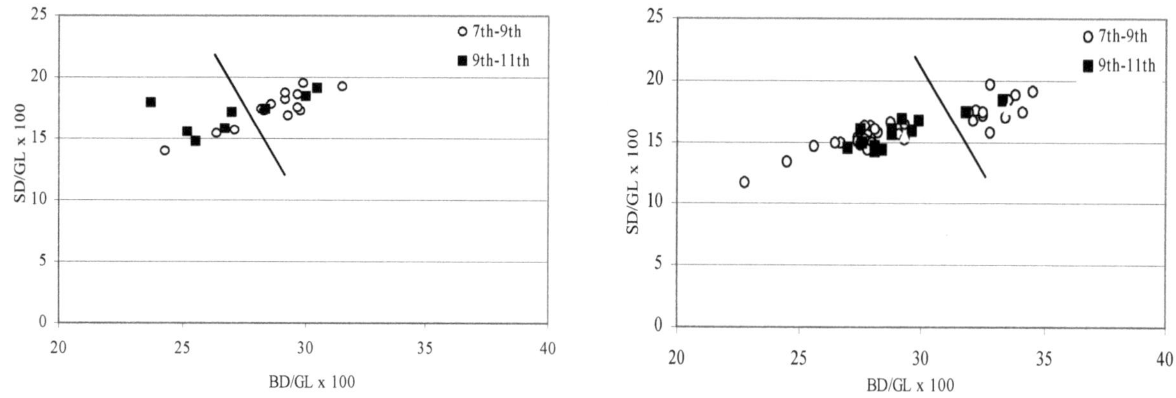

Figure 49: Plots of shape indices (smallest diameter (SD) divided by greatest length (GL) x 100 against the distal breadth (BD)divided by greatest length (GL) x 100) for cattle metacarpals from North Elmham (left) and Southampton (right)

and b show Mid- and Late Saxon cattle from two multi-period settlements: North Elmham and Southampton. By considering the two assemblages separately, both graphs demonstrate clearer bimodality and, if it is accepted that the clusters represent males and females, both suggest a Mid- to Late Saxon shift away from the representation of males towards an increase in females. Taken together the male to female ratio moves from 1:1 to 1:3, certainly a shift that could have contributed to the observed reduction in cattle size.

As far as can be determined from the small data samples for the post-Conquest period, herd composition remained fairly static through to the later medieval period, with a predominance of females. A change in sex structure cannot, therefore, be responsible for the pre- to post-Conquest size decline. Neither can climate be cited as a cause since conditions across Europe are thought to have improved considerably from the tenth century onwards (Lamb 1981; Dark 2000, 28). The agricultural intensification of the post-Conquest period, which reduced further the pasture available, may be responsible for producing the smallest cattle seen since the Iron Age but it is equally possible that social factors played a role. The impact of war and pillage on cattle populations should not be underestimated, especially since the preferred meat of armies was beef, hence 'Beefeaters' (Bökönyi 1995, 42). The Bayeux Tapestry depicts William's men raiding Hastings for cattle, the impact of which could be twofold. First, the indiscriminate slaughter of cattle may have had an adverse affect on stock quality. Second, harrying was often followed by famine and in such periods we might imagine that people would have sacrificed prized breeding animals to feed themselves. The Anglo-Saxon Chronicle records the incidence of famine years and there does appear to be an inverse correlation between increased famine years and decline in cattle wither heights (see Sykes 2001).

When the evidence is viewed together, it would seem that the observed inter-period size decline cannot be attributed to any single cause, more probably occurring in response to a combination of factors. The effects of the Conquest may have contributed to the degeneration of cattle stock but against a background of gradual diminution this is difficult to conclude with confidence. Currently, it is not possible to determine whether new cattle types were introduced to post-Conquest England. The Late Saxon period would appear to be the point at which different metacarpal shapes began to appear, perhaps indicating the emergence of regional types. Far larger samples will be necessary before it is possible to detect the impact of the Normans on cattle stock.

4.2: Sheep

Of the main domesticates, research into the ancestry of the sheep is probably the most complete with historical (Trow-Smith 1957; Ryder 1964, 1983), iconographic (Armitage and Godall 1977), fibre (Ryder 1984) and osteological (Noddle 1978; O'Connor 1982) studies having been undertaken. The general consensus is that across Europe medieval sheep took one of four fairly primitive forms: a small, horned prehistoric type akin to a modern Soay; a slightly improved type of the same form with a coarse-wool coat, perhaps similar to a Shetland; a hairy and horned blackface, the ancestor of modern hill sheep; and a longer woolled polled variety, which may have been a Roman import (Trow Smith 1957, 231). Fine

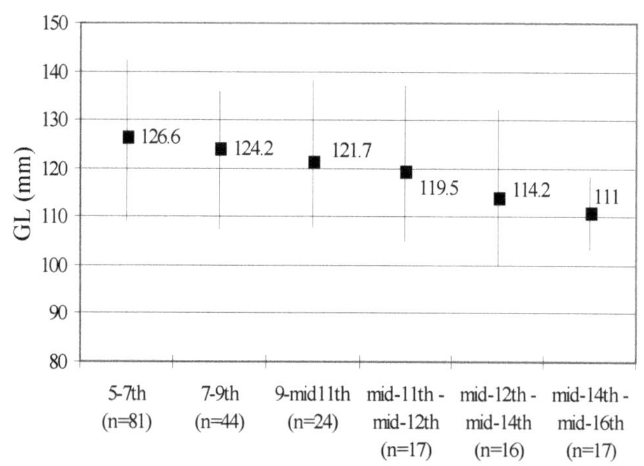

Figure 50: Inter-period variation in the greatest length (GL) – range and mean – of sheep metacarpals

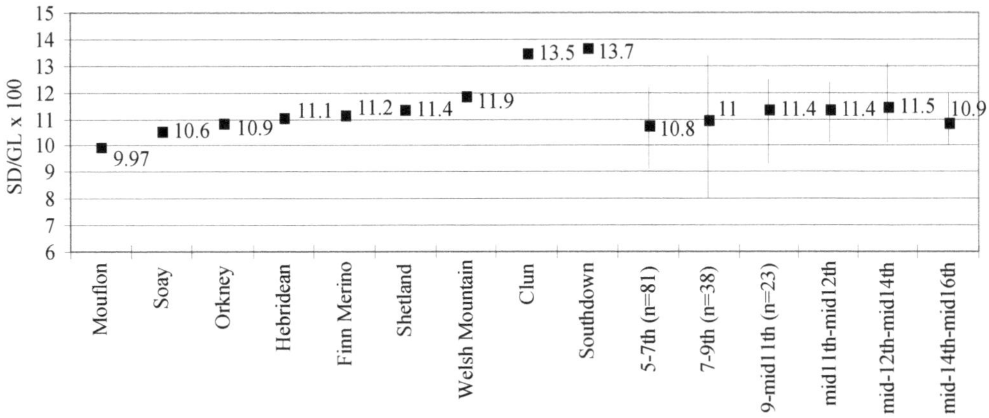

Figure 51: Inter-period variation in the shape indices (smallest diameter (SD) divided by greatest length (GL) x 100) of sheep shown against O'Connor's (1982) indices for modern sheep of known breed.

wool sheep are thought to have developed in the Middle East some time around AD 1000 (Ryder 1984) but by the thirteenth century these types were being imported to Britain from Spain (Trow-Smith 1957, 113). Given the importance of the medieval wool industry, this movement of animals is unsurprising but did the Normans utilise their connections with the Mediterranean to improve the British stock at an earlier date? The zooarchaeological evidence may help to answer this question.

Figure 50 shows the inter-period variations in sheep metacarpal length, demonstrating a steady drop-off in size from the fifth to the mid-sixteenth century. It is tempting to conclude that this diminution represents degeneration in stock quality; however, O'Connor (1982b) has demonstrated that this interpretation may be ill-founded since improved sheep are frequently smaller than their primitive counterparts. He suggests that metacarpal shape may be a better parameter for improvement, slenderness characterising primitive breeds with robustness a trait of developed forms (Figure 51).

Viewed against O'Connor's (1982b) data for modern sheep the archaeological metacarpals demonstrate a gradually increase in robusticity, indicating improvement, through the course of the medieval period. A substantial shift towards robusticity occurred between the Early and Mid-Saxon period, with specimens moving from Orkney-type forms to metacarpal shapes that are seen in modern Hebridean populations. A second more dramatic shift took place between the Mid- and Late Saxon period when sheep attained a level of improvement comparable with the modern Shetland. Because sheep metacarpals are sexually dimorphic, a change in flock structure – with a greater presence of robust male specimens – could feasibly account for the observed change; however, the data (see Sykes 2001) suggest continuity in male-female ratios and up-breeding seems the more likely explanation. From the mid-ninth century there is surprisingly little variation in metacarpal shape until the mid-fourteenth to mid-sixteenth century when they appear to revert to slender forms. This last shift may reflect stock deterioration caused by the murrain that devastated the sheep populations of the fourteenth and fifteenth centuries, although the data (Sykes 2001) indicate a shift in flock structure towards an overwhelming dominance of females, which may be partially responsible.

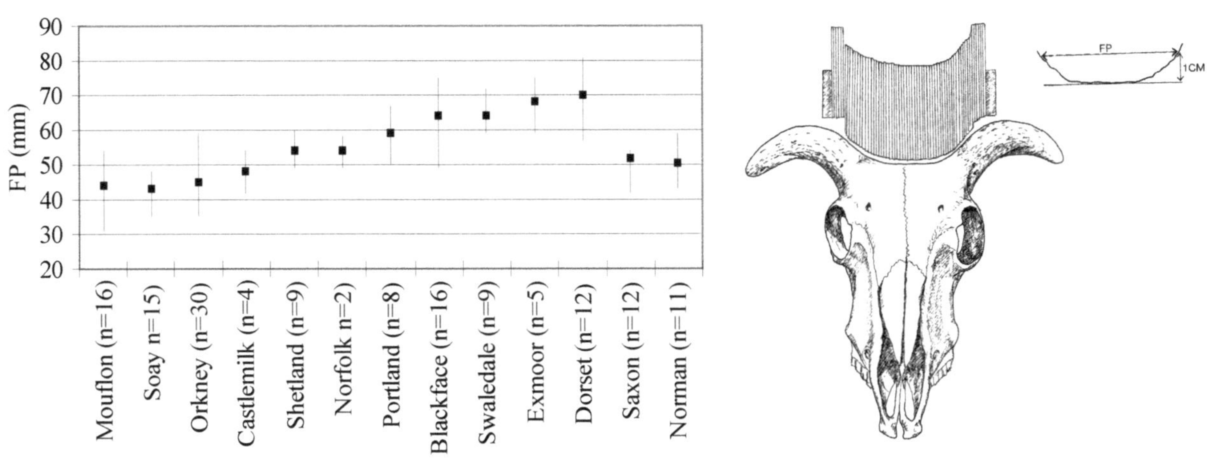

Figure 52: Range and mean plots for the Frontal Profile (FP) measurements for modern sheep breeds compared to those for Saxon and Norman period animals. The FP is taken using a profile gauge and the measurement is taken from a tracing of the profile, 1cm up from the profile base. For further details see Sykes 2001 and Sykes in prep.

It is notable that there is very little difference in the size or shape of Late Saxon and Norman period sheep metacarpals (Figure 50). A similar pattern of continuity is demonstrated by skull/horn-core morphology: studies of frontal profiles (see Figure 52), which are little affected by sex but linked closely to skeletal improvement (Sykes in prep.), indicate that both Late Saxon and Norman period sheep were bred to a level somewhere between modern Castlemilk Moorit and Shetland forms. As such, it would seem that the Norman Conquest had little impact on sheep types; certainly there is no evidence that novel forms were bred or introduced but neither is there any indication that stock was affected adversely.

4.3: Pigs

It is widely accepted that medieval pigs were little different in form to wild boar – long-legged, bristle backed and prick eared – and that it was not until the later medieval period that attempts were made to improve stock (Trow-Smith 1957; Wiseman 1986; Albarella 2006). That said, most researchers would also acknowledge that the origins of modern pig breeds are obscure, the development of lop-eared types and the date at which imports from the East began are unknown. Zooarchaeological studies of the situation have been hampered by the lack of metrical data for post-cranial bones, restricting knowledge about body conformation. On the other hand, lack of measureable bones has encouraged researchers to take dental measurements as standard. These offer an advantage over post-cranial metrics because tooth form, in particular molar width, is little affected by age, sex or residual variation, thus any dramatic variations can be attributed to genetic diversity (Payne and Bull 1988). Work by Sylvia Warman (2000) has demonstrated that modern pig breeds can, to some extent, be differentiated on the basis of their tooth form, molar widths being linked to level of improvement. If the Conquest did impact on pig stock, it seems possible that this may be detectable through dental analysis.

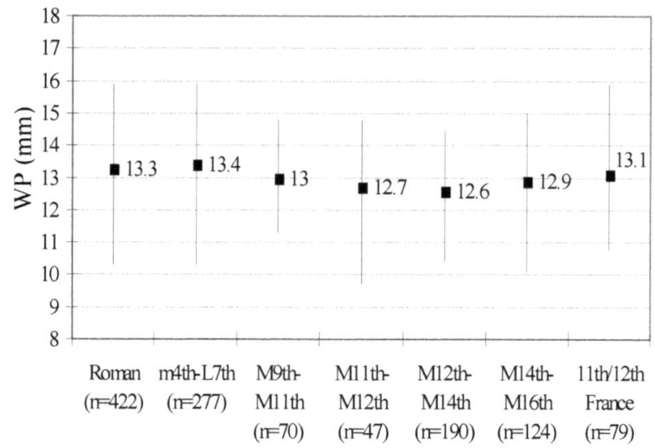

Figure 53: Inter-period variation in the width of posterior (WP) of pig second permanent mandibular molars

For this study, large numbers of dental measurements from sites dating between the first and mid-sixteenth century were acquired, and the range and mean plots for posterior width (WP) measurements for second permanent mandibular molars are shown in Figure 53. The data suggest a steady decline in mean tooth width until the mid-fourteenth to mid-sixteenth century when larger specimens become represented and the average mean increases. Taken at face value this would appear to indicate a gradual improvement of stock. However, it is important to recognise that the samples are likely to include some specimens belonging to the larger wild boar, whose presence may obscure trends concerning domestic animals. It is uncertain when the British wild boar went extinct but Pierce's (n.d.) work on dental measurements suggests that populations were dwindling by the mid-ninth century with few free-living individuals available by the end of the twelfth century. It is probable that much of the size change demonstrated in Figure 53 represents the decline of the wild boar rather than shifts in domestic populations. A better impression of changes

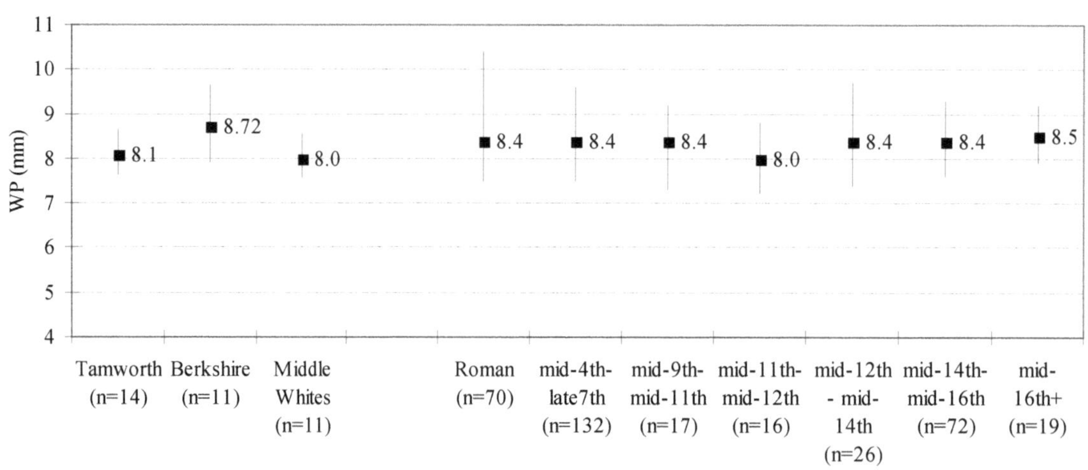

Figure 54: Inter-period variation in the width of posterior (WP) of pig deciduous fourth pre-molars, shown against Warman's (2000) measurements for modern pigs of known breed.

in domestic stock is perhaps provided by Figure 54 which shows posterior width measurements for the fourth deciduous premolar: the theory being that wild boar are less likely to be represented as juvenile animals (see page 34). Here there is considerable uniformity in the width measurements, the samples from all but two periods having a mean posterior breadth of 8.4mm. Interestingly it is the mid-eleventh/mid-twelfth century sample that exhibits a mean breath of 8.0mm and a range that includes specimens smaller than in any preceding or succeeding phase. Because teeth are little affected by environmental or nutritional factors this short-lived decline in tooth size most likely represents a genetic difference. When considered in conjunction with the ageing data, which demonstrate a shift towards earlier slaughter (page 34), it seems possible that the Normans did introduce a new fast maturing pig type. However, the pigs from Vatteville-la-Rue did not exhibit similarly slender tooth forms and other factors, such as a move towards sty husbandry and increased selective breeding, may have been responsible for the short-lived change in tooth size seen in post-Conquest England.

4.4: Conclusion

Whilst the picture remains far from clear, the data presented in this chapter suggest that animal size and conformation did change through the course of the medieval period. Metacarpal measurement for both cattle and sheep demonstrate a gradual decline in bone length but the cause and meaning of the diminution remain uncertain. Traditional explanations would cite poor husbandry or climatic deterioration as responsible and this may well be correct but here other factors – changes in herd composition or increased incidence of human famine – have also been suggested.

It is important to recognise that the reduction in animal size need not, however, reflect a degeneration of stock quality; the data presented here suggest that decline in cattle and sheep size was concomitant with skeletal improvement. In both cases the most substantial shifts in metacarpal morphometry occurred between the Mid- and Late Saxon periods, suggesting that the mid-ninth to mid-eleventh centuries witnessed not only dramatic social reorganisation but also innovative animal husbandry. The events of 1066 seem to have been less influential; sheep types apparently enduring unchanged, although it is noteworthy that cattle metacarpals reached their smallest size in the mid-eleventh/mid-twelfth centuries, perhaps due to Conquest-related factors.

Changes in the morphometry of metacarpals cannot be taken as definite evidence for the change in cattle or sheep types; the development of post-cranial bones is influenced by too many factors other than genotype. Teeth are more reliable indicators of genetic change, and the mid-eleventh/mid-twelfth century reduction in the posterior width of pig Dp^4s is the most persuasive evidence that the Normans impacted upon animal stock. However, it is unwise also to rely on dental metrics in isolation from post-cranial measurements, and far more data from a wider range of skeletal elements will be required before the Norman impact on animal stock, or lack of the same, can be assessed. Nevertheless, on the basis of these pilot studies it seems that there exists real potential for detecting changes in animal type, even where breed historians suggest there are none to be found. Herein lies the benefit of zooarchaeology; the human-animal relationships it charts provide useful insights into the mundane, yet socially telling, day-to-day activities that went unrecorded by medieval historians. With this in mind it seems highly probable that a wealth of information might be obtained when evidence from animal bones and documentary sources can be used in unison. The next two chapters will do precisely this, attempting to marry the large corpus of historical evidence for wild resource exploitation with the zooarchaeological data for fishing, fowling and hunting.

CHAPTER 5: THE NORMAN IMPACT ON WILD RESOURCE EXPLOITATION

'You had heard that the birds of the air and fish of the sea are common to all, but all those that are hunted, wherever they fly, belong to the royal estate' John of Salisbury, *Policraticus* (1.4 I)

Of all the Conquest-related changes, the Norman's impact on wild resources is probably the most notorious, numerous history books detailing how Forest Law was applied to vast areas of common land, denying peasants the right to take wild animals that were now deemed to be the property of the King (Petit-Dutaillais 1914, 173). The conquered population appear to have viewed Forest Law with contempt, the Anglo-Saxon Chronicle devoting approximately one third of William the Conqueror's obituary to complaining about it. Certainly the post-Conquest situation appears to have been fundamentally different to that of pre-Conquest Britain where perceptions of game followed the Roman concept that, until caught, wild animals were *res nullius*, nobody's property. This shift in hunting laws is significant because, as Wickham (1994, 161) has argued, hunting rights reflect socio-political change. Within most farming societies the significance of hunting extends beyond the desire to obtain meat; instead it is a symbolic action, a social and political 'performance' (Marvin 2001; Hamilakis 2003). Whilst different groups may employ hunting for similar reasons, the purpose and meaning of hunting is not cross-culturally uniform: for instance Hamilakis (2003) in his study of Neolithic and Bronze Age Greece has proposed that hunting was linked to perceptions of time and space, and employed as a strategy for negotiating gender and social power; Kessinger (1989) has interpreted hunting in Cashinahua society as embodying masculinity; and Marvin (2001) has explained modern English fox hunting in terms of the social and cultural identity of rural communities. The evolution of medieval hunting law and rituals has recently been examined from a historical perspective by Marvin (2006) but it seems possible that a zooarchaeological study of the wild animals themselves may provide additional information concerning the social impact of 1066.

Several zooarchaeologists have commented upon the presence of wild animals in medieval assemblages and the general consensus is that wild animal exploitation became an important activity only in the post-Conquest period (Noddle 1975; Grant 1981, 1988a). Since the publication of these reports, more data from sites across England have become available. By considering fifth to fourteenth-century developments in wild resource exploitation, this chapter will explore whether the Normans were, indeed, responsible for changing the way in which wild animals were perceived and utilised. Particular attention will be given to site-specific variations in taxa frequency and representation, to see if fishing, fowling and hunting became more exclusive in the post-Conquest period.

5.1: Animals from the Water

Within the last few years, analysis of medieval fishing and fish-eating has undergone something of a renaissance and there exist now several excellent and in-depth surveys of the historical and zooarchaeological evidence (see for instance Barrett *et al.* 2004; Locker 2001; Woolgar 2000; Serjeantson and Woolgar 2006). These studies have demonstrated how the fishing industry developed dramatically around AD 1000 – Barrett *et al.*'s (2004) 'fish-event horizon' – with further intensification during the thirteenth and fourteenth centuries. The Norman Conquest fell centrally within this time frame, a fact that has been noted and addressed variously by different authors: Serjeantson and Woolgar (2006) utilise 1066 as a useful division for considering their data; Woolgar (2000) gives greater weight to the Conquest, suggesting that the Normans introduced a fashion for rigorous lay fasting, but the possibility that this impacted on the fishing industry is rejected outright by Barrett *et al.* (2004, 629) on the basis that the fish event horizon pre-dates 1066 and that there was no precedent for the large scale consumption of marine fish in northern France. Both of Barrett *et al.*'s (2004) arguments are valid; the Conquest cannot be attributed with all economic changes that occurred in medieval England and, as was shown in Chapter 2 (page 24) fish were not a key feature of northern French assemblages before the thirteenth century. However, it seems simplistic to conclude that no Conquest-related change occurred because levels of fishing and fish consumption had already increased before 1066. It is worth critically reviewing the evidence before the hypothesis of Conquest-related change is 'laid to rest' as Barrett *et al.* (2004, 629) suggest it should be.

Figure 55, which is based on presence/diversity data, provides a proxy measure of the inter-period variation in

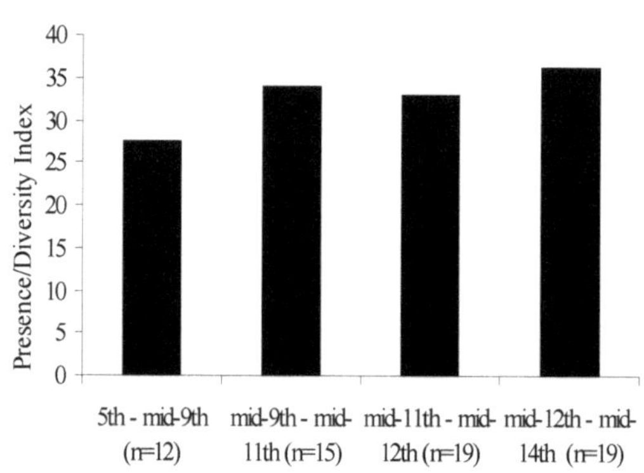

Figure 55: Inter-period variation in the presence/diversity of fish

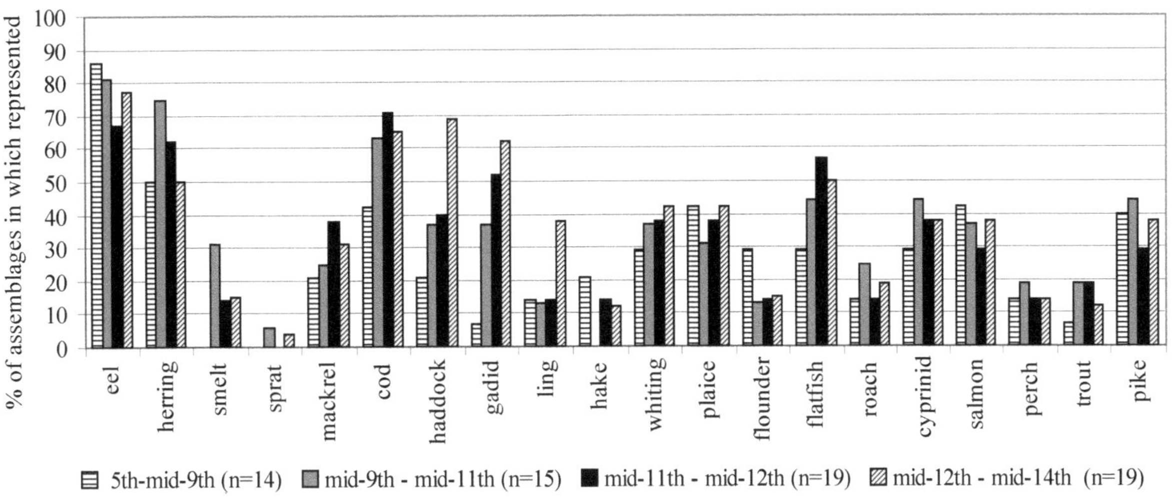

Figure 56: Inter-period variation in fish taxa representation

fish exploitation. It can be seen that, as other researchers have already demonstrated, the major increases in fish exploitation occurred between the mid-ninth and mid-eleventh century, with a second increase after the mid-twelfth century. It is quite clear that no dramatic shift in levels of fish exploitation occurred during the mid-eleventh and mid-twelfth centuries, validating Barrett *et al.*'s (2004) suggestion that the Normans cannot be held responsible for the 'fish event horizon'. Previous work by other authors is also supported by Figure 56 which shows how the exploitation of the different taxa changed through time, these data are summarised in Figure 57 where the fish are grouped into freshwater, migratory and marine species. During the mid-fifth to mid-ninth centuries marine fish were the least well represented, whereas migratory and freshwater species – notably eel (*Anguilla anguilla*), plaice/flounder (pleuronectid), salmon (*Salmo salar*) and pike (*Esox lucius*) – were more numerous. This taxa range is suggestive of localised exploitation, whereby traps, such as those reviewed by O'Sullivan (2003, and see also Salisbury 1981), were constructed in tidal areas to obtain estuarine species. As such, the zooarchaeological evidence supports Bede's *Life of St. Wilfrid* which claimed that he arrived among the south Saxons to find them starving because, although they caught eels, they did not know how to fish off-shore. According to the story, St Wilfrid taught the people of Sussex how to use nets in the sea and 'they soon took three hundred fish of all kinds' (Miller 1890, 304).

Whether or not St Wilfrid was responsible, the mid-ninth to mid-eleventh centuries did see increased exploitation of marine species, the rise in herring (*Clupea harengus*) representation being a particular characteristic for the period (Figure 56). But the increased presence of freshwater and migratory species (shown in Figure 57) suggests that fishing intensified across the board: it isworth noting here that because the graphs are not relative frequencies (each bar is an independent value) the increase in freshwater and migratory species is not masked by the greater rise in marine fish, a problem noted by Barrett *et al.* (2004, 628). There is certainly more evidence, in the form of place-names and Saxon charters, for large-scale freshwater fishing. A survey of Tidenham, Gloucestershire, refers to the *cytweras* (basket weirs) and *haecweras* (hackle weirs) of the Severn and Wye estuaries. Other Saxon charters and place-names also commonly mention *haec(c)*, meaning hatch/sluice or fish trap (Bond 1988, 86). It is also from the Late Saxon period that fishing begins to be mentioned in Old English literature; *Aelfric's Colloquy* providing a charming insight into the thoughts of an early eleventh-century fisherman: incidentally the range of species he cites are very similar to those found in the archaeological record (Swanton 1975, 110). One species that Aelfric's fisherman does not mention is cod (*Gadus morhua*), a species that in pre-Conquest England was yet to be named (Sayers 2002, cited in Barrett *et al.* 2004, 623). Figure 56 shows that some cod was obtained in the Late (and even Mid-) Saxon period but it seems likely that many of these fish were imported: the *Saga of Egil* records that dried fish, probably cod, was imported to England from Norway by the tenth century (Seebohm 1952, 109).

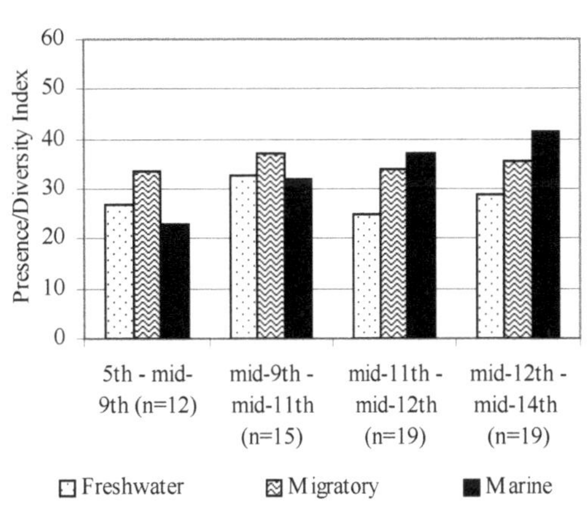

Figure 57: Inter-period variation in the presence/ diversity of freshwater, migratory and marine fish

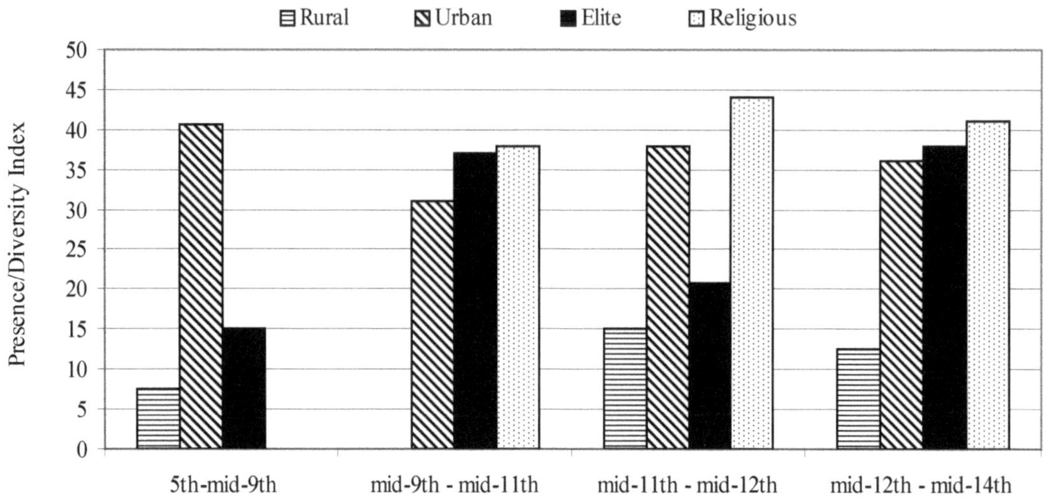

Figure 58: Inter-period and inter-site variation in fish Presence/Diversity

According to Serjeantson and Locker (1997), references to the capture and trading of cod increase from the mid-eleventh century, also the point that cod becomes the best represented species archaeologically (Figure 56). The fishing technology used to obtain gadids differed from that used to procure other marine fish. Rather than employing large nets, used to catch shoal fish such as herring, cod were obtained using hooks and long lines. According to Vince (1994, 114), large fishing hooks appear at the Billingsgate site for the first time in deposits post-dating 1050. Although this is roughly coincident with the Norman Conquest, it is clear that line and hook fishing developed in Norway and that it was the Vikings, who settled on the Orkney and Shetland Islands, who were responsible for introducing the technology to England (Serjeantson and Locker 1997).

Between the mid-twelfth and mid-fourteenth century the representation of cod declines whereas the presence of haddock (*Melanogrammus aeglefinus*), whiting (*Merlangius merlangus*) and ling (*Molva molva*) increases dramatically, boosting the overall representation of marine fish shown in Figure 57. Coincidentally, it is in the texts of the fourteenth century that the first references to ling appear (Burnley 1992, 438). In line with the growth of the marine fishing industry, professional fishmongers also rose in abundance: one is recorded for Winchester in 1148 with many more being present by the early fourteenth century (Keene 1985, 259).

This brief review of the evidence supports Barrett *et al.*'s (2004) theory that the Normans were not responsible for the 'fish event horizon'; however, it remains a possibility that the Conquest caused subtle changes to fish distribution and consumption, with different sections of society eating different fish species in different quantities. To consider the social, rather than overarching economic, factors that governed fish exploitation requires that the zooarchaeological data are considered at the level of site type. Social analyses of fish data have traditionally been limited due to lack of samples, and indeed sample sizes in this study are woefully small, but it is hoped that this attempt to consider Conquest-related changes in fish procurement and consumption may offer hypotheses that can be tested as more information becomes available.

Inter-period and inter-site variations in the representation of 'fish' are shown in Figure 58; again each bar is independent and depicts presence/diversity data rather than relative frequencies based on absolute numbers. It shows that levels of fish representation were far from uniform, the variation between sites apparently more dramatic than that between periods. As might be predicted, rural assemblages always demonstrate low levels of fish representation whereas urban assemblages, regardless of period, are characterised by an abundance of fish. These patterns fit with evidence from stable isotope analysis of human bones, which suggests that the diet of urban populations contained a significantly greater marine component than that of the rural population (Müldner and Richards 2006, 235).

Surprisingly the urban assemblages dating between the mid-ninth and mid-eleventh centuries appear to show a decline in fish representation and herein lies the problem of using a presence/diversity index, it takes no account of absolute numbers. Barrett *et al.* (2004) have shown convincingly that the consumption of marine fish increased dramatically during the Late Saxon period, with a contemporaneous decline in the representation of freshwater and migratory species. Figure 59 suggests that the representation of these last groups did drop on urban sites and it is probably due to their decline that the level of overall 'fish' representation appears low according to the presence/diversity data. Nevertheless, it is interesting that freshwater and migratory species became less abundant on urban sites of this period, especially since they are particularly well represented in mid-ninth to mid-eleventh century assemblages from elite sites and religious houses (Figures 60 and 61). It is tempting to see this inter-site variation as evidence for elite-imposed

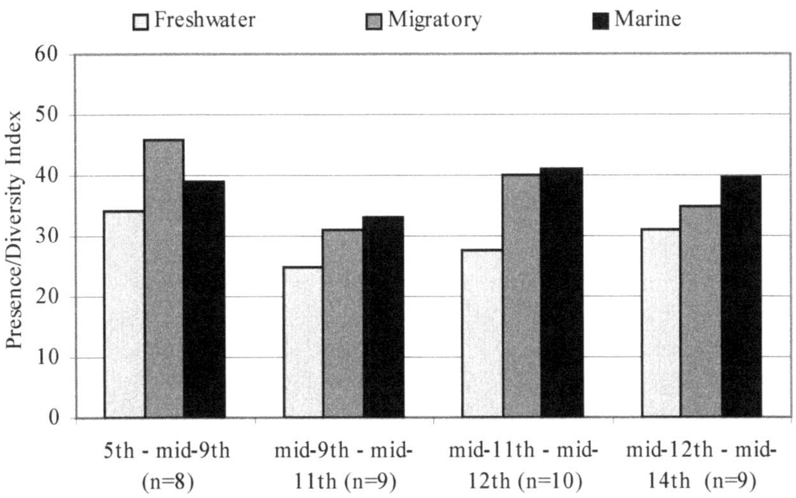

Figure 59: Inter-period variation in the presence/diversity indices of freshwater, migratory and marine fish in urban assemblages

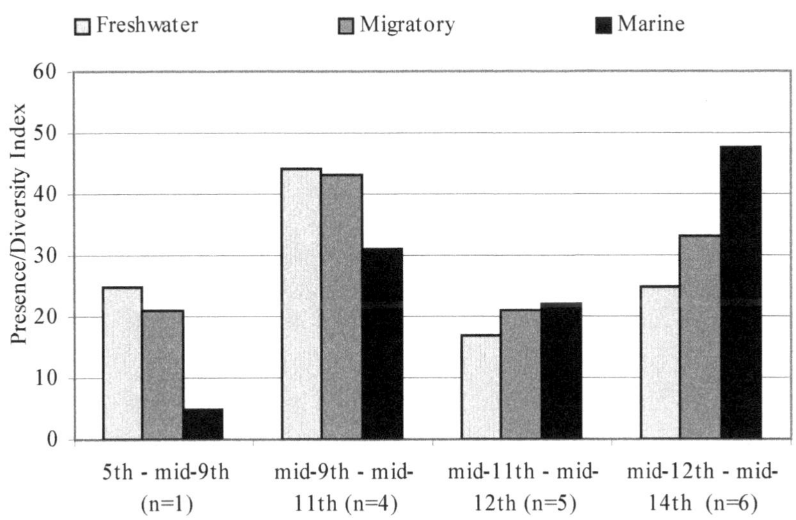

Figure 60: Inter-period variation in the presence/diversity indices of freshwater, migratory and marine fish in elite assemblages

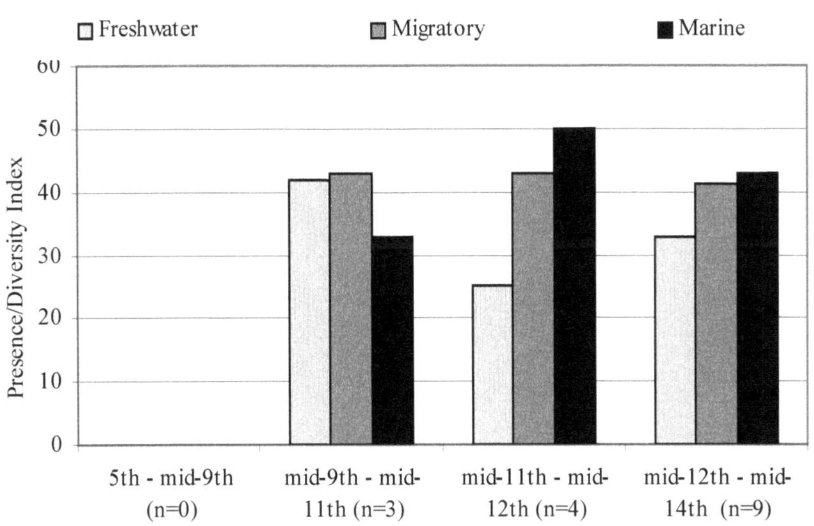

Figure 61: Inter-period variation in the presence/diversity indices of freshwater, migratory and marine fish in assemblages from religious houses

restrictions, the new thegnly class attempting to create and secure luxury resources by taking control of freshwater fisheries (Fleming 2000, 8; Hoffman 1996, 653). This possibility is born out by the historical evidence, the eleventh-century charter for Tidenham, Gloucestershire, insisting that that 'every rare fish' belonged to the lord and could not be sold without his permission, whereas every alternate fish caught within 30 hides of the whole estate had to be given to the abbot of Bath Abbey (Robertson 1939, 109).

Data for mid-ninth to mid-eleventh-century religious houses (Figure 61) suggest that the ecclesiastics consumed more fish than any other section of society, a fact most plausibly attributed to the Benedictine Rule which forbad ecclesiastics to consume quadruped flesh during periods of fast (for monastics between 190-200 days of the year) but allowed fish to be eaten (Knowles 1940, 458). The apparent absence of fish on fifth to mid-ninth-century religious sites should not be taken literally, since neither the report for Wicken Bonhunt (Crabtree, 1996) nor that for North Elmham (Noddle 1980) provided details of fish remains. Hagen (1998, 84), however, has argued that ecclesiastics did not consume fish in large quantities until the Late Saxon period, when monasteries began to acquire grants of fisheries. The earliest documents defining monastic fishing rights date to the late seventh century but they become more common from the tenth century onwards: for instance in his will Theodred, bishop of London (942-951) left fishing ground in Norfolk to St Pauls (Whitlock 1955, 553), and Thorney Abbey, founded in 973, was endowed with 'a quarter of the lake called Wytlesmere and two fisheries' (Bond 1988, 80). By 1086 religious communities were also receiving supplies of fish through food rents. The Domesday Book records numerous herring tolls that were paid and received in Sussex, Kent and Suffolk, with higher frequencies of eels and salmon noted for Yorkshire and Lincolnshire (Dulley 1969, 36; Lennard 1959, 250). There is even some evidence from early charters, such as those for Rye Abbey, to suggest that fish rents were retained where food renders were otherwise relinquished (Round 1899, 73). Certainly Figure 58 suggests that monastic fish consumption reached its peak in the mid-eleventh to mid-twelfth centuries: was this as a result of the Conquest? It was, after all, only under their first Norman prior that the monks of Winchester were weaned from their diet of meat on to one of fish (Knowles 1963, 459). It seems possible that, in an attempt to demonstrate superior piety, the Norman monks wanted to be seen eating more fish than their Saxon counterparts; this will be discussed further in chapter 8.

If, as Serjeantson and Woolgar (2006, 104) have suggested, the Norman monks had adopted 'a new zeal for clerical abstinence' the same cannot be said for the lay aristocracy. According to Figure 58, fish presence/diversity crashes on elite sites dating to the mid-eleventh to mid-twelfth century. This does not appear to be an artifice of sample size since Serjeantson and Woolgar (2006, 129) have shown that the ratios of fish to birds on several elite sites – Carisbrooke Castle, Middleton Stoney and Faccombe Netherton – are particularly low suggesting a real lack of fish consumption. The drop in fish representation is less of a surprise when it is remembered that levels of fish diversity/representation were equally low (with an index of just 13) in ninth to twelfth-century France, and it seems possible that the apparent post-Conquest drop in elite fish consumption reflects the dietary preferences of the new Norman lords.

Norman dietary preferences may equally account for the post-Conquest increase in cetacean representation. Gardiner (1997) has demonstrated that although porpoise and dolphin bones are found frequently in assemblages pre-dating 1066, the remains of whales become common only from the mid-eleventh century. It is clear from historical evidence that whaling was not customary in Anglo-Saxon England, Aelfric's fisherman rejecting the idea of whaling as too dangerous (Swanton 1975, 110). By contrast, whaling was widespread in northern France, from where salted whale meat (*craspois*) was regularly imported to England: the laws of Aethelred laid down tolls for the importation of *craspois* and wine from Rouen to London (Robertson 1939, 73).

More commonly associated with the Normans are the establishment of artificial fishponds and the restriction of fishing rights (Rowley 1997, 132). From the eleventh century onwards many areas, such as Holderness, seem to have had their common rights to fishing revoked (English 1979, 209). It is known that ponds were first created by damming natural water bodies but it is uncertain when the concept of artificial fishponds was introduced. Physical remains of fish ponds are notoriously difficult to date and historical evidence for their establishment is scarce before the twelfth century (Aston 1988, 4). At least six natural ponds were in existence in Britain prior to 1066, and two tenth-century charters referring to *Strigan pole* and *Strygan pol*, meaning 'fishpond', have been linked to artificial ponds in Oxford (Bond and Chambers, 1988, 356). The Domesday Book records numerous '*vivaria vel piscinae*' (Bond 1988, 92) and by the end of the twelfth century at least ten royal artificial ponds had been established (Steane 1988, 42).

If the Normans were responsible for increasing the management of, and restrictions concerning, freshwater fish it seems possible that this would be reflected in the zooarchaeological record; elite sites showing an increased consumption of freshwater species with a corresponding decline on sites of lower status. Figures 60 and 61, however, show no such pre to post-Conquest shift, in fact the reverse is true with the representation/diversity of fish increasing for mid-eleventh to mid-twelfth-century urban sites and declining in assemblages from religious houses and elite settlements. It is not until the period between the mid-twelfth and mid-fourteenth century that high-status sites (both lay and ecclesiastical) demonstrate a return to freshwater fish, with cyprinids, salmon and pike becoming well represented. It is also in this period that eels become the dominant taxon on elite sites, perhaps suggesting new aristocratic investment in fish traps: O'Sullivan (2003, 463) has highlighted a significant increase in the construction of fish weirs during the late twelfth and thirteenth centuries.

5.1i: Summary

From the evidence provided above it would appear that fishing techniques developed throughout the medieval period, with localised trapping of freshwater taxa being replaced by wide-scale exploitation and trading of both freshwater and, in particular, marine fish. By the Late Saxon period, the fishing industry was already well established. Provisioning systems must also have been advanced because there is evidence that most sections of society, perhaps with the exception of the rural population, were receiving quantities of fish.

Since most developments had occurred prior to 1066 they cannot, as Barrett *et al.* (2004) have already argued, be attributed to Norman influence. The possibility that the Conquest resulted in the imposition of tighter fishing controls is difficult to ascertain, the zooarchaeological evidence does not support historical sources, showing no obvious shifts to suggest the introduction of fish ponds. If anything, better evidence for resource restrictions is apparent for the Late Saxon period when the representation/diversity of freshwater fish on elite and ecclesiastical sites reached its zenith, whereas levels declined on urban sites. Current zooarchaeological data actually indicate that the post-Conquest aristocracy had little interest in fish consumption. Considering that levels of fish representation/diversity were also low in eleventh/twelfth century France, this together with the increased representation of cetacean remains is perhaps the most credible evidence that the Norman aristocracy did impose their own systems of fish exploitation on Anglo-Saxon England.

5.2: Wild Birds

It is necessary from the outset to be clear about to what the term 'wild' refers, especially since the medieval period saw several wild taxa, for instance swans, doves/pigeons and partridges, become semi-domestic. These managed species have been included in the 'wild' count for both logistical and theoretic reasons. Practically speaking it is either difficult or impossible to separate the remains of wild and managed species, few researchers attempt to do so and thus the information is simply not available. The decision to include these species was also made on the basis that their move from wild to semi-domestic is highly significant, indicating a change in the human-animal relationship. Whilst these 'wild' species have been included in the study, other true wild species have been omitted: members of the crow family and birds of prey have been excluded from the count because, although clearly symbolic creatures, it is uncertain whether they were considered food during the medieval period.

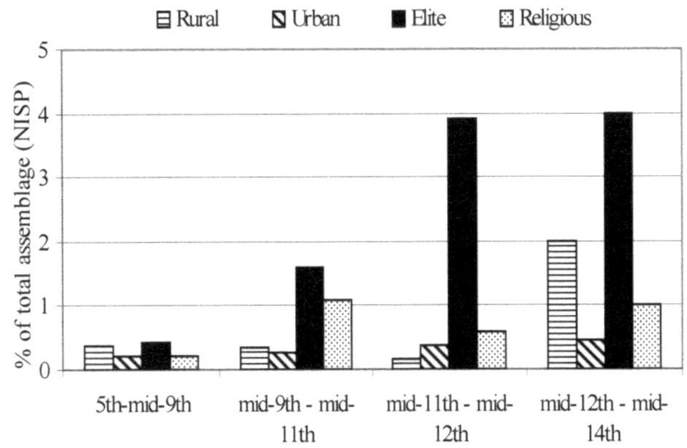

Figure 62: Inter-period and inter-site variation in the representation of wild birds, expressed as a percentage of the total assemblage (NISP)

Few assemblages dating to between the mid-fifth and mid-ninth century contain high frequencies of wildfowl (Figure 62), and game birds can have made little contribution to the diet of any section of society. This is not to suggest, however, that their capture and consumption lacked social meaning, especially since early medieval iconography (Hawkes 1997, 319-23) and funerary deposits (Bond 1994, 134) show an association between hunting birds and the social elite. The origins of hawking are uncertain but it is generally assumed that the technique developed in Asia and spread into Europe during the third and fourth centuries AD (Prummel 1997, 335). By the sixth century, hawking appears to have been an established continental practice and Burgundian law codes describe penalties for the theft or damage of someone else's hawk (Drew 1972: code 84). Historical evidence for hawking in England begins to appear around the eighth century: in 748-752 King Aethelbert of Kent had to ask Boniface to send two hawks from Germany (Kylie 1911, 157). This is coincident with the appearance of hunting birds in high-status assemblages: a near complete skeleton of a sparrow hawk (*Accipiter nisus*) was recovered from the eighth-century levels at Hartlepool monastery (Allison 1988) and a peregrine falcon (*Falco peregrinus*) was recovered from the seventh to tenth-century site of Brandon, Suffolk (Crabtree, 1996). The strong implication must be that the symbolic value of hunting birds was far greater than their role in food acquisition. Whilst no section of fifth to mid-ninth century society ate large numbers of wild birds, the consumption of fowl obtained through falconry, but perhaps not other methods, would have been a powerful sign of social identity. This is indicated zooarchaeologically by the fact that wildfowl are slightly better represented on elite settlements, in particular those on which raptorial birds are present – such as Brandon in Suffolk (Crabtree 1996) and Flixborough in Lincolnshire (Dobney and Jaques 2002) – than in assemblages from sites of lower status, for instance Pennylands, Buckinghamshire (Williams 1993) and Quarrington in Lincolnshire (Rackham 2003).

Clearer status-based variation in wild bird consumption becomes apparent between the mid-ninth and mid-eleventh centuries. All site-types demonstrate a rise in

Site	Date	Information	Reference
Brandon, Suffolk	650-900AD	Partial peregrine falcon skeleton	Crabtree 1996
Hartlepool Monastery	8th century	Sparrowhawk skeleton	Allison 1988
Flaxengate, Lincolnshire	10th-E 11th century	Partial goshawk skeleton from pit F685	O'Connor 1982a
Castle Mall, Norwich	L 9th - 11th century	Four goshawk bones	Albarella *et al.* 1997
Faccombe Netherton, Hants.	980-1070	Partial skeleton of a goshawk	Sadler 1990
Canterbury Cathedral, Kent	late 11th century	6 sparrow hawk bones	Driver *et al.* 1990
Exeter, Devon	12th century	2 sparrow hawk skeletons	Maltby 1979a
Portchester Castle, Hants.	1100-1200	10 goshawk bones from a single individual.	Eastham 1977
Middleton Stony, Oxford	12th-early 13th century	Sparrow hawk skeleton	Levitan 1984a

Table 10: Archaeological representation of partial/complete hunting-bird skeletons (after Cherryson 2002)

wildfowl representation but increases are particularly marked for assemblages from elite settlements. It can be no coincidence that, by comparison to the earlier period there is substantially more evidence for falconry. Several mid-ninth to mid-eleventh-century sites have yielded complete skeletons of hunting birds (see Table 10) and historical references are also more common: there are several Mercian charters which mention the maintenance of trained birds (Sawyer 1968, S 1271); William of Malmesbury recorded that the Welsh paid a tribute of trained hawks to Aethelstan (Whitlock 1955, 474) and the eleventh-century *Aelfric's Colloquy* provides detailed accounts of falconry, suggesting that the technique was now commonplace. Aelfric's Fowler makes it clear, however, that hawking was not the only method he used to catch birds:

> 'I trap birds in many ways; sometimes with nets, sometimes snares, sometimes with lime, by whistling, with a hawk or with a trap'
> (Swanton 1975, 111).

The rise of such professional fowlers, who would have serviced elite and monastic households but also supplied the urban population, must be partly responsible for the Late Saxon increase in wild bird consumption. Nevertheless, such professions would not have become established without incentive and, considering that hunting practices chart social change, these moves in wildfowl representation can be seen as symptomatic of the widespread shifts – the growth in commercialisation and, in particular, the increase in social hierarchy – that characterise Late Saxon England. Senecal (2001, 252) coined the term 'thegnly culture' to describe the composite habits and activities that emerged as the new aristocratic class sought to jockey for social position. If, as Senecal argues, conspicuous displays of wealth were key to securing social position, it seems likely that falconry and the consumption of wildfowl were important components of the package. That the ownership of hunting birds became a badge of thegnly identity is clear from contemporary iconography, the Bayeux Tapestry being amongst a suite of artworks that employ falcons to denote aristocratic figures.

It is clear, then, that on the eve of the 1066 the aristocracy were already procuring and consuming proportionally more wild birds than other social groups. At the point of the Conquest, however, the situation appears to have become particularly polarised (Figure 62). Unlike the preceding centuries, which had witnessed a rise in wildfowl frequencies across the board, the mid-eleventh to mid-twelfth centuries saw their representation increase only (and dramatically) on elite sites, whereas wild bird frequencies decline for villages and religious houses. By the mid-twelfth to mid-fourteenth centuries the gap between high and low-status sites appears to have narrowed, suggesting that the pre- to post-Conquest shift is not part of a long-term trend. Instead the pattern for the Norman period hints at the type of unequal access that might be anticipated with the aggressive introduction of Forest Law.

Whilst this study in wild bird representation has revealed some interesting and apparently Conquest-related changes, the clustering of data into a 'wild bird' group masks greater complexity in species composition. To remedy this Figure 63 shows representation data, by period, for some of the main taxa found in medieval assemblages. Here it becomes clear just how varied the exploitation of different birds was, each taxa exhibiting its own unique pattern of representation. Some species such as the curlew (*Numenius arquata*), godwit (*Limosa* spp.), lapwing (*Vanellus vanellus*) and snipe (*Gallinago gallinago*) show little variation in their representation, being present in low levels throughout the medieval period. The pheasant (*Phasianus colchicus*) falls within this group which is interesting given that the Normans are popularly attributed with its introduction: several scholars have suggested that it was imported, perhaps from Norman Sicily, at some point during the late eleventh or twelfth century (Lever 1979; Rackham 1997, 50, and Rowley 1999, 104). This assumption has largely been based on the historical evidence where the

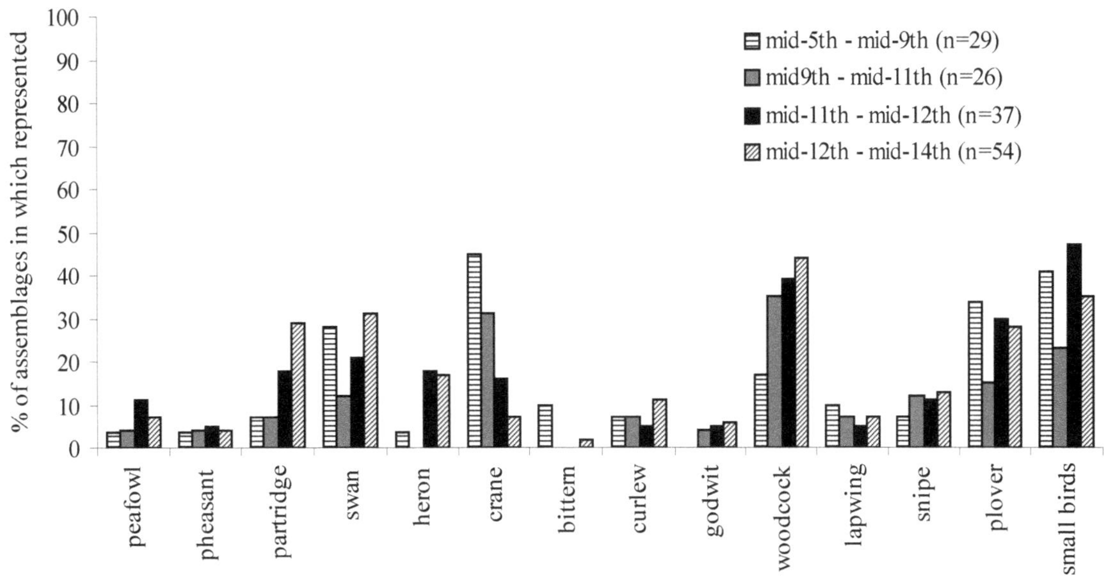

Figure 63: Inter-period variation in the representation of the main wild bird taxa

appearance of the word *fasianus* in medieval Latin texts had been interpreted as meaning *Phasianus colchicus*. According to Rackham (1997), pheasants are first recorded in the 1098 monastic charters of Rochester; although Yapp (1983) has argued that all pre-thirteenth century references to *fasianus* actually denote the capercaille (*Tetrao urogallus*).

Clutton-Brock (1984) has pointed out that pheasants are mentioned in the Old Welsh texts on the Nine Huntings (AD945) which state that:

> 'A pheasant is called a barking hunt, because when the pointers come upon it, and chase it, it takes to a tree where it is hunted by baiting.' (Probert 1823 cited in Clutton Brock 1984, 168)

If not a mistranslation, this reference is interesting when viewed in conjunction with the zooarchaeological data. Figure 63 indicates that a small number of pheasant have been recovered from pre-Conquest sites: Appendix Ie shows that specimens have been recovered from the Saxon town sites of Fishergate in York (O'Connor 1991) and Lincoln (Dobney *et al.* 1995). This would seem to undermine the theory that the Normans were responsible for introducing the pheasant; however it is noteworthy that these early examples are all from trading settlements where pheasants may have arrived from the continent. Were the Normans responsible for the establishment of the pheasant, the dearth of their remains in mid-eleventh to mid-twelfth-century deposits is interesting. It can probably be explained by issues of identification; pheasant remains are very difficult to separate from those of other galliformes. Even in the later medieval period, when documentary references to pheasant are common, their remains are still archaeologically scarce and far more research will be necessary before the question of pheasant introduction can be answered satisfactorily.

Similar problems exist for the identification of domestic doves and pigeons, rendering the study of their management problematic. According to Serjeantson (2006), dovecotes and the associated remains of adult pigeons and squabs become common from the eleventh and twelfth century onwards. The timing of their appearance has frequently been linked with the Conquest and it may be that the concept of pigeon/dove management was a Norman import. The tradition does seem to have been known in pre-Conquest Normandy and a probable dovecote is illustrated on the Bayeux Tapestry, next to the clerk in the Aefgyva scene, where the doves in the margin may suggest either a religious or erotic significance (Thorpe 1983).

Peafowl (*Pavo cristatus*) are also considered widely to be a Norman introduction, and Figure 63 demonstrates that their representation does increase in the post-Conquest period. With the exception of the specimens from Mid-Saxon Wicken Bonhunt (Crabtree 1996) and Late Saxon Thetford (Jones 1984), peafowl are largely absent from pre-Conquest assemblages. By contrast, *Pavo cristatus* remains have been identified in several elite assemblages dating to the Norman period: for instance Faccombe Netherton (Sadler 1990, 505), Carisbrooke Castle (Serjeantson 2000) and Ludgershall Castle (Poole n.d.). A similar situation exists in Ireland where McCormick (1991, 49) has identified peafowl only from post-Conquest deposits. Historical evidence also points towards a post-Conquest introduction: by the twelfth century Alexander Neckham was able to describe the peacock as a bird of the English court (Holmes 1952, 195). Taken together, the evidence does suggest the Normans as responsible for the peafowl's introduction, a case made stronger when it is remembered that peafowl are well represented on elite sites from eleventh/twelfth-century France (Chapter 2).

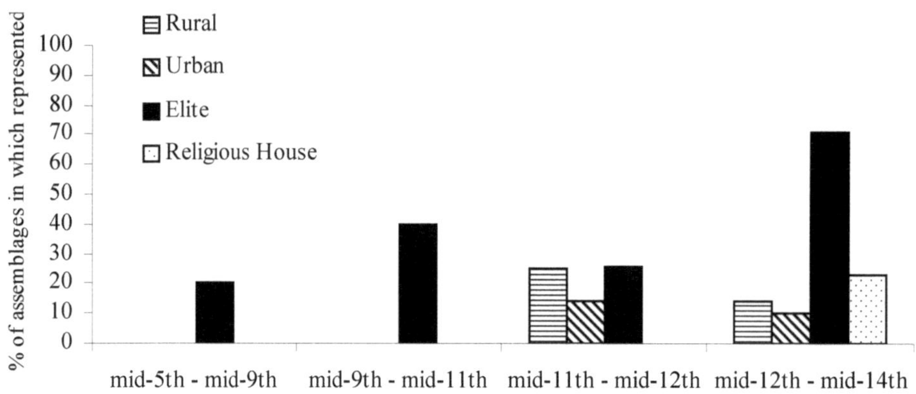

Figure 64: Inter-period and inter-site variation in the representation of the grey partridge (Perdix perdix)

The heron (*Ardea* sp) is also well represented on elite French sites (Chapter 2) and it seems likely that Norman traditions of hunting and consuming this species were transported to England following 1066. It is certainly noticeable how heron representation changes from the pre- to post-Conquest period. They are entirely absent from mid-ninth to mid-eleventh-century assemblages and their presence is recorded on only three fifth to mid-ninth-century sites: West Stow (Crabtree 1996), Flixborough (Dobney and Jaques 2002) and Wroxeter (Hammon *pers. comm.*). Interestingly there is also a dearth of heron place-names (Boisseau and Yalden 1998) which is surprising considering that the heron is, and must have been, the most common of all long-legged marsh birds. The situation contrasts markedly with that for crane (*Grus grus*), which are regularly mentioned in place-names and are also the best represented wild bird species between the mid-fifth to mid-eleventh centuries (Figure 63). After the mid-eleventh century, herons become increasingly well represented in assemblages from elite sites and religious houses (Appendix Ie). It is also at this point that they begin to appear in manuscripts: according to Yapp (1981, 15) the earliest depiction of a heron is in a Psalter dated to *c.*1170. It may be significant that the representation of herons increases just as the frequency of crane (*Grus* sp) declines, perhaps due to environmental change and over-hunting (Morrison n.d.; Rackham 1997, 37). Reduced availability of crane could feasibly have encouraged the uptake of heron hunting. Alternatively, the fact that the piscivorous heron begins to be hunted in a period when the first artificial fish ponds are documented historically, may reflect new concerns with preserving fish stocks: if this is the case, heron representation may be a better indication of pond construction than can be gleaned from the remains of the fish themselves (see page 60). Ponds and stews would have been easy hunting grounds for the heron whose population must have benefited from their establishment. At the same time, the heron's attraction to these artificial ponds would have made them predictable targets for wildfowlers and it is no surprise that by the later medieval period herons were being raised within parks as falconry targets: for example, in 1501 the Duke of Buckingham was sent 17 live herons from his park at Newton Blossomville (Woolgar 1999, 114).

One avian species whose representation can be linked directly with parks, and the elite's desire to control luxury resources, is the partridge (*Perdix perdix*) whose abundance increases in line with the proliferation of parks. In Section 2.5 (page 24) it was noted that partridge were the best represented wildfowl in French medieval assemblages and, as with the heron, it seems possible that Norman preferences are behind the post-Conquest increase in partridge representation. However, when the data are shown by site type (Figure 64) it can be seen that partridges had status associations from the beginning of the medieval period, when it may be assumed that they were taken occasionally through falconry. Their representation on elite sites from the Norman period is actually lower than that for Late Saxon high-status sites and it was not until the mid-twelfth century that the aristocracy invested considerable sums raising partridges within the confines of their pleasure parks and warrens (Wilson 2003, 119; Lasdun 1991, 6). Historical evidence indicates that private enclosures were sometimes breached by peasant poachers who carried off birds for their own consumption (Liddiard 2005, 103). This is confirmed by the zooarchaeological evidence which shows that partridges begin to be represented on lower-status sites of the post-Conquest period.

Considering that the Normans are attributed with imposing restrictions on the utilisation of wild resources, it seems possible that they were responsible for the well-known regulations pertaining to the consumption of swans (*Cygnus* spp); especially since swan representation begins to increase in the post-Conquest period (Figure 63). Popular conception is that swans are royal birds and, whilst this is not entirely true, the Crown has asserted prerogative rights over swans since the twelfth century, with governance over swan ownership first recorded in the mid-thirteenth century document *De legibus et consuetudinibus Angliae* (Henry de Bracton, cited by MacGregor 1996, 40). In all probability, the restrictions antecede this earliest record and the zooarchaeological

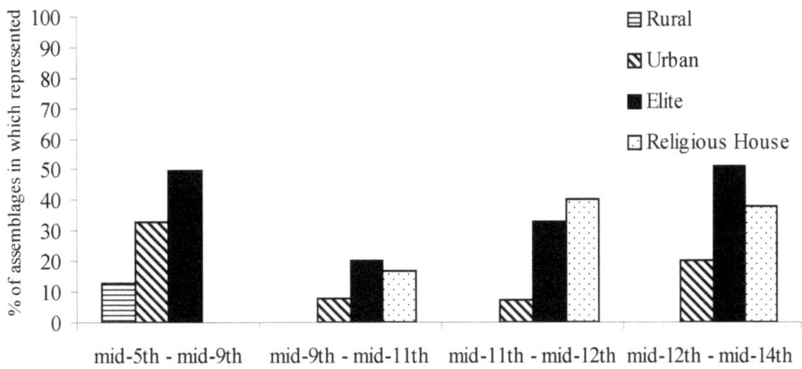

Figure 65: Inter-period and inter-site variation in the representation of Swan (Cygnus sp.)

evidence affords an insight into the development of the tradition. Figure 65 shows that even during the mid-fifth to mid-ninth centuries swans were best represented on elite sites. That said, swan remains have also been found on lower-status settlements, notably at rural West Stow and on the trading sites of Ipswich (Crabtree 1996) and Sandtun (Murray and Hamilton-Dyer 2001). The very presence of swan remains on these sites suggests that this taxon had not yet become the preserve of the elite. Between the mid-ninth and mid-eleventh centuries, patterns of swan representation appear to have changed: they are absent from rural sites, present on a small number of urban sites but more widespread amongst elite assemblages. Whilst this pattern becomes more defined in the post-Conquest period, the status-based restrictions would seem to be of Late Saxon origin. Such legislation would not be out of place in this period when, as has already be seen (page 60) certain wild animals were beginning to be viewed as the lord's property – their possession and consumption a mark of aristocratic identity. As demonstrated in chapter 2, there was no similar tradition in medieval France and in this case the Normans appear to have exploited, rather than introduced, existing restrictions.

5.2i: Summary

It is clear that patterns of wildfowl exploitation changed during the Norman period, levels of game bird consumption increasing dramatically on elite sites. The types of species targeted also shifted from the mid-eleventh century, with herons and peafowl being consumed more regularly than was the case in the early medieval England. These changes can, perhaps, be attributed to the arrival of the Normans who introduced new exotic species (peafowl and possibly the pheasant and domestic doves), encouraged the hunting of their favoured quarry (the heron), and strengthened controls over wildfowl utilisation. But these shifts must be viewed against a backdrop of increasing wildfowling and resource restriction. Even between the mid-fifth and mid-ninth-centuries there is evidence that status association were attached to wild birds, in particular those that might be caught through hawking. By the mid-ninth to mid-eleventh centuries these associations had become further defined with some species, notably the swan, becoming the preserve of the elite. If there was a precedent of restrictions over wildfowl prior to the Conquest, it seems possible that the same was also true concerning the exploitation of wild mammals. This possibility will now be considered.

5.3: Wild Mammals

In many respects the wild mammal data exhibit similar trends to those for wild birds. Figure 66 shows that few assemblages dating from the fifth to mid-ninth century contain high frequencies of wild mammals. Grant (1981) has suggested that in a period when attention focused on

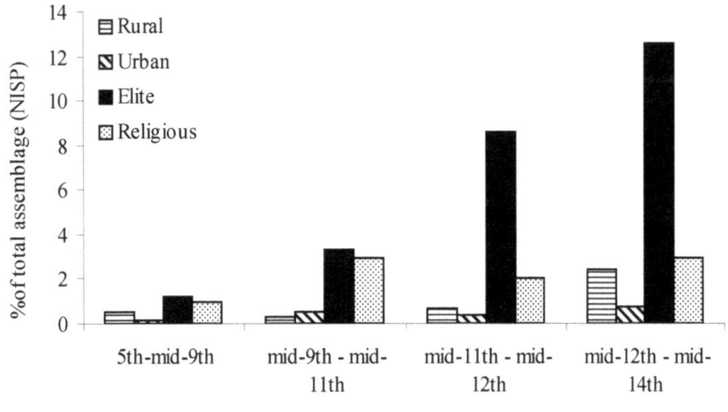

Figure 66: Inter-period and inter-site variation in representation of the main wild mammals, expressed as a percentage of the total assemblage (NISP)

agricultural production, farmers may have dedicated little time to hunting, perhaps even driving away deer in order to protect crops from their ravages. Whilst this idea is plausible it is more probable that, as with wild birds, the significance of game mammals extended beyond their value as providers of meat. This finds support in Old English literature; for instance in *Beowulf*, Hrothgar names his great hall 'Heorot', the 'Hart' (Marvin 2006, 17). Early medieval iconography and funerary deposits also indicate clear associations between hunting and the social elite: weapons and armour often carry motives of wild animals, such as the boar crested helmets mentioned by Lucy (2000, 51); a bronze figurine of a red deer was recovered from the seventh-century Sutton Hoo (Clutton-Brock 1976); and the remains of the animals themselves are frequently represented, along with hunting paraphernalia, in high-status male burials and cremations (Crabtree 1995; Lucy 2000, 90-4, 1112-13). Zooarchaeologically, these status-associations are indicated by the fact that elite and ecclesiastical assemblages contain marginally more game than those from other site types.

Wild mammals are better represented on all mid-ninth to mid-eleventh-century sites, except villages. Towns show an increase from 0.2% to 0.6% whereas the frequency of game on elite sites rise from 1.2% to 3%. Growth in elite game consumption coincides with the social reorganisation of the ninth century. With the break up of the old estates, the newly-landed thegnly class was able to establish its own hunting reserves, a bold statement of power and authority, and there is even some evidence that woodland was allowed to regenerate for this purpose (Hooke 1998, 139, 157). Biddick (1984a) has argued that these ninth-century changes in landscape organisation, may have curbed peasant hunting rights, perhaps explaining the reduced frequency of game on rural sites. It can be no coincidence that as an aristocratic hunting culture began to emerge, words relating to animals and hunting began to change: around 1000 AD the word *deor*, which had originally meant 'animal' in general narrowed to mean 'deer' in particular (Cartmill 1993, 67).

Terminology surrounding wild animals changed again after 1066, the meat obtained by hunting being labelled according to the Norman '*venesoun*', from which the modern word 'venison' is derived. The zooarchaeological record for the mid-eleventh to mid-twelfth centuries demonstrates an unparalleled increase in the representation of wild mammals on elite sites (Figure 66). Rural sites also demonstrated a slight rise in the presence of wild mammals but towns and religious houses both show a reduced frequency compared to the preceding period. Overall, division between high- and low-status sites became particularly pronounced during the Norman period, the status based inequality even exceeding that demonstrated by French assemblages (Chapter 2, page 20). This supports Gilbert's (1979, 11) assertion that the Forest Law introduced to England was more severe than it had been in pre-Conquest Normandy.

Game consumption increased at all social levels between the mid-twelfth and mid-fourteenth centuries. Patterns of inequality were retained but were not accentuated further, confirming the mid-eleventh to mid-twelfth centuries as the period during which differential access became established. It is encouraging that the zooarchaeological data are consistent with traditional beliefs that the Normans both delighted in hunting and restricted rights to it; however, to advance our understanding of Norman influence on hunting traditions, it is necessary to examine the data at the more detailed level of species frequency.

Representation data for the main wild species – red deer (*Cervus elaphus*), roe deer (*Capreolus capreolus*), fallow deer (*Dama dama*), wild boar (*Sus scrofa*), rabbit (*Oryctolagus cuniculus*) and hare (*Lepus* spp) – are shown in Figures 67-70. They reveal a considerable discrepancy between the zooarchaeological and historical record: whilst documents regularly cite wild boar as one of the most common game animals, faunal assemblages suggest that this species was rarely hunted in medieval England. The zooarchaeological scarcity of wild boar could be the result of identification problems, as it can be difficult to separate the remains of wild and domestic pig, especially where the two have interbred. However, even where researchers have studied tooth size, a more reliable criterion for distinguishing species, evidence for wild boar has remained slight: Pierce (n.d) suggested that wild boar must have been exceptionally uncommon by the thirteenth century, if not before (see Chapter 4). It must be assumed that the 'wild boar' recorded in many medieval documents, as for instance those relating to a 1251 Christmas dinner held by Henry III (Rackham 1997, 36, 119), were in fact domestic pigs. Emphasis on wild boar in English hunting manuals can most probably be attributed to the fact that most of these texts were based on continental originals, in which the centrality of wild boar is unsurprising: it was seen in Chapter 2 that wild boar were frequently hunted by the elite of France.

Perhaps the most obvious shifts in taxa representation are those for the three cervid species; the native red deer and roe deer and the imported fallow deer. Figures 67 and 68 show clearly that, throughout the Saxon period, both village and town assemblages comprised higher frequencies of red than roe deer. To some extent this might be accounted for by antler working, with artisans preferentially collecting and working the shed antler of the larger red deer. Where possible I have, however, attempted to exclude antler fragments from this synthesis and, therefore, the data set should reflect the 'real' situation. If so, it contrasts markedly with the representation data for religious houses, assemblages from the mid-fifth to mid-twelfth century all showing a preponderance of roe deer (Figure 70). Why ecclesiastics should have consumed the venison from roe deer in preference to that from red deer is unclear but it is interesting to note that the same is true on the continent (Yvinec 1993 and see Figure 15). It may be significant that in both England and France ecclesiastics were often

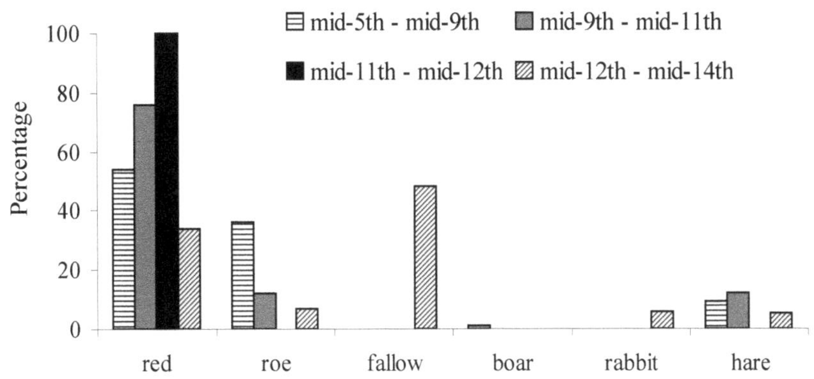

Figure 67: Inter-period variation in the relative percentages of the main wild mammals recovered from rural assemblages

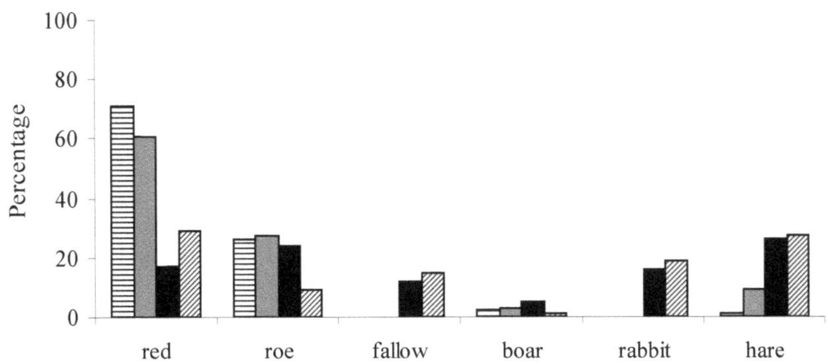

Figure 68: Inter-period variation in the relative percentages of the main wild mammals recovered from urban assemblages

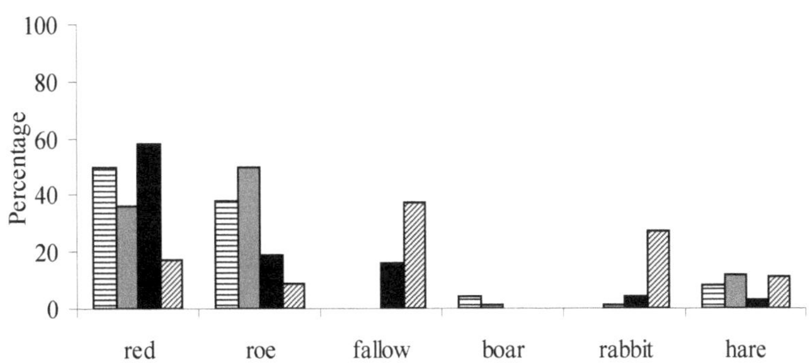

Figure 69: Inter-period variation in the relative percentages of the main wild mammals recovered from elite assemblages

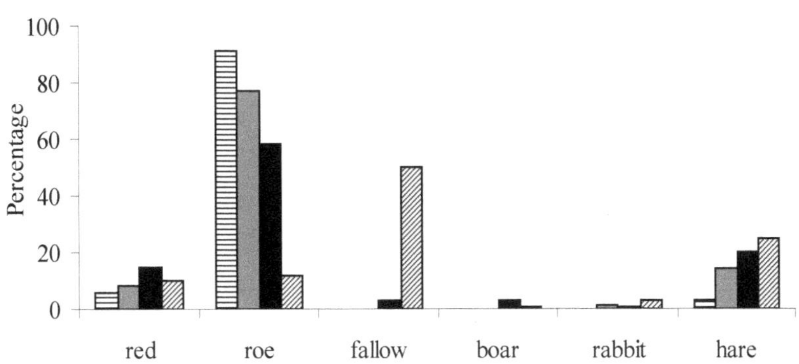

Figure 70: Inter-period variation in the relative percentages of the main wild mammals recovered from religious house assemblages

granted the rights of chase, that is, the freedom to hunt lesser quarry such as roe deer and hare. Another explanation for the roe-dominated assemblages could lie in the symbolism of roe deer: in later medieval art and literature roe deer are depicted as faithful, chaste and abstemious. In a period when it was believed that the character of an animal could be acquired through the consumption of its flesh (see for instance Salisbury 1994, 44; Scully 1995, 42), it seems possible that roe deer venison was considered suitable for men of the cloth.

Ecclesiastics were not the only section of society who procured and consumed roe deer, this species is also well represented on elite sites dating from the fifth to mid-ninth centuries but more particularly those from the mid-ninth to mid-eleventh centuries (Figure 69). Indeed, wild assemblages from several Late Saxon elite sites – Faccombe Netherton (Sadler 1990), Goltho (Jones and Ruben 1987) and Bishopstone (Ingrem n.d.) – are dominated by roe deer remains. By the Norman period, however, high-status sites demonstrate a dramatic shift away from roe and towards red deer, a trend even apparent for religious houses. This shift has been recorded for many multi-period sites, including Trowbridge Castle (Bourdillon, 1993) and Eynsham Abbey (Ayres *et al.* 2003) and may represent a decline in the roe population (Rackham 1997, 39). Alternatively, it could reflect changing preferences and hunting strategies. In pre-Conquest England most aristocratic hunting appears to have centred on enclosed wooded areas, such as the *siluis uenationibus* 'woods for hunting' mentioned in the Late Saxon charter for Grimley, Worcester (Sawyer 1968, S 1370), or the *haia in qua capiebant ferae* 'enclosure in which they used to capture wild animals' referred to in the Domesday accounts for Worcestershire (Hagen 2002, 135). Hunting in wooded enclosures would have been weighted towards the capture of roe deer, which are predominantly a woodland species, adapted to negotiating thick undergrowth (Darling 1937). By contrast red deer are less suited to woodland and are better adapted to open landscapes (Clutton-Brock *et al.* 1982). The Norman introduction of the 'forest' concept changed perceptions of the hunting grounds towards unbounded tracts of moor, heath and agricultural land, the type of environment where red deer will form herds. I suggest that it was this change in hunting landscape that promoted red deer to premier beast of the hunt. Animal bone evidence from eleventh/twelfth century France, where red deer were the dominant cervid species by far, indicate that the pre- to post-Conquest shifts in deer representation can be attributed to Norman influence. Certainly, the Anglo-Saxon Chronicle mentions William the Conqueror's love of red deer and the controls he placed on hunting them. Imposition of restrictions may explain the substantial post-Conquest decline in red deer on urban sites. No similar decrease is apparent for rural assemblages of the mid-eleventh to mid-twelfth century; indeed, their frequency actually increases. This may reflect peasant defiance of the new game laws; that they continued to hunt regardless of the restrictions.

Changes in the ratio of red to roe deer are not, however, the only pre- to post-Conquest shifts in cervid representation. Figures 67-70 show that fallow deer became more abundant in the Norman period (the circumstances of their introduction are considered in Chapter 7). Rackham (1997, 123) has argued that the post-Conquest increase in fallow deer was connected with the growing fashion for emparkment; certainly the frequencies of both appear to rise in parallel. That the two are connected is unsurprising since fallow deer, an exotic species, would not have been set free to wander the landscape but jealously guarded in enclosures. The difficulties of poaching fallow deer from within these physical barriers may explain why they are poorly represented on lower-status sites; although Figure 67 suggests that fallow deer are the dominant species on rural sites, this pattern is due entirely to the assemblage from Lyveden in Oxfordshire, Appendix Ia demonstrates that they are absent from all but one other rural assemblage. Quantities of fallow deer venison did percolate into post-Conquest towns. Despite restrictions on the sale of venison, some may have been purchased from hunters or parkers, but documents record that venison also arrived through a system of organised trafficking: court rolls from the thirteenth century, record the illicit trade running between the Midland forests and Bristol (Birrell 1982, 20). The roll for Cannock, Staffordshire, for example, mentions one man who had hidden deer carcasses in a cart-load of timber (Birrell, 1982, 18). Problems of carcass concealment may explain why, in the post-Conquest period, towns began to receive a greater frequency of rabbits (whose introduction is examined in Chapter 8) and hares, animals that could have been smuggled with less risk of discovery.

5.3i: Summary

Few history books neglect to mention the Normans' 'love of hunting' and the zooarchaeological evidence confirms that the Norman elite both utilised and restricted resources of wild mammal more heavily than their Saxon predecessors. Comparison with elite assemblages from France, however, demonstrates that the celebrated Norman hunting obsession was not so much a Norman tradition but more a passion that could be indulged only with the occupation of England and the personal wealth that this brought. Zooarchaeological evidence also confirms Gilbert's (1979) suggestion that the Forest Law introduced on this side of the Channel was more strictly applied than was the case in Normandy, since wild animal assemblages from France demonstrate less status-based inequality than those from post-Conquest England.

5.4: Discussion and Conclusion

Traditionally, Anglo-Saxon society has been viewed as drawing little from wild resources and the zooarchaeological evidence for fifth to ninth-century England supports this perception. The animal bone data demonstrate that no section of society expended much time in gathering wild birds, mammals or fish and where these resources were targeted, exploitation focused on animals that were locally available.

Rural dependence on domestic resources appears to have continued into the Late Saxon period with village assemblages demonstrating the lowest levels of fish, wildfowl and wild mammal representation. Other assemblage types, however, in particular those from urban sites, demonstrate a general widening of the resource base, the limited taxa ranges of the Early/Mid-Saxon period superseded by a greater variety of animals. O'Connor (1994) argued that these changes, which he noted from medieval York, reflect market development. Similar conclusions have been drawn by Barrett *et al.* (2004) and it may be assumed that the Late Saxon rise in wild resources reflects the growing urbanisation and commercialisation noted in Chapter 3.

Aside from economic change, the Late Saxon period was also a time of increased social hierarchy and gender definition, with emphasis being placed on military activity and aristocratic authority. An upsurge in hunting and hawking, activities predominantly of the male elite at this time, must be symbolic of these changes but it can also be seen as reflecting the concerns of the newly-landed thegnly class to set themselves apart from their social inferiors, as well as each other. Increased participation in flamboyant hunting and hawking forays, or the consumption of fish obtained from private fisheries would fit Senecal's (2001) suggestion that competitive display amongst the aristocracy grew significantly during the Late Saxon period. The desire of the elite to maintain their authority and prevent the lower classes aping their lifestyle seems to have been the impetus for the privatization of wild resources, for which we see the beginnings in this period.

So, the transition towards increased but restricted exploitation of wild resources was already underway by 1066 but it was not until after the Conquest that dramatic changes occurred. Zooarchaeological data demonstrate that the frequency of wild mammals and birds on high-status and monastic sites increased considerably, whereas low-status sites were receiving little more game than in the preceding period. By contrast, levels of aristocratic fish consumption dropped during the Norman period, although cetacean remains become more abundant on elite sites. All these changes occurred with rapidity indicative of external influence and the fact that they coincide with the date of the Conquest is highly suggestive of Norman involvement. Credence is added to this conclusion when it is recognised that post-Conquest assemblages from elite sites show traits typical of those from eleventh/twelfth century France: the apparently low levels of fish consumption, the dominance of red deer and the representation of herons and peafowl.

The introduction of exotic species – peafowl, fallow deer and rabbits – changed fundamentally the methods by which wild resources were managed, with the establishment of numerous warrens and parks that came to be stocked with a wide range of species. The extent to which the Normans can be held responsible for these new landscape features is debateable and will be considered over the course of the next two chapters, but their introduction to post-Conquest England raises an important point. The popular perception has always been that the Normans restricted access to animals that, prior to the Conquest, were free to all; however, this study has shown that many of the animals exploited in Norman England were neither utilised by the lower classes of Saxon England (for instance herons, partridges and swans) nor physically available to them (such as the introduced peafowl, pheasants, fallow deer and rabbits). That restrictions grew up around the enclosures, and animals maintained within them, is unsurprising given that they were maintained at great expense. But other than removing areas of land from free access, the establishment of parks and warrens did not alter significantly the lot of the native peasant.

The same cannot be said for the introduction of the forest concept, which undoubtedly restrict commoner's rights to previously available wild animals. In this way the decades following 1066 were undoubtedly a watershed in the development of English wild resource exploitation, a shift that reflects a new social order. The status bestowed upon red deer, roe deer and wild boar created instantly a luxury resource but, by definition, it also increased their desirability. Despite the controls placed upon them, occupants of low-status sites succeeded in obtaining and trafficking supplies of game. In situations when luxury resources trickle down the social scale, the upper classes often construct new ways in which to differentiate themselves. The possibility that the Normans introduced new hunting rituals and etiquette as a way of maintaining the status boundary will now be considered.

Chapter 6: Deer Hunting: Methods and Rituals

'Stop in God's name! What are you at? Whoever saw a hart broken up in this fashion?'
The huntsmen fell back a pace... 'We know of no better way in these parts than to split it clean down from the head and then into four, so that none of the quarters is much bigger than another. That is the custom of this country'
'The usage is different in the land where I was reared, there they excoriate a hart'.
(Gottfried von Strassburg 1960, 78)

In the previous chapter it was seen that Norman England witnessed an upsurge in aristocratic hunting, with attention shifting from the chase of woodland roe deer to the pursuit of red deer across open land. With a change in the hunting landscape, it might be expected that post-Conquest hunters not only targeted different species of deer but also used different techniques to catch them. Certainly this has been suggested by several historians who have argued that a new and highly ritualised form of hunting, the *chase par force*, was introduced to England by the Normans (Gilbert 1979, 58; Cummins 1988, 51; Marvin 2006). These conclusions are, however, based on very slim evidence, since few texts on hunting date to before the thirteenth century. Zooarchaeological analysis offers the opportunity to examine how patterns of deer exploitation may have changed through time; were deer carcasses treated similarly in the pre- and post-Conquest period or is it possible to detect shifts that might represent the introduction of new rituals? Given that hunting is a social action, any change in hunting traditions is significant and requires explanation. In this chapter the possibility that the Normans did introduce new hunting methods will be explored and the motivation and meaning of any changes considered.

6.1: Hunting Techniques

By the later medieval period there existed a large corpus of literature, both practical and literary, detailing the manner by which different quarry should be taken (Cummins 1988; Rooney 1993; Almond 2003). Whilst the medieval mind clearly appreciated the nuances that separated different hunts, in effect there were, and still are, only two principal hunting methods: animals can either be driven towards hunters or hunters can chase after them. This is, of course, a great simplification; the choice of hunting paraphernalia and the inclusion/exclusion of dogs or horses add variety to the methods. And before the zooarchaeological data are examined, it is worth considering briefly the evidence for hunting weapons because this has some bearing on the interpretation of pre- and post-Conquest hunting techniques.

It may be assumed, not unreasonably, that since hunting and warfare were linked ideologically, both employed the same weapons. Several good surveys of medieval weapons have been undertaken (notably Bradbury 1997) and they all highlight the same factors that set apart the Anglo-Saxon and Norman arsenals: the use of bows and arrows. This difference is expressed iconographically by the Bayeux Tapestry, which depicts the Norman use of archers but their almost complete absence within Harold's army. Lewis (in press) has argued that the association of Normans with archery may have been a device on the part of the Tapestry's designer to identify the players of the different sides; although he also accepts that the depictions may have been grounded in truth. The Anglo-Saxons did know of archery: one Saxon archer is depicted in the Bayeux Tapestry, there are a few other illustrations of archery in Anglo-Saxon art, and the bow is thought to be the answer to two (17 and 23) of the Exeter Riddles (Bradbury 1997, 18). These examples are, however, insufficient to suggest that the use of bows and arrows was widespread. According to Henry of Huntington (*Historia Anglorum*, cited in Bradbury 1997, 21), William the Conqueror believed that the English were 'not even in the possession of arrows', a belief that can only have been strengthened when he ran out of arrows whilst hunting and was required to direct a local blacksmith in the art of arrow-making (*Chronicle of Battle Abbey*, cited in Bradbury 1997, 23). Archaeologically, finds of arrows from pre-Conquest England are scarce, spears and swords being by far the dominant weapon, but their frequency increased considerably in the post-Conquest period (Jessop 1996).

If archery was not a technique employed in pre-Conquest hunting, what methods did the aristocracy of the Late Saxon period use to catch the roe deer that they consumed in such high numbers? Gilbert (1979) and Cummins (1988) have both suggested that the 'drive' was the most common pre-Conquest hunting method, whereby deer were chased into some sort of enclosure and then killed with knives or hunting spears. Such practices are in accord with Aelfric's Hunter who states:

> 'I weave myself nets and set them in a suitable place, and urge on my dogs so that they chase the wild animals until they come into the nets unawares and are thus ensnared; and I kill them in the nets.' (Swanton 1975, 109)

The nets and enclosures into which wild animals were chased are referred to variously as *haga* or *haia*. They were often clearly-defined features within the landscape and the *Rectitudines Singularum Personarum* specifies that geneats and cottagers were required to maintain them on behalf of the lord or king (Lemanski 2005, 32-33). By 1086, the Domesday Book was able to record numerous *haiae* across the country, and several entries refer specifically to *haiae capreolis* – hays for taking roe deer (Thorn and Thorn

1986, 6.14). It can be no coincidence that roe deer are the dominant deer species in the Saxon levels at Faccombe Netherton in Hampshire, a site for which a 'white haga' was recorded in 961 (Hooke 1989, 128).

Drives would have been highly effective methods of obtaining large quantities of venison in a single event, although to be successful it would have required the participation of many people. The Domesday Book mentions that citizens of Hereford, Shrewsbury and Berkshire were obliged legally to act as drivers (Loyn 1970, 366) and it seems likely that this ability to muster manpower was used as a conspicuous display of royal or *thegnly* resources (Cummins 1988, 51). Driving may have continued as the preferred hunting method into the post-Conquest period but several historians have suggested that the classic form of medieval hunting, the *chase par force*, came to dominance after 1066 (Gilbert 1979, 58; Cummins 1988, 51; Marvin 2006).

Details of the *par force* hunt are given in all of the later medieval hunting manuals. It should be noted from the outset that all of these texts date to at least 200 years after the Norman period thus their relevance to Saxo-Norman England is questionable. Nevertheless, they provide a useful insight into medieval traditions of venery, details of which have been summarized by scholars such as Thiébaux (1967), Cummins (1988), Almond (2003) and Marvin (2006). According to the manuals, the *chase par force*, was a long and wide-ranging hunt in which a single deer, the strongest of the group, would be selected as the quarry with the intention that it would provide sport for the whole day. Pursued by hounds, hunt servants and the mounted aristocratic hunters it would be chased to exhaustion and, if the hunt was successful, be killed by a sword to the heart. Then came the crescendo of the hunt, the 'unmaking' or 'excoriation', where the deer was skinned, disembowelled, butchered and the meat distributed according to set protocol. The earliest account of the unmaking process is provided by Gottfried von Strasburg's *Tristan* (written *c.*1210) but the fullest description is given by the fifteenth-century *Boke of St Albans* (see Brewer 1992). These texts suggest that, in the case of a stag, its testicles and penis were first removed and hung on a forked stick (the *forchée*) which was used to collect together all the prized organs and titbits: the *forchée* was later carried at the front of the homeward procession. The animal was then skinned, first split from the chin down to the genitals and out to each leg before being flayed down to the spine. At this point the feet were removed from the carcass but left attached to the skin, which was now used as a convenient blanket upon which the butchery was undertaken. The shoulders and haunches were removed and the rest of the carcass de-fleshed with the meat and the antlers being carried home in the skin, presumably using the feet as handles. Certain parts of the carcass did not, however, return with the lord to his residence: the pelvis was left at the kill-site as an offering to the crow or raven, whilst the right and left shoulders were presented to the best hunter and forester or parker as his fee. By the mid-twelfth century, knowledge of these unmaking rituals, and in particular the Gallic terminology surrounding them, was deemed to be a mark of nobility, a fact condemned by John of Salisbury who remonstrated that 'the scholarship of the aristocracy consists in hunting jargon' (*Policraticus* 1.4).

If carcasses were treated in the way suggested by the texts, one might expect that the introduction of the unmaking rituals might be indicated in the zooarchaeological record by a shift in body-part representation, and indeed this does seem to be the case. Unusual deer body-part patterns have been recognised from several late medieval sites. Reports for the castle sites of Launceston (Albarella and Davis 1996), Sandal (Griffith 1983), Okehampton (Maltby 1982), Prudhoe (Davis 1987) Barnard (Jones *et al.* 1985a) and Castle Rising

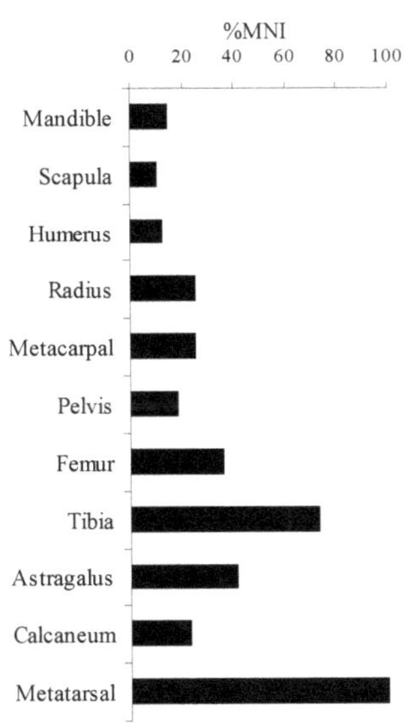

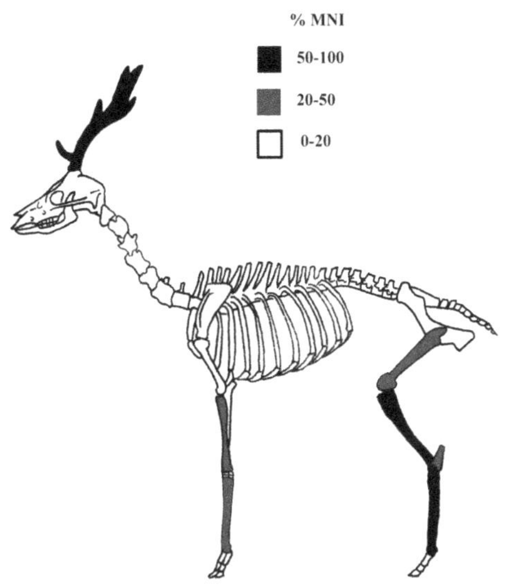

Figure 71: Anatomical representation patterns, shown as a percentage MNI, for deer (red, roe and fallow) from mid-twelfth to mid-fourteenth-century assemblages from elite sites

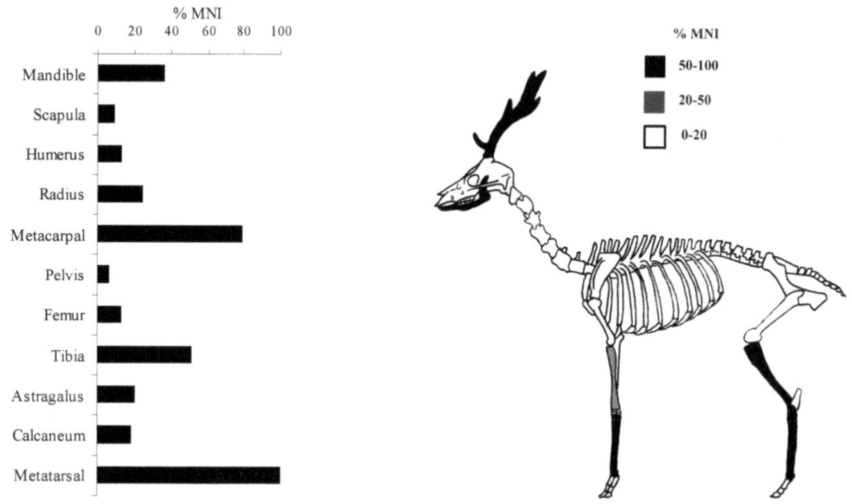

Figure 72: Anatomical representation patterns, shown as a percentage MNI, for deer (red, roe and fallow) from mid-eleventh to mid-twelfth-century assemblages from elite sites

(Jones *et al.*, 1997) have all mentioned a considerable over-representation of foot elements. Data for these assemblages are shown synthesised in Figure 71 and it is clear that the metapodia, in particular metatarsals, greatly outnumber all other elements. Many theories have been put forward to explain this patterning – that they are an artifice of inter-element variation in identification or preservation (Grant 1988a), that they represent butchery waste (Sadler 1990), or the importation of salted haunches of venison (Albarella and Davis 1996) – but I would suggest that they reflect the unmaking rituals, hence the absence of the pelvis, and elements from the upper forelimb. That shoulder joints did not arrive back at high-status settlements is reinforced by the evidence from later medieval household accounts, which commonly record haunches of venison but seldom mention the presence of the forequarters (Birrell 2006, 184). The final destination of the shoulders has recently been confirmed by zooarchaeological studies of forest workers's residences which show, as might be expected from the historical evidence, a preponderance of forelimb elements, indicating that foresters and parkers regularly received their allotted portions of venison (Sykes 2007).

Figure 72 demonstrates that Norman deer assemblages from elite sites are also skewed towards the representation of the hindlimbs, indicating that *par force* hunting was, indeed, practised in the years following the Conquest. Compared with the later medieval data, however, the Norman assemblages show less of a discrepancy between the metacarpal and metatarsals and it seems possible that, at this point, the forefeet were not yet considered trophies: the practice of retaining forefeet is not recorded until the fifteenth century when, according to Thiébaux (1967, 267), the hunt became even more heavily ritualised. If it is accepted that these body-part patterns reflect the unmaking rituals, their appearance in the archaeological record ought to pinpoint the arrival date of *par force* hunting and confirm whether or not the Normans were responsible for introducing this technique.

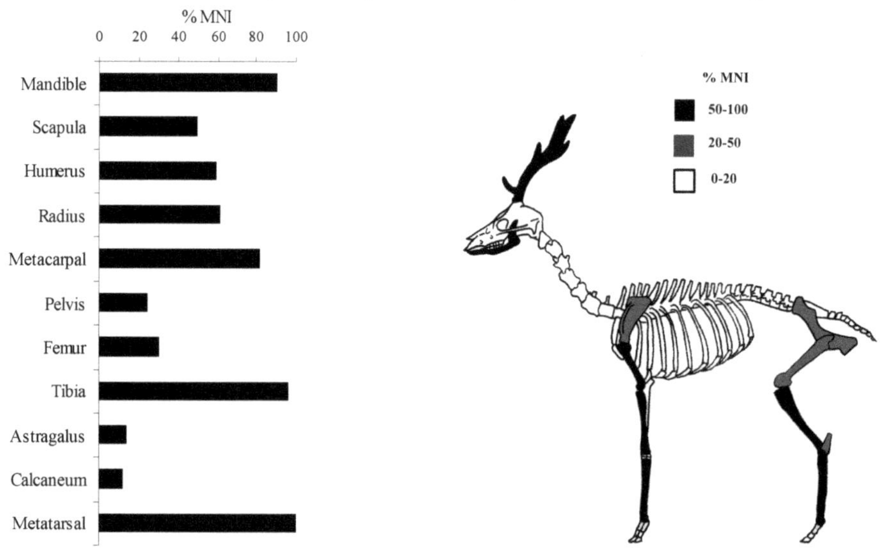

Figure 73: Anatomical representation patterns for deer (red and roe) from mid-seventh to mid-eleventh-century assemblages from elite sites

Data for pre-Conquest deer are shown in Figure 73. It is immediately obvious that the patterns are not similar to those of the succeeding period. All parts of the skeleton, in particular the shoulders, are well represented and even the pelvis is present in considerable numbers. Overall, the patterns for the Saxon period would suggest a situation where the deer were being brought back to the settlement complete, or at least with minimal butchery having taken place. There is no indication of patterns that could be attributed to the unmaking rituals. Consequentially it would seem that, if the body-part patterns are associated with the *par force* rituals, this method of hunting arrived in Britain at some point after the Conquest. Studies of multi-phase assemblages, including Eynsham Abbey (Ayres *pers. comm.*), Faccombe Netherton (Sadler *pers. comm.*), Goltho (Jones and Ruben 1987), Portchester Castle and the Cheddar Palaces (IIId) indicate that the distinctive skeletal patterning appears for the first time in the late eleventh century: where dating permits they first become apparent shortly after 1066.

Butchery data also demonstrate a pre- to post-Conquest shift in carcass processing. Whereas Saxon deer were butchered in the same way as sheep and cattle, with the majority of marks deriving from marrow extraction, post-Conquest deer butchery is different in style to that of domesticates. One butchery mark becomes particularly common and widespread in post-Conquest deer – the shaving of the crest and tuberosity of the tibia, which was apparent on 38% of those Norman period tibiae that I examined. This pattern is reminiscent of that observed for the Vatteville assemblage, where a high percentage of deer tibiae exhibited the same mark (Figure 16, page 21). Skeletal representations for Vatteville's deer are also akin to the situation in post-Conquest England. In Chapter 2, I suggested that the butchery and body-part patterns might be related to the de-boning and transport of venison but now it seems more likely that they reflect *par force* hunting techniques. If this is the case, it would seem that, as the historians suggest, the Normans were responsible for the introduction of *par force* hunting to England, and

Table 11: Anatomical data (MNE) for red, roe and fallow deer from Vateville-la-Rue

	11th/12th century	13th-15th century
Antler	4	45
Mandible		7
Scapula	4	
Humerus	5	1
Radius	4	6
Metacarpal	2	4
Pelvis	3	1
Femur	6	4
Tibia	9	8
Astragalus	3	5
Calcaneum	2	6
Metatarsal	9	6
Phalanges	2	6

that this hunting method did form part of a 'Norman Package'. Before this is concluded absolutely, it is worth re-considering the evidence from France.

In Chapter 2, the deer anatomical representations were shown synthesised for all periods but sample sizes are just sufficient to permit the data to be split into two groups: eleventh/twelfth century and the thirteenth to fifteenth century. Skeletal frequencies for these two date groups are shown in Table 11 and, surprisingly, they are not the same. Whilst those for thirteenth-fifteenth century deer are similar to the patterns for post-Conquest England, those for the twelfth-century material are closer to the pattern for Anglo-Saxon England: all body parts, including the scapula, humerus and pelvis were well represented. If the Vatteville-la-Rue data are representative of the wider continental situation, they indicate that *par force* rituals did not appear in France until at least the thirteenth century. Lack of detailed zooarchaeological information for medieval France makes this theory difficult to prove but it is supported by the evidence from medieval Germany. Figure 74 shows the body-part data for German deer and, similar to twelfth-

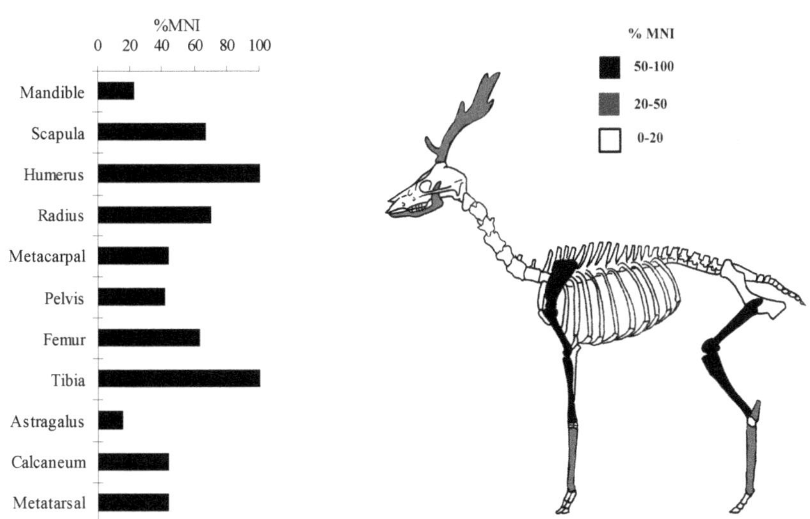

Figure 74: Anatomical representation patterns for deer (red and roe) from elite sites in medieval Germany

century Vatteville, there is no patterning suggestive of the unmaking rituals. These findings have considerable implications concerning the theory that *par force* hunting spread to Britain from the continent. Rather than being imported from Normandy, it would seem that the unmaking rituals either evolved in post-Conquest England or followed a different path of introduction. To determine which explanation is the most likely, it was decided to extend the analysis of deer body-part patterns to sites in southern Europe.

Baker and Clark's (1993) survey of zooarchaeological work in northern Italy provided some useful information concerning hunting. In particular, Baker and Clark noted that, where present, deer assemblages were usually dominated by foot bones and antler. Hunting does not, however, appear to have been a frequent activity and few sites have produced quantities of deer remains. Whilst this is the case in northern Italy, hunting may have been more important in the south of the country and on the island of Sicily. Little zooarchaeological work has been undertaken in these regions (Gillian Clark *pers. comm.*) but one report for the eleventh to fourteenth-century site of Brucato, Sicily (Bossard-Beck, 1984a and 1984b) proved exceptionally informative. A number of red, roe and, in particular, fallow deer, bones were recovered from this site and their skeletal representations are shown in Table 12.

Anatomical Element	Red	Roe	Fallow
Antler		1	153
Mandible			1
Scapula	Not Recorded		
Humerus		1	3
Radius	2	2	9
Pelvis	2	4	4
Femur	1		12
Tibia	2	3	11
Astragalus			2
Calcaneum		1	1
Metacarpal		1	1
Metatarsal	3	1	10
Phalanges	2	5	4

Table 12: Anatomical data for deer (red, roe and fallow) from medieval Brucato, Sicily (Bossard- Beck 1984a)

It is interesting to note that red and roe deer, although present only in small numbers, were represented equally by all body parts. They certainly give no indication of the effects of unmaking rituals; indeed, the results for roe deer show the pelvis to be one of the best represented elements. By contrast fallow deer demonstrate a more selective body-part pattern. Most obvious is the high frequency of antler, which seems to have been inflated both by the importation of shed antlers and fragmentation caused by working this element. More significant are the patterns for the post-cranial skeleton, which show an over-representation of hind limb elements, similar to that observed for post-Conquest England. Scapulae were not recorded for this site but the dearth of humeri and pelves is indicative of the *par force* rituals.

Whilst data from a single site need not be representative of the wider situation, they may suggest that the unmaking rituals were of southern, rather than northern, European origin. This does not exclude Norman involvement. Sicily was conquered by the Normans between 1061 and 1091 and strong links existed between the two kingdoms during the late eleventh century, providing ample opportunity for the exchange of ideas. Hunting practice no doubt formed part of this; Henry II's park at Woodstock is thought to have been based on the model *parco* at Palermo of King Roger II of Sicily (Rowley 1999, 181).

The data from Brucato suggest that the unmaking traditions were specific to fallow deer and, if this is the case, it has implications for the connection of the unmaking rituals to the *chase par force*. The practice of maintaining fallow deer within parks would have precluded this form of wide-ranging hunt, a method better suited to the open forest. Furthermore, the physiology and behaviour of fallow deer is not conducive to *par force* hunting because they have poor stamina and tend to maintain a herd structure during flight (Recarte *et al.* 1998). Fallow deer hunting must have been limited to the drive or 'bow and stable', whereby deer were chased into nets or towards archers who were positioned on a platform, or 'stand'. This is indicated not only by historical records but also by the zooarchaeological evidence: arrow fragments have been found embedded in fallow deer remains (Sadler 1990, 487). It must, therefore, be assumed that not only were the unmaking rituals employed regardless of hunting method but that, after their introduction, they were applied rapidly to all cervid species, not just fallow deer.

That it was the unmaking rituals, rather than the total *par force* hunting package, that were imported to post-Conquest England appears to find literary support in Aelfric's Colloquy: the hunter explains that, aside from driving animals, he sometimes used fast running hounds to chase deer (Swanton 1975, 109). Similarly, in the hunting scene of Gottfried von Strassburg's (1960) *Tristan*, which is set in Cornwall, the royal party had already caught their stag using the *par force* technique when Tristan arrived. Tristan did not object to the way in which the animal had been hunted but took issue when the English huntsmen attempted to quarter the deer following the customs of the country. At this point Tristan demonstrates how to unmake the deer according, he says, to the traditions of his homeland (Brittany). It is interesting to note that although much of the terminology he uses is French, the English party refer to some of it as Arabic: Arabic influence was very strong in Sicily, again indicating this island as a potential source for the unmaking rituals.

Even if the unmaking rituals were of Sicilian origin, it seems improbable that their Sicilian meaning was retained after their uptake in England. This is supported by the fact that none of the Middle English hunting texts clarify the symbolism of, or reason for, the rituals; even Gottfried's Tristan does not explain *why* his unmaking methods are superior to those of his Cornish hosts (Rooney 1993, 88). I would suggest that the Sicilian unmaking rituals were adopted and modified by the Anglo-Normans for two main reasons. First, it conformed to a wider pattern of cultural imitation; the Normans of Sicily were held in great esteem by the Anglo-Normans, who seem to have used the achievements of their Mediterranean kinsmen to bolster their own identity (Jameson 1938; Loud 2003). Secondly, and most importantly, the new flamboyant hunting rituals offered the Normans an opportunity to symbolically act out their dominance over the native people and their land. Even where Anglo-Saxon lords had retained their land after 1066 and continued to hunt, the new and complex rituals that were steeped in Gallic terminology would have acted as a suitable linguistic hurdle to exclude the English. In much the same way as is depicted in Gottfried's story of Tristan, it might be assumed that the new unmaking rituals were suggested to be more 'correct' than, even morally superior to, the English techniques.

6.2: Conclusion

While zooarchaeological analysis cannot prove that the *par force* hunt was imported to post-Conquest England, it is clear from the skeletal representation data that novel butchery practices associated with unmaking rituals were introduced shortly after 1066. By ritualising what must previously have been a relatively straightforward drive or *par force* hunt, the post-Conquest elite were responsible for the creation of a brand new tradition. Customs are seldom fabricated or modified without reason and one motivation for their creation is to legitimize change, such as that which would have accompanied the Conquest.

Combined with the introduction of Forest Law and the controls it placed on wild animal exploitation (Chapter 5), hunting provided the Normans with an outlet through which they could project everything they wished to emphasise about themselves: their 'Frenchness', their social superiority and, because hunting is linked to warfare, their martial supremacy. By embedding social status and ethnicity in law, landscape and language, hunting would have been a powerful statement of Norman identity and it is perhaps no wonder that the 'love of the chase' has come to be a character trait so frequently attributed to the Norman elite.

Another introduction frequently attributed to the Normans is the fallow deer. We have seen in this chapter that the unmaking rituals may, originally, have been linked to this species. The possibility that fallow deer arrived at the same time as the rituals will be considered in the next chapter.

CHAPTER 7: BIOGEOGRAPHY OF THE ANGLO-NORMAN TRANSITION - FALLOW DEER AND RABBITS

The 'received wisdom' is that the Norman Conquest impacted dramatically on British wildlife, with the introduction of several new animal species (Henderson 1997; Rackham 1997; Rowley 1999; Yalden 1999; Bartlett 2000; Langbein and Chapman 2003). As was seen in Chapter 5, there is some good zooarchaeological evidence to suggest that peafowl were re-introduced and managed from the late eleventh century and that pheasant may also have become established during this timeframe. Yet, the belief that the Normans introduced fallow deer (*Dama dama*) and rabbits (*Oryctolagus cuniculus*) to England has never been tested critically, 1066 has simply been used as a convenient peg on which to hang their arrival. In truth, the situation was probably more complex and it is likely that these species were introduced on multiple occasions, hence the suggestion by several authors that the Romans were also responsible for their importation (Sheail 1971; Lever 1979; Chapman and Chapman 1975; Lister 1984).

The uncertainty surrounding the introduction of fallow deer and rabbits is due largely to the ambiguous nature of the historical and archaeological evidence. Repeated changes to the 'English' language have meant that etymologists find it difficult to confirm the identity of animals mentioned in early texts and place-names (see for instance Yapp 1981; Hough 2001). Even where references are indisputably to fallow deer and rabbits, the earliest records can be viewed only as a *terminus ante quem*, the animals most probably arriving prior to the first accounts. Similar problems exist for the zooarchaeological data. Our understanding of the past biogeography of these species has been confused by 1) the ancient trade in fallow deer body parts, in particular antler, and 2) the propensity of modern rabbits to burrow down into archaeological deposits. Both of these factors have meant that fallow deer and rabbits have, in death, been represented in areas and contexts in which they were absent during life. The situation is complicated further by problems of misidentification by zooarchaeological researchers: the skeletal remains of fallow deer are close in appearance to red deer and roe deer, and the bones of rabbits may easily be confused with those from hare or birds, especially where reference collections are not consulted. Misidentification is particularly acute for assemblages that date to periods where, conventionally, these species are deemed to be absent. Reliance upon historically-based introduction dates has meant that fallow deer and rabbits found in deposits pre-dating 1066 are often either overlooked or dismissed as 'intrusive' (Crabtree 1990; Miles 1984). On other occasions, early deposits containing their remains have had their dating questioned, or even re-assigned to the post-Conquest period (Jones 1993; Sutermeister 1976). As a consequence, the belief that the Normans introduced fallow deer and rabbits has persisted whilst potential evidence for alternative introduction dates is frequently disregarded. In order to clarify the circumstances surrounding the arrival of fallow deer and rabbits, this chapter presents evidence derived from zooarchaeological literature surveys, the re-analyses of pivotal assemblages and metrical studies of purported early fallow deer and rabbit specimens.

7.1: Fallow deer

*Figure 75: Fallow deer (*Dama dama*) buck. Photo © R. Ford Digital Wildlife*

The prehistory of the European fallow deer (*Dama dama*) is examined in detail by Chapman and Chapman (1975) and Lister (1984). In brief, current evidence suggests that although represented in assemblages dating to interglacial phases from the last 400k years BP, fallow deer seem to have become extinct across north-west Europe during the Devensian cold phase (70-10ka years BP), surviving only in areas of Turkey (Uerpmann 1987; Masseti 1996). Resulting restrictions in breeding population may have caused the 'genetic bottleneck' in which the species lost its biochemical polymorphism (Pemberton and Smith 1985). Lack of genetic variation retards ability to adapt to altered environments, which may explain why fallow deer did not re-colonise north Europe naturally after the last glaciation. It has been suggested that a relict population also survived in the Balkans, on the basis that a small number of specimens have been recovered from Neolithic sites in Bulgaria, Greece and on several Greek islands (Bökönyi 1971, 1989). It seems more likely, however, that these specimens derive from fallow deer populations imported to these areas during the early Neolithic period (Yannouli and Trantalidou 1999).

By the classical period, fallow deer had become incorporated into Greek culture; they are represented iconographically, mythologically, and zooarchaeologically within votive deposits (Boessneck 1973; Poplin 1984; Yannouli and Trantalidou, 1999). From the Balkans, the Phoenicians are thought to have transported fallow deer around the Mediterranean coast (Millais, 1906); indeed, a small number of *Dama dama* specimens have

Site	Date of Site	Description of Specimens	Reference
Westward Ho!, Devon	Mesolithic	misidentified	Churchill 1965
The Mound, Somerset	Iron Age	6 fragments (intrusive)	Darvill and Coy 1985
Lydney, Gloucester	Iron Age	shed antler	Wheeler 1932
War Ditches, Cambridgeshire	Iron Age	two pieces of worked antler	White 1964
Monkton, Kent	Roman	antler and several post cranial bones	Bendrey 2003
Canterbury Kent	Roman	1 metatarsal	Bendrey *pers. comm*
Fishbourne Roman Palace, Sussex	Roman	evidence of breeding population	Sykes *et al.* 2006
Catterick Bridge, Yorkshire	Roman	no details	Meddens 1990
Wroxeter, Shropshire	Roman	tibia shaft fragment	Meddens 2000
Redlands Farm, Northamptonshire	Roman	1 calcaneum	Davis 1997b
Barnsley Park, Gloucestershire	Roman	6 fragments	Noddle 1985c
Portchester Castle, Hampshire	Roman	11 fragments	Grant 1975
Scole Dickleburgh, Norfolk	Roman	almost complete shed antler	Baker 1998
St Albans, Hertfordshire	Roman	shed antler	O'Neil 1945
Cowdery's Down, Hampshire	Roman	radius (superior bone preservation)	Maltby 1983
Wraysbury, Hampshire	Roman	2 fragments (from mixed context)	Coy 1989a
Ipswich, Suffolk	Saxon	misidentified	Jones and Serjeantson 1983
Portchester Castle, Hampshire	Saxon	28 fragments	Grant 1976
Barking Abbey, London	Saxon	no details provided	Rackham 1994
Hereford, Herefordshire	Saxon	3 fragments	Noddle 1985a
Cheddar Palaces, Somerset	Pre-930	1 fragment	Higgs and Greenwood 1979
Brandon Road, Norfolk	10th century	distal humerus Bt 36.3mm	Jones 1993
Goltho, Lincolnshire	850-1000	antler fragments	Jones and Ruben 1987
Cheddar Palaces, Somerset	L10th/E11th	3 fragments	Higgs and Greenwood 1979
St Ebbes, Oxfordshire	10th/11th	present	Wilson *et al.* 1989
Castle Rising Castle, Norfolk	Saxo-Norman	7 fragments	Jones *et al* 1997
North Elmham, Norfolk	Saxo-Norman	13 fragments	Noddle 1980
Castle Acre, Norfolk	Saxo-Norman	1.75% of the assemblage	Lawrance 1982
Portchester Castle, Hants	Saxo-Norman	1 fragment	Grant 1977
Dudley Castle, West Midlands	Pre-1070	1 tibia	Thomas 2005
Goltho, Lincolnshire	1000-1080	25 fragments	Jones and Ruben 1987
Bath, Avon	10th/11th	4 fragments	Grant 1979
Trowbridge, Wiltshire	950-1139	4 fragments	Bourdillon 1993
Exeter, Devon	1000-1150	1 fragment	Maltby 1979
Guildford Castle, Surrey	1000-1170	3 fragments	Sykes 2005
Launceston Castle, Cornwall	1104-1175	1 fragment, many by 13th century	Albarella and Davis 1996
Dudley Castle, West Midlands	1071-1175	5 fragments	Thomas 2005
Sandal Castle, Yorkshire	1106-1188	5 fragments	Griffith *et al.* 1983
Lurk Lane, Yorkshire	1070-1188	1 fragment, many by 1190-1400	Scott 1991
St Martins-at-Palace-plain, Norfolk	11th/12th	2 fragments	Cartledge 1987
Bramber Castle, Sussex	12th century	10 fragments	Westley 1977
Goltho, Lincolnshire	1080-1150	5 fragments	Jones and Ruben 1987
Exeter, Devon	1100-1200	5 fragments	Maltby 1979
Trowbridge, Wiltshire	1139-1200	3 fragments	Bourdillon 1993
Faccombe Netherton, Hampshire	980-1204	91 fragments	Sadler 1990
Guildford Castle, Surrey	1170-1230	4 fragments	Sykes *et al.* 2005
Kings Lynn, Norfolk	1050-1250	1 fragment	Noddle 1977
Alms Lane, Norfolk	1150-1275	1 fragment	Cartledge 1985
Carisbrooke Castle, Isle of Wight	1100-1293	27 fragments	Smith 1994
Portchester Castle, Hampshire	12th/13th	7 fragments	Grant 1977
Lewes Castle, Sussex	12th/13th	19 fragments	Sykes 1997
Hereford, Herefordshire	12th/13th	2 fragments	Noddle 1985a
Eynsham Abbey, Oxford	12th/13th	6 fragments	Ayres *et al.* 2003
Copt Hay, Oxfordshire	medieval	present	Pernetta 1974
Middleton Stony, Oxfordshire	medieval	present	Levitan 1984
North Elmham, Norfolk	medieval	22 fragments	Noddle 1980
Oakhampton Castle, Devon	pre-1300	48 fragments	Maltby 1982
Castle Rising Castle, Norfolk	medieval	152 fragments	Jones *et al.* 1997

*Table 13: Principal early examples of fallow deer (*Dama dama*) as reported in the zooarchaeological literature*

been recovered from a Phoenician settlement on the island of Motoya, near Sicily (Ryder, 1975). But it was during the Roman period that the movement of fallow deer increased in scale. There are several historical records of the capture, taming, emparkment and exhibition of deer, some mentioning fallow deer by name:

> 'wild creatures, such as roe deer, chamois and also scimitar-horned orynx, **fallow deer** and wild boars sometimes serve to enhance the splendour and pleasure of their owners...Those who keep game shut up for their own pleasure are content to construct a park, on any suitable site in the neighbourhood' (Columella *De Re Rustica* Book IX, I, 1, my emphasis).

Roman texts suggest that game parks, or *vivaria*, were common in Italy from the late Republic onwards (Pliny *Historia Naturalis* Book VIII 78. 211; Varro *Rerum Rusticarum* Book III.12.1-2; see Toynbee 1973, Anderson 1985 and Starr 1992 for an overview). Interestingly, a *vivarium* has been identified archaeologically at the Settefinestre Villa where fallow deer remains have also been noted (Carandini 1985; King 1985). Several other rural sites in southern Italy, and also Portugal, have yielded small numbers of fallow deer bones but there is little evidence to suggest that there were established free-living populations (King 2002; Mackinnon 2004; Davis 2005). According to Columella (*Res rustica* IX. I, 4), the chieftans of Gaul also set up game parks, a reference which has led some researchers (for example, Lever 1979) to suggest that the Gauls themselves were responsible for the introduction of fallow deer to Britain. Similarities between the material culture from pre-Roman sites in Gaul and south-east Britain indicate considerable cross-channel contact during the pre-Roman period (Millett 1998). There is, however, little archaeological evidence to support a pre-Roman introduction.

Table 13 presents the results of a zooarchaeological literature survey for references to archaeological fallow deer and only three Iron Age assemblages are claimed to contain *Dama* specimens. Of these, the remains from The Mound are said to have been medieval intrusions (Darvill and Coy 1985), whilst the material from the Lydney (Wheeler and Wheeler 1932) and War Ditches (White 1964) were both shed antler and, as such, cannot be taken as positive evidence for the presence of living animals.

Fallow deer are better represented on sites dating to the Roman period; nevertheless, they are rare considering the large number of Roman sites that have been excavated. The specimens from Roman Portchester Castle have always been the lynch-pin in the argument that the Romans introduced fallow deer but re-analysis of the assemblage suggested that the purported *Dama* specimens were actually misidentified (Sykes 2004). More recently, however, evidence for a Roman introduction has been amassing. Bendrey's (2003) finds of post-cranial fallow deer bones from the late-second to fifth-century site of Monkton in Kent provided the first clear hint that fallow deer were established in Britain whilst it was under Roman control. This possibility was proved conclusively when fallow deer remains from Fishbourne Roman Palace, Sussex, were AMS dated to the first century AD. Teeth from two mandibles were submitted for strontium isotope analysis, a geochemical provenancing technique, which revealed that one individual was imported to Fishbourne as a fawn whilst the other was born and raised at the site, indicating that captive breeding populations were established in Romano-British parks (Sykes *et al.* 2006). There is, however, a big step from the localised breeding of exotica to their establishment as a naturalised fauna and, at the moment, the evidence suggest that the Romans did not succeed in bridging this gap; fallow deer appear to have died out with the withdrawal of the Roman Empire.

Although several historians have suggested that fallow deer were present in the landscape of Saxon England, the evidence upon which their assumptions have been based is questionable. For instance, *Aelfric's Colloquy* is often cited as referring to fallow deer, or *dammas*, but it is clear that this is a mistranslation – the Saxon word for roe buck '*rann*' having been confused with the Latin *dam* (Yalden 1999,153). Similarly, Hough's (2001) argument that the place-name elements *fealu*, *pohha* or *pocca* refer to 'fallow deer' is unconvincing (Cullen and Parsons *pers. comm.*). Nor is there any substantive zooarchaeological evidence to suggest that fallow deer were present in Saxon England: a few site reports do mention the possible presence of fallow deer but nearly all the specimens are either antler fragments (Jones and Ruben 1987), dubiously identified (for example Jones and Serjeantson, 1983) or poorly dated (Jones 1995). Two assemblages – those from Portchester Castle and the Cheddar Palaces – have been responsible for maintaining the case for a Saxon establishment of fallow deer; however, again re-examination of the material was unable find any evidence to corroborate the presence of fallow deer on these sites (Sykes 2004). During analysis all deer bones were measured and comparison with the archived metrical records confirmed that the specimens originally recorded as fallow deer had been misidentified: Figure 76 shows that a putative *Dama* humerus from Portchester Castle falls more comfortably within the size range for roe deer.

Before it is concluded that fallow deer were not present in pre-Conquest Britain, the specimens recorded for Barking Abbey (Rackham 1994, 131) and Hereford (Noddle 1985a) should be mentioned. Neither report provided details of the specimens or mentioned the significance of their presence. The material from these sites would clearly benefit from re-examination, yet, even if these specimens prove to be correctly identified and dated, the case for fallow deer living free in the Saxon landscape would remain weak.

Today most scholars assume that it was the Normans who successfully brought, and established, fallow deer from across the Channel. The Anglo-Saxon Chronicle's obituary

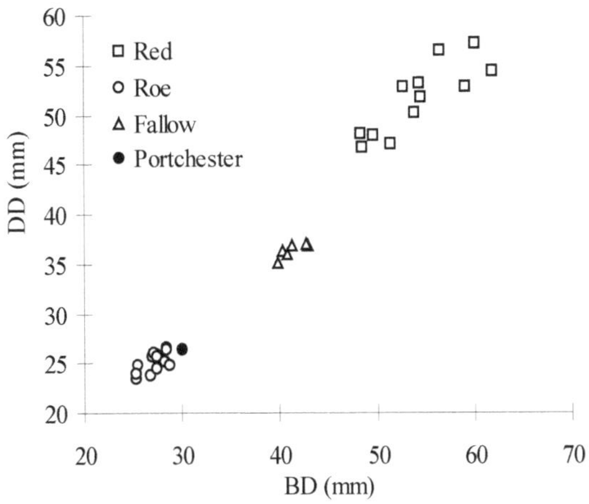

Figure 76: Measurement of the Portchester Castle humerus – breadth of distal end (BD) plotted against depth of distal end – against the distributions for adult red, roe and fallow deer from reference collections and securely identified archaeological specimens

of William the Conqueror does not mention fallow deer, even though the native red and roe deer are both recorded as the King's favourites (Garmonsway 1967); however, correctly identified specimens from Goltho (1000-1080) and Dudley Castle (pre-1070) indicate that they were present in England at this time, and I would propose a late eleventh-century date for the introduction of this species. Table 13 shows that from the mid-eleventh century fallow deer are represented increasingly, and by 1180 Alexander Neckham was able to list fallow deer amongst the animals commonly available in Britain (Bartlett 2000, 673).

Motivation for the introduction of fallow deer has been linked to the Norman's 'love of the chase' and their desire to introduce the hunting parks that were, apparently, common in pre-Conquest Normandy (Gilbert 1979; Rackham 1997, 123; Yalden 1999, 153; Gautier forthcoming). Henry I (1100-1135) seems a likely candidate for their importation. He devoted much of his time to the study of venery and his dedication to the sport was well known amongst his contemporaries, William of Warenne nicknaming him 'Stagfoot' because of it (Mayr-Harting and Moore 1985, 30). According to William of Malmesbury, Henry was also intrigued by exotic animals, collecting lions, leopards and camels in his park at Woodstock (Bartlett 2000, 672). The menagerie is also said to have included deer and it seems probable that fallow deer formed part of the collection (Clark 2006, 18). Trade in exotic animals was commonplace amongst medieval royalty, animals either exchanged between leaders to cement political relationships or taken from conquered lands as a metaphor for empire. As early as the ninth century Charlemagne was sent an elephant, called Abu l'Abbas, by the Abbasid caliph (Hodges 2000, 36).

It makes sense that fallow deer were originally maintained in menageries because, despite their capacity for rapid breeding, it would have taken many generations for them to become established as the premier park species. For the first few decades after their introduction, fallow deer would have been a great rarity, prized exotica to be observed and admired rather than chased and eaten. As such, the link between fallow deer and hunting parks may have developed at a later date, perhaps supporting Liddiard's (2003; forthcoming) contention that parks were not, as is so frequently suggested, a Norman introduction.

Whilst it may be acceptable to remove parks from the Norman Package, the possibility that fallow deer were imported from established populations in pre-Conquest Normandy deserves consideration. A quick survey of the zooarchaeological literature for northern France reveals that fallow deer cannot have been imported from Normandy: they simply were not present. According to Clavel (2001) and Lepetz and Yvinec (2002) fallow deer are absent from French assemblages pre-dating the thirteenth century, a situation which Audoin-Rouzeau's (1993) survey suggests is echoed across northern Europe. Rather than fallow deer being imported to Britain from Normandy, French zooarchaeologists believe that fallow deer were introduced to France from England (Yvinec *pers. comm.*). Whilst this suggestion contradicts long-standing British perceptions, there is historical evidence to suggest that the movement of fallow deer went in directions opposite to those we might expect: the Close Rolls for 1238 mention that King Henry III ordered 120 live fallow deer to be boxed up and sent to Flanders, no doubt, as Rackham (1997, 131) suggests, to set up a game park. It would seem therefore that, as with other elements deemed to be part of the 'Norman package', the introduction of fallow deer to England was more complex than would first appear.

Zooarchaeologically, the eleventh-century distribution of fallow deer is little different to the Classical period, with populations confined to southwest Asia and southern Europe (Sykes 2004). Because the distribution of fallow deer was restricted, the possible sources of British fallow deer are also limited; Anatolia, Greece or Sicily being the most likely areas of origin. Rowley (1999, 104) suggested that the Normans might have discovered fallow deer in the Levant whilst on the first crusade. Normans were certainly based in Antioch until it fell to Moslem armies in 1268; however, the only *Dama* species available in this area was the Persian variety *D. d. mesopotamica* so this scenario seems unlikely (Massetti 1996).

That fallow deer were present in England but absent across the rest of northern Europe, is reminiscent of the pattern of occurrence noted for the 'unmaking' rituals: is it possible that fallow deer too were imported from the Norman Kingdom of Sicily? It is a strong possibility since fallow deer were established on the island at this point (Bossard-Beck 1984). As was mentioned in the previous chapter, the late eleventh century saw stronger connections between the Normans of England and Sicily than between those of England and Normandy, perhaps explaining why fallow deer were established in Britain long before they arrived in northern France.

7.1i: Summary

For many years the introduction of fallow deer into Britain has been discussed and various theories concerning the peoples responsible have been proposed. By undertaking a survey of European site reports, and discounting dubiously-identified and poorly-dated specimens, it has been possible to reconstruct the ancient biogeography of fallow deer. It is clear that fallow deer have long held a cultural significance, their reintroduction to Europe after the last Ice Age being due entirely to human transportation.

Their desirability inspired many peoples to attempt their emparkment, breeding populations even being established in Roman Britain. It was not, however, until the eleventh century that fallow deer were successfully re-established in northern Europe when the Normans of England took receipt of animals sent, most probably, from the Norman Kingdom of Sicily. It was the transportation and subsequent establishment of this beautiful species that is perhaps the greatest legacy of 1066, the Norman Empire achieving where the Roman Empire failed.

7.2: Rabbits

In 2003 Callou published her excellent research, which utilised osteometrics, osteomorphology, genetics, iconography and historical evidence to chart the natural history of the rabbit. Her work focused on mainland Europe, greatly enhancing our knowledge of the species' biogeography in this area, but she was also able to touch on the evidence from Britain, concluding that rabbits were brought to the island some time during the late eleventh or twelfth century. Whilst in accord with much of the existing British zooarchaeological and historical literature, the proposed date range offers no resolution to the question of a Norman introduction. This section seeks to build on Callou's work and refine the dating evidence for the rabbit's arrival in Britain.

Rabbits have only been referred to as such for the last two hundred years. Prior to this they were known as 'conys' and only juvenile animals were called 'rabettes'. Cony derives from the Latin '*cuniculus*', which, like the animal itself, is of Iberian origin: butchered rabbit remains have been recovered from many prehistoric and historic sites across Portugal and Spain (Davis 2005). Their presence on the peninsula was first recorded by the Phoenicians who colonised the coastal regions around 1100 B.C. (Zeuner 1963, 410). Outside Iberia, the earliest reference to rabbits dates to the second century B.C., the Greek historian Polybius mentioning their presence on the island of Corsica: although Callou (2003, 230) suggests that the text is actually referring to another lagomorph *Prolagus sardus*. By the first century B.C. rabbits were being managed in Italy: Varro (*Rerum Rusticarum* 12.7) recommended that they be kept in *leporaria*. Within two hundred years they had been taken to south-west France, and breeding populations were apparently established by the sixth century: in 590 Bishop Gregory of Tours is said to have scorned the Lenten consumption of foetal rabbits ('laurices'), which were not considered to be meat and so could be eaten on days of fast (Zeuner 1963, 413). The desire of monks to acquire foetal rabbits is seen as one of the main driving forces of rabbit domestication (Nachstein 1949). It has been suggested that between the sixth and tenth centuries, rabbits were bred within the walls of French religious houses and that the practice gradually spread from monastery to monastery (Zeuner 1963; Van Damme and Ervynck 1988). Surely it can be no coincidence that the earliest archaeological rabbits have been recovered from monastic sites: those from the eleventh- to twelfth-century site of Charité-sur-Loire in France, and the late twelfth/thirteenth century specimens from the Belgian sites of Ename Abbey (Ervynck *et al.* 1999) and Dune Abbey (Gautier 1984). Nevertheless, rabbit breeding can have been common only in southern and central France because in 1149 the abbot of the Benedictine monastery of Carvey, on the Wesser (Germany) asked the abbot of Solignac, in south-west France, for two pairs of rabbits (Nachstein, 1949). Even by the late thirteenth century, Peter de Crescentiis of Bologna wrote that rabbit distribution was restricted primarily to Spain, Lombardy and Provence (cited in Veale 1957). The work of both Clavel (2001) and Callou (2003) suggests that, in northern France, rabbit remains are scarce in assemblages dating before the twelfth century.

The earliest British references to rabbits are presented below in Table 14. There is one reference to rabbits in Anglo-Saxon England which comes from the Marksbury charter, dated nominally to A.D. 936; however, this document is an unreliable source, possibly a thirteenth-century forgery. Alexander Nequam's list of English wildlife, written in the 1180s, makes no mention of rabbits on mainland England, although fallow deer are recorded (cited in Bartlett 2000, 672). At this point, rabbits appear to have been kept on small off-shore islands such as Drake's Island (recorded for 1135), Scilly (1176) and Lundy (1183-1219). The earliest reference to rabbits on the mainland comes from a deed, dated to between 1135 and 1272, relating to the Dartmoor warren (Henderson 1997, 105). Rabbits could, of course, have been established in England for some time prior to these first records. This theory was taken to the extreme by Warry (1988) who argued that the rabbit was native to Britain and that its absence in the historical record is due simply to taxonomy: that before the twelfth century, rabbits and hares were known collectively as '*lepus*' and that the introduction of the word 'coney' has, mistakenly, become associated with the arrival of the animal itself. Whilst this theory rather flies in the face of popular opinion, it does highlight a significant flaw of modern zooarchaeology: that researchers tend to project current meanings and systems of classification, namely Linnaean, back on to past societies. As Serjeantson (2000a) has shown, there are many ways, other than on the basis of genetic similarity, in which animals can and have been classified: age, sex, size, appearance, manageability, the list goes on. In ancient Mesopotamia, for instance, sheep

Date	Reference
936	Survey attached to a charter of Marksbury (Somerset) mentions *Conigrave* 'Rabbit Grove' (cited in Grundy 1932)
1087	The Domesday Book records one *Warenna leporum* at Gelston in Lincolnshire (cited in Darby 1977)
1135	Ownership of Drake's Island '*cum cuniculi*' (with rabbits) recorded as transferring to Plympton Priory (cited in Henderson 1997, 105)
1176	Richard de Wyka granted 'all my tithes from Scilly, and expressly those from the rabbits' to Tavistock Abbey (cited in Finberg 1947, 365).
1183-1219	At some point during this period the tenant on Lundy Island was given entitlement to take 50 rabbits from certain *chovis* on the island (cited in Veale 1957)
1187-94	A grant of land made by Simon le Bret to the canons of Waltham (Essex) mentions that he had a *cunicularium* there (cited in Bartlett 2000)
1204	Rights '*in warennis et cunigariis*' are mentioned in a charter granting land in Connaught to Hugh de Lacy (cited in Veale 1957)
1209	King John gave permission to William Picot to take roe deer, hare, rabbit, pheasant and partridge on his land in Lincolnshire (cited in Veale 1957)
1225	A *custod' cuniculorum* was mentioned on the manor of Bowcombe, Carisbrooke, on the Isle of Wight (cited in Veale 1957)
1235	The *Close Rolls* record that Henry III presented a gift of *decem cuninos vivos* from his park at Guildford (cited inVeale 1957)
1240	King Henry III ordered a supply of rabbits for his Christmas feast. From this date the Liberate Rolls (for the years 1240-5 and 1245-51) record numerous orders for rabbits (cited in Veale, 1957)
1340	The *Inquisitiones nonarum* records 100 acres of land in Ovingdean (Sussex) which were devastated by William de Warrene's rabbits (cited in Blaauw 1847)

Table 14: Early historical references to rabbits

and goats were viewed as the same taxon, '*senu*' (Hesse and Wapnish 1985, 11). Likewise, the Anglo-Saxons gave porcupines the same name as hedgehogs '*igil*' (Crossley-Holland 1993, 91). Even today the animal named the 'Belgian Hare' is actually a rabbit (Robinson 1984). Perhaps the strongest indication that rabbits could have been classified together with hares is provided by the Roman author Varro (*De re rustica*, Book XII, 5,) who stated that 'there are three kinds of *Lepus*, of which *cuniculus* is one'. If this were the case it may account for the Domesday reference to the '*warenna leporum*' at Gelston (Lincolnshire). Bond (1988) argues that this should be regarded as an enclosure specifically for hares but this seems unlikely on the basis of their solitary behavioural ecology. If rabbits were present in pre-Norman Britain, it might be expected that their remains would be represented zooarchaeologically. Indeed, considering the uncertainty of the historical evidence, a study of the rabbit's archaeological presence may provide the best opportunity for determining the date of, and Norman involvement in, their introduction.

The main sites on which rabbits have been recorded are listed in Table 15. Their presence in some prehistoric assemblages could be taken as corroborative evidence for Warry's (1988) theory; however, most of these early specimens or the deposits from which they were recovered are of doubtful date. The specimen from the Mesolithic site of Thatcham, Berkshire, is a classic example of the effects of bioturbation, radiocarbon dating confirming the specimen as an eighteenth century intrusion (Cowan 1991). Similarly, the Bronze Age deposits from Brean Down, Somerset, were noted as being disturbed by rabbit burrows (Levitan 1990) and the rabbit remains from Barton Court, Oxfordshire (Miles 1984) and West Stow, Suffolk (Crabtree 1990) are also thought to be intrusive. In most cases these conclusions have been based on the abundance of partial, often juvenile, skeletons, suggestive of a situation where infant rabbits have died in their burrows. The assemblage from the Roman site of Carne's Seat in Sussex also contained large number of juvenile lagomorph bones, again recorded as intrusive (Beech 1986).

For many years the only Roman specimen that was not obviously intrusive or misidentified was that from Latimer in Buckinghamshire (Hamilton 1971); however, as with fallow deer, new evidence supporting a Roman introduction has recently come to light, with butchered rabbit bones being recovered from sealed Romano-British deposits from Lynford in Norfolk and Beddingham in Sussex. Neither set of remains have, as yet, been radiocarbon dated but the associated pottery suggests first and third century dates respectively (Curl *pers. comm.*). Interestingly, all the specimens are said to be noticeably smaller than modern British animals, closer to the size exhibited by Mediterranean populations: it has been demonstrated repeatedly that the geographic variation in rabbit size obeys Bergman's (1847) Law concerning the relationship of environmental temperature

Site	Date	Description	Reference
Thatcham, Berkshire	M	Intrusive, C14 dated to 1710	King, 1962
Brean Down, Somerset	BA	A number of specimens noted as intrusive	Levitan 1990
Croft Ambrey, Hereford	IA		Whitehouse 1974
Barton Court, Oxford	RB	Small number of bones thought to be	Miles 1984
Latimer, Buckingham	RB	Single bone - status uncertain	Hamilton 1971
Gorhambury, Hertfordshire	RB	2 specimens	Locker 1990
Carne's Seat, Sussex *	RB	278 bones of adult and juvenile 'hare' classified as intrusive	Beech 1986
Catsgore, Somerset	RB	Small number of bones - status uncertain	Leech 1982
Hambledon, Buckingham	RB		Cocks 1921
Beddingham, Sussex	RB	butchered bones from 3rd century context	Curl *pers. comm.*
Lynford, Norfolk	RB	Butchered bones from 1st century pit	Curl *pers. comm.*
West Stow, Suffolk	MS	197 'probably intrusive' rabbit specimens recovered as partial skeletons	Crabtree 1990
Brandon, Suffolk	MS	32 specimens	Crabtree 1996
Abbots Worthy, Hampshire	MS	Single rabbit bone from pit F7341	Coy 1987
Ribblehead, Yorkshire	LS		Rackham nd
Burpham, Sussex *	LS	Small number of bones recovered from a pit	Sutermeister 1976
Steyning Village, Sussex*	LS	11 of 17 pits said to contain rabbit bones	O'Shea 1993
Cheddar Palaces, Somerset*	LS	Tibia (proximal end unfused) from pre 930	Collected personally
16-22 Coppergate, Yorkshire	LS	1 specimen from late 9th-mid 11th -century	O'Connor 1987
North Elmham, Norfolk	SN	L9th/10th century 1 specimen; 6 'Saxo-Norman' bones and 3 from 'Medieval' deposits	Noddle 1980
Goltho, Lincolnshire	SN	2 specimens dating to 1000-1080 and 1 fragment 1080-1150	Jones and Ruben 1987
Faccombe Netherton, Hampshire *	SN	24 bones recovered from Saxon layers but believed to be intrusive, although some are possibly contemporary.	Sadler 1990
St Martin-at-Palace, Norwich	SN	1 specimen	Cartledge 1987
Rayleigh Castle, Essex	SN		Hinton 1912-13
Redcastle Furze, Norfolk	SN	E/M 11th century 1 tibia L 11th century 1 mandible, 1 pelvis	Wilson 1995
Thetford, Norfolk	SN	10th century 1 specimen 11th/12th century 1 specimen	Jones 1993
St Nicholas' Street, Norfolk	SN	11th and 12th centuries - 8 specimens	Hutton MacDonald 1999
Brighton Hill South, Hampshire	SN	M-L 11th-12th century 165 rabbit bones	Coy 1995
Eynsham Abbey, Oxford	SN	Single bone from L11th/12th century deposit	Ayres *et al.* 2003
Castle Rising, Norfolk	SN	10 specimens dating to the Saxo-Norman period and 333 from 'Medieval' deposits. Bone fusion and element representation data suggest that they may be intrusive.	Jones *et al.* 1997

*Table 15: Principal early examples of rabbits (*Oryctolagus cuniculus*) as reported in the zooarchaeological literature.*

M= Mesolithic, BA = Bronze Age, IA = Iron Age, RB = Romano-British, MS = Mid-Saxon, LS = Late Saxon, SN = Saxo-Norman

to an animal's surface-area/body-volume ratio, with animal size increasing with latitude (Vanderford 1997; in particular see Callou 2003, 78). This suggests that the Roman specimens were imported directly from southern Europe, again indicating that no populations existed in the northern part of the Roman Empire.

Whilst the Lynford and Beddingham evidence seems (at present) insufficient to conclude that *Oryctolagus cuniculus* was established in Roman Britain, the studies do demonstrate that detailed examination of rabbit bones can provide substantive new information about ancient animal biogeography. Unfortunately, few zooarchaeological reports provide sufficient details of the context, taphonomy and metrics of rabbit bones to allow a comprehensive study of the medieval situation. With help from the zooarchaeological community, it has been possible to gather a small number of measurements from butchered, gnawed and burnt specimens from thirteenth- to eighteenth-century contexts, which provide a useful dataset (see Appendix IVb); however, I found it necessary to re-analyse several of the assemblages, those marked with an asterisk, shown on Table 15, to collect data pertaining to the more pivotal ninth to twelfth centuries.

Within Sussex there are two Late Saxon sites – Burpham (Sutermeister 1976) and Steyning (O'Shea 1990) – that purportedly produced rabbit bones from well-stratified pits. During re-analysis, I was unable to locate the rabbit mandible mentioned in Sutermeister's (1976) report and it seems possible that, because he was aware of the specimen's significance, it was removed from the assemblage and has not since been reintegrated. Without being able to confirm the identification and status of the Burpham specimen it cannot be used as positive evidence for a pre-Norman rabbit introduction. By contrast, the material from Late Saxon Steyning, where rabbits are said to have been recovered from '11 of the 17 sealed pits' (O'Shea 1993), proved especially informative. O'Shea's exemplarily detailed inventory made it possible to target the apparently rabbit-yielding contexts for analysis. One deposit in particular was described as containing just fifteen specimens, including two rabbit femora 'of different sizes, one from the left and one from the right'. On re-analysis, the material conformed exactly to the catalogue except the two femora belonged to domestic fowl rather than rabbit. Indeed, I was unable to identify any of the rabbits noted in O'Shea's records, although a number of bird and hare bones did fit his descriptions. On this basis it seems fairly clear that the specimens from Steyning were originally misidentified, again weakening the case for the existence of a rabbit population in Late Saxon England. If these two sites are representative of the situation further a field, the evidence would seem to support Bond's (1988) argument that rabbits were absent from pre-Conquest Britain.

Iconographic evidence is frequently cited in support of a Norman introduction date for rabbits: lagomorphs are famously depicted in the Bayeux Tapestry and also on

Figure 77: Norman rabbit or 'hare and the hound'? Corbel from Kilpeck Church, Herefordshire. Photo R. Jones

one of the corbels of Kilpeck Church in Herefordshire (Figure 77). It is debateable, however, whether these representations are actually rabbit or hare; I would argue that their association with hounds is more suggestive of the latter – the traditional 'hound and the hare' – chiefly because rabbits were not usually caught through coursing. Even if these depictions are rabbit, they need not indicate the presence of this species in England, after all, lions, camels, centaurs and griffins are also illustrated on the Bayeux Tapestry. Zooarchaeologically there is, apparently, more evidence for the presence of rabbits in Norman period assemblages (Table 15) but this picture may simply be the creation of researchers who, understandably, accept misidentifications or intrusive specimens because the pattern fits the school-book dictum. Many of these 'Norman' examples – for example those from Brighton Hill South, Hampshire (Coy 1988) – are clearly intrusive and are documented as such, whilst others – notably those from Castle Acre (Lawrance 1982) – look suspicious and warrant re-analysis.

My attempts to re-examine the Norman period assemblage from Bramber Castle in Sussex proved problematic because many of the bones had been discarded after the original analysis (Westly 1977) and it was not possible to draw any conclusions on the basis of the archived material. Re-analysis of the assemblage from Faccombe Netherton, Hampshire, was more fruitful. Whilst a large number of the rabbit remains from the Norman contexts are thought, with good reason, to be intrusive, Sadler's (1990) report mentions that a few specimens were retrieved from undisturbed contexts within sealed features dating to between AD 980 and 1180. Examination of these specimens revealed that their preservation and appearance was consistent with the wider assemblage and, although none exhibited butchery, gnawing or burning marks, one specimen – a femur – was measurable having a smallest diameter (SD) of 6.1mm

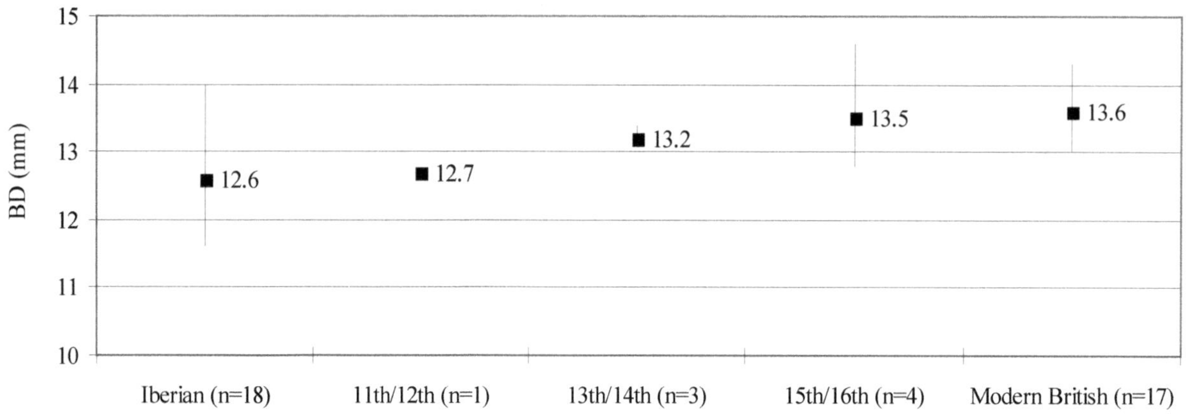

Figure 78: Measurement of eleventh/twelfth century rabbit femur – breadth of distal (BD) – from Faccombe Netherton shown against the range and mean for securely dated archaeological specimens and modern reference material

and a maximum distal breadth (Bd) of 12.7mm. Figure 78 shows that this specimen is rather small compared to the later medieval rabbits and, although the data are extremely limited, it is tempting to suggest that it represents an early generation import from southern Europe. The same may be true for a specimen recovered from a late twelfth-century context at Clay Hill Castle, Sussex (Sykes n.d.). In this case the specimen was a distal right humerus which came from an undisturbed level and was confirmed as archaeological on the basis that it had been burnt. Measurement showed it to be small (SD3.5mm and Bd 8.2mm) relative to later medieval and modern rabbits, its size again being more comparable with south European animals (Figure 79). The dating of the Clay Hill example puts it beyond the Norman period and whilst the date range for the Faccombe Netherton assemblage spans both the Late Saxon and Norman periods it seems most probable that the bone dates to the later part of the range, the late twelfth-century.

When these zooarchaeological data are combined with the textual, iconographic and landscape evidence, which suggest rabbits and warrens began to appear only from the thirteenth century (Henderson 1997), it is difficult to substantiate the claim of a Norman introduction. Instead it seems more likely that rabbits formed part of a late twelfth-century package of ideological reform, which saw the emergence of new attitudes to nature and a revival of classical manicured parks which were founded to contain, amongst other animals, rabbits (Clark 2006, 18; Sykes in press).

7.2i: Summary

It has long been thought that the Normans were responsible for introducing rabbits but when all the data are considered it seems likely that the species was successfully established in Britain, perhaps after a failed Roman attempt, during the Angevin period, most probably during the 1170s or 1180s. There is little evidence to suggest that rabbits were present in northern France much before they arrived in England, appearing in both regions at approximately the same date. If this is the case, the idea that rabbits were introduced by the Normans, as part of a pre-existing 'Norman package' is difficult to uphold. Debate concerning the introduction of the rabbit will, no doubt, continue until a comprehensive radiocarbon dating programme has been undertaken (Lauwerier and Zeiler 2001) but when this is achieved I suspect that it will prove conclusively that attributing the Normans with the introduction of rabbits has been the most enduring of all their 'Myths' (RCH Davis 1979).

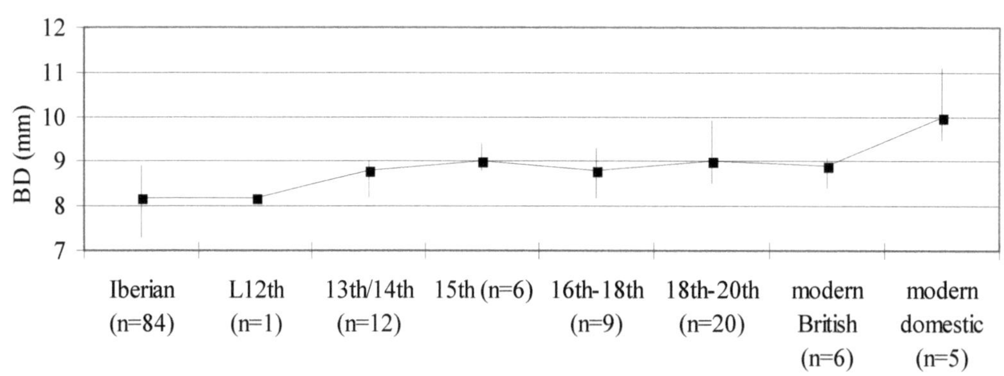

Figure 79: Measurement of the late twelfth-century rabbit humerus – breadth of distal (BD) – from Clay Hill, Sussex shown against the range and mean for securely dated archaeological specimens and modern reference material

7.3: Conclusion

As an island, most animal introductions to Britain have been anthropogenic, purposeful and socially meaningful. Yet research into the establishment of non-native fauna has traditionally been the preserve of zoologists, conservation biologists and natural historians, interested predominantly in the ecological impact of their arrival (Lever 1979, Yalden 1999). Whilst these are important issues in their own right, this chapter has shown that there exists considerable potential for viewing the 'ecological' data from an archaeological perspective, so long as the existing data are examined critically.

The evidence for fallow deer has hinted that such approaches can inform on patterns of trade, exemplified by the apparent connection with Sicily, but also broader questions of culture and ideology: that the movement and maintenance of fallow deer not only demonstrated power and authority but, in the early years of their arrival, must also have served as a metaphor for the Norman Empire.

Motivation for the introduction of rabbits is likely to have been equally significant but circumstances surrounding their arrival are still far from clear. It is possible to conclude, with some confidence, that rabbits were not a Norman import, more probably arriving in the later part of the twelfth century. Their eventual establishment can be linked to the growing interest in nature and exotic animals that this period witnessed, a movement also responsible for the proliferation of parks and fallow deer populations. This serves to remind us that whilst the introduction of new faunal species is culturally driven, their importation and establishment may be caused by factors other than 'invading peoples'.

Whilst mirroring cultural change, the introduction and management of exotic animals also impacted on human society, culture, lifestyles and whole landscapes. Together these changes are reflected in cuisine and the possibility that Conquest-related shifts in consumption practices are detectable zooarchaeologically is the subject of the next chapter.

Chapter 8: Cooking, Class and Cultural Identity

Foodways are culturally determined, developing through complex social, political and economic interactions. As a result, diet and consumption practices reflect not only a group's social structure but can also, potentially, provide substantial information concerning the collective beliefs of the consumers. For this reason food has, for many decades, been the focus of anthropological work that has sought to explain diet as a socio-cultural system and understand the impetus behind changing food habits (for instance Lévi-Strauss 1966; Douglas 1975; Goody 1982; Counihan and Van Esterik 1997; Caplan 1997).

Mennell (1985) demonstrated that the cuisines of England and France developed in tandem. He was unable, however, to find historical evidence for ethnic-based dietary difference and argued that variation was greater between the rich and poor than between the two nations. Medieval France and England were socially, politically and religiously very similar and overall differences between their foodways may not have been particularly dramatic. In general, it is only when groups become inter-dependent that the differences between them become apparent and group identities, which often include dietary signifiers, are formed. The Norman Conquest and its ensuing changes in social stratification, economy and resource access must have produced precisely this kind of climate and it seems likely that dietary shifts, resulting from both internal and external influences, occurred during the period. Food terminology certainly changed after the Conquest, at least at the aristocratic level. For instance, the earliest English cookery book, attributed to the twelfth-century monk Alexander Neckham, used little English terminology and most of the text was written in Norman-French and Latin (cited in Goody 1982, 136). More famously, choice joints of meat were re-named with French titles – 'beef' (*boeuf*), 'veal' (*veau*), 'mutton' (*mouton*) and 'pork' (*porc*), whilst the poorer cuts retained their Saxon names, such as ox tail (Goody 1982, 136; Davidson 1999, 67, 604). These changes in terminology may reflect wider changes in food distribution and consumption, and it seems possible that studies of Saxon to Norman diet may have much to reveal about the Conquest.

Despite this potential, the Norman period has been neglected in terms of dietary studies; especially when compared to the well researched earlier (Hagen 1998; 2002; Magennis 1999; Banham 2004; Gautier 2006) and, in particular, later medieval periods (Dyer 1988 and 1994; Hammond 1995; Mennell 1985; Woolgar 1999). Availability of historical evidence is clearly the reason for this gap but whilst the lack of eleventh and twelfth-century documents has restricted studies of Norman period diet, the historical silence actually speaks volumes about attitudes to food: clearly it was not done to be preoccupied with dining. Indeed, from the limited historical evidence that we do have, the overall impression provided is one of Norman abstinence – a sort of dietary piety. It is clear from historical evidence that the Norman *Gens* deliberately cultivated a pious cultural image, attempting to show themselves as abstemious in both their lifestyle and dining traditions (Potts 1995, 149; Bliese 1991). The Bayeux Tapestry, for instance (see Figure 80) portrays William and his men in a pious light, staged in a 'Last Supper' scene eating food blessed by the bishop. By contrast Harold's retinue is shown quaffing alcohol from vast drinking horns. Although clearly dramatic devices, the depictions do appear to reflect contemporary perceptions of the two ethnic groups. As Thomas (2003, 299-302) and Gautier (n.d.) have shown, the overwhelming stereotype presented in the historical texts is that the English were preoccupied with eating and drinking, and that few other *Gens* could compete with their levels of feasting.

Communal consumption was a feature of Anglo-Saxon society and a tradition common to all Germanic groups (Brown 1988; Thompson 1995; Magennis 1999; Effros 2002; Pollington 2003). The sense of belonging provided by the feast hall is demonstrated by the tenth-century poem *The Wanderer*, which describes the despair of those forced to leave it (Leslie 1966). In some historical texts the Saxon devotion to feasting was reported in a positive light,

Figure 80: Dining scenes from the Bayeux Tapestry: Harold and his men drinking in a hall at Bosham, Sussex (left) and William and his men on the eve of the battle (right)

as a trait of wealth, nobility and generosity (Thomas 2003, 299). But more often it was linked to sinfulness, and through the course of the medieval period the Germanic feast hall traditions saw the English gradually gain a reputation as irreligious gluttons (Hagen 1998, 103; Short 1995). According to Thomas (2003, 302), this image of the English may have fuelled Norman feelings of superiority, especially since their dining habits were seen as 'delicate but not to excess' (William of Malmesbury *Gesta Regum Anglorum* 245 cited in Handcock 1996, 316).

The degree to which these stereotypes reflect actual cultural practices would appear, superficially, to be beyond our grasp. However, this chapter seeks to explore whether the Norman aristocracy deployed dietary piety as a mechanism for displaying their moral superiority over the conquered Anglo-Saxon population. Zooarchaeological analysis does not enable consumption rates to be calculated, thus the degree of 'gluttony' in pre- and post-Conquest England cannot be ascertained. Nevertheless, Christian texts mention the dietary restrictions imposed by the Church and indicate the type of symbolic foods that might be eaten by 'good Christians'. By using animal bone assemblages from religious houses as a benchmark for pious consumption, the data for elite Saxon and Norman sites will be considered to see which group, if either, was more observant of religious dietary restrictions.

For the medieval period there is every reason to suspect that evidence for group identity might be gleaned from zooarchaeological studies of diet because, from at least the time of Galen (AD 129-*c*.210), the phrase 'you are what you eat' had more than a proverbial significance. It was widely accepted that every living organism was composed, in varying proportions, of the four elements (earth, air, water, fire) which determined the plant's/animal's 'humour'. This humour, or character, could be transferred to any person who consumed it – visually, audibly or orally (Salisbury 1994, 44; Scully 1995, 41; Grant 2000). Because it was believed that human behaviour was fundamentally influenced by diet, it really *mattered* what people ate, and particular ingredients, especially flesh foods, were deliberately consumed or avoided by different gender, social and ethnic groups.

Galenic philosophy meant that medieval cooks were required to know how to amend a food's natural temperament to suit particular individuals (Scully 1995). For example, if a meat was perceived as excessively cold and moist, it had to be corrected by the drying and heating process of roasting. Baking was a method of moderately drying and heating food, whereas boiling warmed moderately but also added moisture to dangerously dry foods. Salt was considered to be very dry and warm, and could, therefore, be used to counteract problems of decay, which were associated with all things moist. Since this thinking seemingly encompassed all Christian cookery in medieval Europe, it is debatable whether the Normans could have introduced novel cooking techniques to post-conquest England; however, the possibility that the Conquest resulted in the introduced new styles of food preparation and consumption will be considered.

Essentially, this chapter draws together data from the previous seven chapters to examine whether the Norman Conquest influenced what and how food was eaten; whether dietary studies can provide evidence for changes in social structure (in particular the emergence of social hierarchy); and to investigate the possibility that the Normans used foodways, in particular expressions of 'dietary piety', to construct and reproduce their identity. To best achieve these aims, this study will focus on the upper echelons of society, where Norman influence was most marked.

8.1: A Colonial Diet?

Diet is often invoked, sometimes simplistically, in archaeological discourse concerning population migration; studies of transitional periods frequently suggesting that the arrival of novel dietary elements, the uptake of new food preferences/taboos or the adoption of different processing techniques reflect the importation of colonial foodways: for instance see Richards (2003) for the Mesolithic/Neolithic transition, Price and Getsdóttir 2006 for the Viking colonisation of Iceland, and King (1978 and 1991) or Dobney (2001) for the Iron Age/Roman transition. Studies of modern cuisines suggest that colonial and migrant diets are seldom so easily separable from those of the wider population (Caplan 1997, 13). Differences do exist, however, and there is potential for archaeological examples to be detected through animal bone studies (O'Connor 2007, 5).

In the preceding chapters it has been shown that new food animals were introduced to post-Conquest England whilst other native species began to be exploited in different quantities. In particular, elite sites demonstrate an increased representation of pigs and herons, whilst the frequency of domestic birds rises dramatically from 9.6% to 19.5% (Figure 39). It was suggested in Chapter 3 that these last shifts do reflect continental influence, elite sites in northern France having similarly high levels of domestic bird representation. Linguistic evidence supports the hypothesis, the word *capun* (capon) appears in England for the first time around the date of the Conquest, which Burnley (1992, 429) suggests reflects the preferences of the conquering aristocracy. Whilst dietary traditions in Normandy may be responsible for some of the changes in post-Conquest England, Norman Sicily was perhaps even more influential, the island probably supplying pheasants, fallow deer, and the 'unmaking' traditions to late eleventh-century England. Interestingly, similar Anglo-Sicilian links have been identified by the culinary historian Constance Hieatt (2002) who has argued convincingly that Anglo-Norman recipes are closer to Sicilian than medieval French cooking. Archaeobotanical evidence also supports these connections, Moffett (1989) has suggested that a new free-threshing tetraploid wheat (*Triticum turgidum/durum*), apparently absent from northern France, may have been imported to Norman England from Sicily. Considering Sicily's history of occupation, it seems probable that most of these elements had Arabic origins (see page 74 and Hieatt 2002, 26). That they were adopted by the Normans of Sicily and later transported to Norman England demonstrates the fluidity of food-culture associations and highlights the problem of linking particular cuisines to specific peoples: different groups are quite capable of eating the same things. What the evidence does show, however, is that by being incorporated into the Norman Empire, England became exposed to new culinary influences and it seems possible that their uptake was accompanied by different techniques of food preparation and cooking.

Methods of cooking are notoriously difficult to detect zooarchaeologically, boiling and baking leave few perceptible traces on bones (Montón Subías 2003). Roasting has greater potential for leaving archaeological traces but quantification of burning can be problematic (Kent 1993; Jones 1993). Nevertheless, taphonomic studies are not without value and it is noteworthy that, of the ninth to twelfth-century assemblages studied (Sykes 2001), the incidence of charred specimens does rise in the post-Conquest period. This may indicate an increase in the preparation of roasted meats, although data are currently too limited to draw any firm conclusions.

Potentially, butchery marks provide a better indicator of food processing and cooking, and studies of pre- and post-Conquest butchery patterns (for the raw data see Sykes 2001) do indicate that some changes do occur around the late eleventh century. It is very clear that bones were more heavily processed in the pre-Conquest period, with an average of 41% of long bones being split longitudinally, presumably for marrow extraction (Figure 81a). This butchery pattern is ubiquitous for Saxon assemblages and even the bones of deer were utilised in this way (for example, Higgs and Jarman 1977; O'Connor 1984; Crabtree 1983). Both Grant (1976, 272) and Hagen (1998, 58) have argued that incidence of bone splitting reflects a reliance upon stews and soups. *Aelfric's Colloquy* (Swanton 1975, 112) illustrates that cooks in the Late Saxon period were familiar with both boiling and roasting techniques but there is little to suggest that methods were particularly elaborate. Magennis (1999, 37-8) has even suggested that cooking skills may have been little valued.

Ceramic evidence suggests that methods may have become more complex after the Conquest; by the mid-twelfth century roasting dishes (or dripping pans) begin to appear regularly in archaeological contexts (Brown 2002, 135-7). A move away from the boiling of meat may also be indicated by the butchery evidence: a much lower percentage (6%) of the butchered post-Conquest bones showed signs of marrow extraction and where evidence is apparent the methods are different, with bones being chopped through at the mid-shaft rather than down the length of the shaft (Figure 81b). These butchery patterns were also demonstrated by the Vatteville assemblage where the incidence of bone splitting was similarly low. On this basis it seems plausible that the shift reflects continental influence. Both William the Conqueror and Charlemagne are documented as having a penchant for roast meat (Groundes-Peace 1971, 12; Thorpe 1969) and the assemblage from Vatteville-la-Rue yielded a number of burnt and charred bird bones, perhaps evidence that birds were spit-roasted in the fashion depicted on the Bayeux Tapestry (Thorpe 1973). Alternatively, increased popularity of roast meats may simply be tied to the emergence of an aristocratic sumptuary cuisine (see below): roasting is not only expensive in terms of fuel and labour but the valuable fat from the carcass is also lost. Woolgar (1999, 137) mentions that, by the later medieval period, consumption of roast meat was a sign of status and it seems probable that this was also the case in the earlier medieval period.

It is unlikely that stews became unfashionable in Norman England, given that stewing is both a practical and economic method of cooking because all of the juices are retained within the pot. Furthermore, since post-Conquest cattle and sheep were, on average, older than those eaten by the pre-Conquest population (see page 36) their tough meat may have required considerable boiling before consumption; this was probably achieved with the use of large cooking pots like that shown on the Bayeux Tapestry (Thorpe 1973). Indeed, studies of medieval ceramics suggest a twelfth-century proliferation of sagging-based cook-pots, perhaps indicating a rise in the consumption of stews (Hinton 1990, 130).

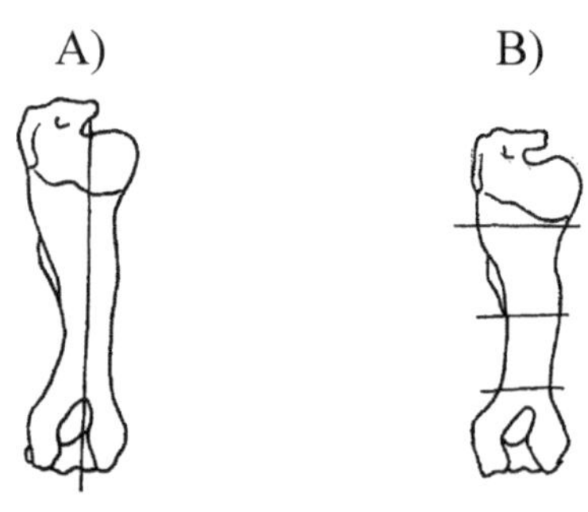

Figure 81: Methods of splitting long-bones in a) pre-Conquest and b) post-Conquest England

	Total No Butchered	Brining Indicators		Smoking Indicators	
		n	%	n	%
Pre-Conquest	98	20	20	12	12
Post-Conquest	129	5	4	20	16
Vatteville	28	2	7	6	21

Table 16: Frequency of butchery marks suggestive of different meat preservation techniques: brining and smoking

Other butchery data suggestive of pre- to post-Conquest changes in meat processing are provided by the patterns for the shoulder blade (Table 16). A considerable number of cattle scapulae from the pre-Conquest assemblages showed signs that their spines had been chopped or sliced off and their glenoid cavity trimmed (Figure 82a). Similar patterns have been noted for Roman assemblages, in particular those from Lincoln where it was suggested that they resulted from preserving meat in brine (Dobney *et al.*

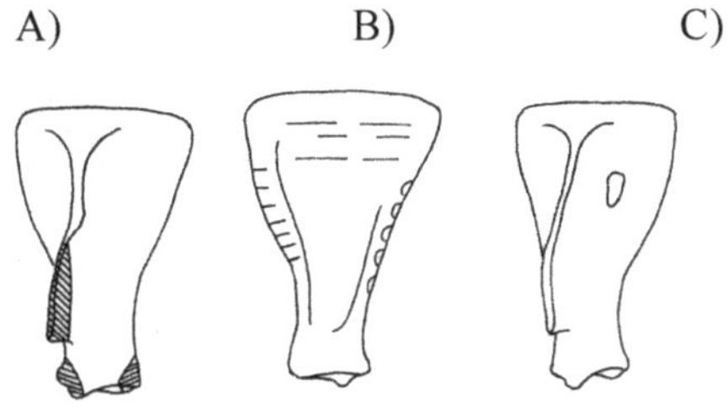

Figure 82: Butchery marks commonly found on scapulae from a) pre-Conquest and b-c) post-Conquest assemblages

1995, 26-7). Dobney *et al.*, argued that by chopping through the scapula at these points the salt water would permeate the meat better than through simple immersion. The Norman period data provided less evidence for this practice, with greater number of scapulae exhibiting cut and shaving marks on their medial surface (Figure 82b). Lauwerier (1988) interpreted these marks as produced when meat, which had become firmly attached to the bone as a result of the smoking process, was cut away. That meat was smoked is suggested by the scapulae which displayed meat-hook holes in their blades (see Figure 82c). Only one example of a meat-hook hole was found in the Saxon period deposits and the cut marks of the type shown in Figure 95a were also less frequent. This might suggest that techniques of meat preservation shifted from the Saxon to Norman period, with smoking rather than brining becoming the preferred method. Again the data for Vatteville are more comparable with the evidence from the Norman rather than Saxon period assemblages.

These changes in butchery technique appear to have been accompanied by a shift either in the way meat was supplied to high-status sites or how it was consumed at them. As was seen in Chapter 3, anatomical representation data for cattle, caprines and pigs (Figures 41-43) all indicate pre- to post-Conquest shifts away from head-dominated patterns towards the over-representation of meat-bearing elements, the latter being very similar to the patterns seen for assemblages from elite French sites. Regardless of why these shifts occurred, they may explain the linguistic evidence – that choice cuts of meat were renamed with French labels – and suggest that the consumption of particular joints became linked to social and ethnic identity.

8.2: Dietary Hierarchy

Compared with other types of identity, zooarchaeology has made the greatest contribution to investigations of socio-economic status (for instance Crabtree 1991; Albarella and Davis 1996; Ashby 2002; Grant 2002; Ervynck *et al.* 2003). For medieval England, evidence for dietary variety and in particular the consumption of animals obtained through hunting and hawking have become the traditional marker of elite identity. Whilst there are problems with applying such basic criteria uncritically to the whole period (Sykes 2005), it is clear from the evidence presented in Chapter 5 that they have some validity for Saxo-Norman England.

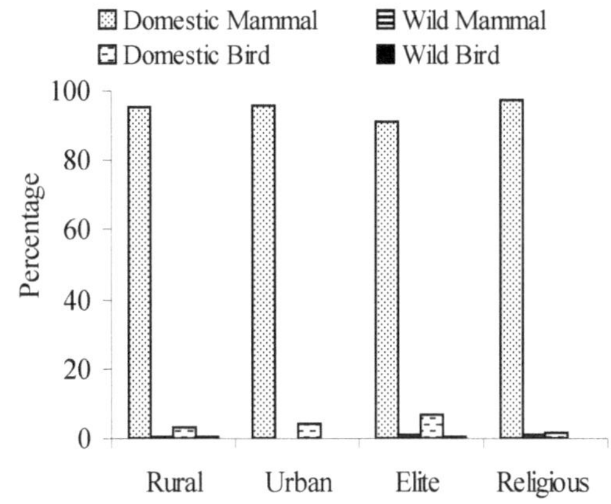

Figure 83: Composition of assemblages from sites dating to the fifth to mid-ninth-century

Throughout the previous chapters, the overriding impression provided of the Conquest is that it increased social differentiation. As Figure 83 shows, there is little separating fifth to mid-ninth-century assemblages from rural, urban, elite and religious houses; all are characterised by high frequencies of domesticates with elite assemblages marked only by a marginally higher representation of domestic birds and wild mammals. This is not to suggest that Early and Mid-Saxon society was egalitarian, clearly it was not, but dietary expression of social status must have taken a different form: van der Veen (2003, 412) has argued in societies lacking overt social stratification, status differentiation may be expressed in terms of 'quantity' of quotidian fare rather than through variety and refinement of foodstuffs.

Status-based differences in assemblage composition are more apparent for the mid-ninth to mid-eleventh century, with elite and ecclesiastical assemblages demonstrating domestic birds and game animal in higher frequencies than those from rural and urban settlements (Figure 84). It is also in this period that the consumption of particular taxa – some fish, sea mammals and swans – seemingly became restricted to the elite. These dietary changes conform to wider social patterns noted by Fleming (2001) and Senecal (2001) who suggest that acts of competitive display became prominent in the Late Saxon period as thegns

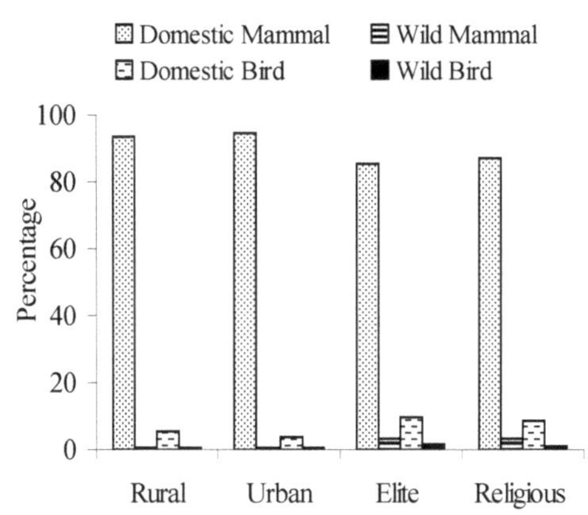

Figure 84: Composition of assemblages from sites dating to the mid-ninth to mid-eleventh century

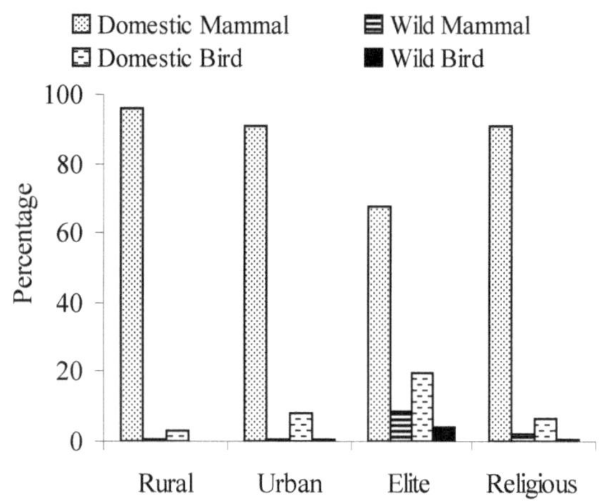

Figure 85: Composition of assemblages from sites dating to between the mid-eleventh and mid-twelfth centuries

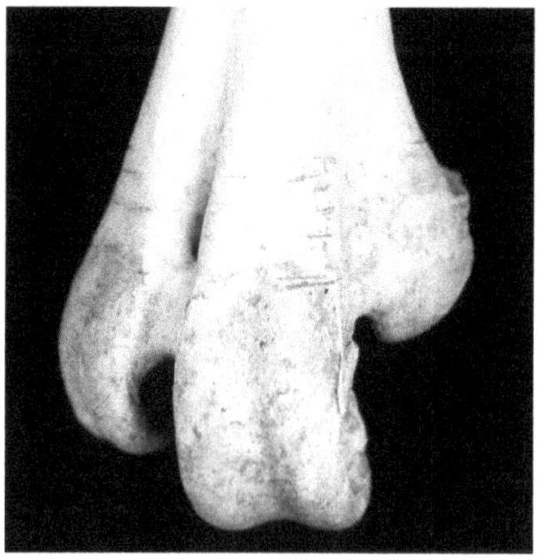

*Figure 86: Eleventh-century crane (*Grus grus*) tarsometatarsus showing fine cut marks on its distal end – evidence of dressing? Photo courtesy of T.P. O'Connor*

attempted to assert their aristocratic identity. Against this background, the changes that followed 1066 – in particular the blanket restriction on low-status utilisation of wild resources – seem even more dramatic. The Norman period marks the point at which diet apparently became strongly differentiated, with the elite assemblages showing far greater variety in species representation than any other site: this is demonstrated by a marked decline in the representation of domestic mammals (from 85.2% in the Late Saxon period to just 67.6%) and an unparalleled increase in the abundance of domestic fowl and wild animals (Figure 85).

These broad changes mask more subtle shifts, such as the post-Conquest introduction of animal new species, exotic luxuries whose ownership and consumption would have been a strong marker of elite identity. The arrival of these new animal ingredients, plus the novel or increased consumption of native wildlife (see above) must have been accompanied by new dining practices, social graces employed to further enhance feelings of exclusivity. The first book on etiquette – Daniel of Beccles' *Book of the Civilized Man* – is thought to have been written during the reign of Henry II (Bartlett 2000, 583) but it seems likely that table manners came into existence earlier. Zooarchaeological evidence for new dining etiquette has been provided by finds of bird remains with butchery marks suggestive of unnecessarily complex methods of carcass preparation and presentation. For instance, eleventh-century deposits (most probably dating to the post-Conquest period – O'Connor *pers. comm.*) from Lincoln produced a crane (*Grus grus*) tarsometatarsus that exhibited fine cut marks on its distal end where the toes had been removed from the leg (Figure 86, also see O'Connor 2007, 7). This practice cannot be linked to the logistics of butchery – it would make more 'sense' to remove the legs, which carry no meat, at the knee – but it is consistent with the methods of 'dressing' crane carcasses, which later became documented in medieval recipe books (Furnivall 1868, 162). A second example comes from eleventh/twelfth-century Carisbrooke Castle, Isle of Wight, where another tarsometatarsus, this time from a peacock (*Pavo cristatus*), was found with similar cut marks on its distal end (Serjeantson 2006, 142-3). The careful preparation of this bird may suggest that the later medieval practice of serving cooked peacocks re-skinned and in full plumage was, in fact, a Norman period innovation. Post-conquest England certainly saw the emergence of new customs surrounding venison consumption, whereby diners were provided with 'meaty symbols' of their social position; the lords eating the prized portions (liver and testicles), the best hunters receiving the shoulders and the lower-ranking members of the retinue given the intestines or 'numbles': indeed the saying 'to eat humble [numble] pie' is derived from the low-status and social humiliation associated with the consumption of the intestine (Goody 1982, 142).

That dietary restrictions and new table manners start to become apparent shortly after the Conquest suggests that the Norman period witnessed the development of what Goody (1982, 134) calls a 'sumptuary cuisine' – the conspicuous consumption of rare, restricted, and complexly prepared foods by elite groups – which represents an extreme form of dietary inequality. If it is accepted that diet is a direct reflection of social differentiation, the zooarchaeological data would suggest that the Normans were responsible for the development of dietary, and thus social, hierarchy. The remaining question is whether this new dietary hierarchy was imported from pre-Conquest Normandy or developed *in situ*. When the French and English data are compared they suggest that the dietary inequality of post-Conquest England exceeded that seen in France. The evidence suggests that rather than resulting in the importation of new systems of dietary/social hierarchy, the Conquest produced a situation where the newly-landed Norman aristocrats suddenly attained excessive wealth, allowing them a higher standard of living and a greater capacity to engage in conspicuous consumption than had been available to them in Normandy. In all probability the level of luxury enjoyed by the Norman elite would have separated them from the native population. Even where Saxon aristocrats (those who retained their land) were able to engage in similar levels of consumption, new forms of social etiquette, courtly language and consumption practices may have served to separate the two ethnic groups. How this 'sumptuary cuisine' was reconciled with the abstemious and pious character attributed to the Normans will be considered in the next section.

8.3: Dietary Piety and the *Gens Normanorum*

Whilst social hierarchy was an important element influencing medieval diet, consumption practices were structured, above all, by Christian beliefs and concepts of morality (Woolgar 2006, 191). By the seventh century, when the first religious houses were founded in England, monastic diet was already governed by the sixth-century Rule of St Benedict of Nursia (*c.*480-*c.*550) which stated that all monks, except the weak and sick, had to abstain from the flesh of quadrupeds but were provided with a daily ration of two dishes of *pulmentum* – essentially pottage, a loaf of bread and a measure of wine (Harvey 2006, 215). The Benedictine Rule was established in the Mediterranean, where a vegetarian diet was not restricted by problems of seasonal production, as it is temperate climates. Fortunately for north European monks (and for zooarchaeologists) the Rule was suitably vague to allow fish, birds, eggs and dairy products to be incorporated into the monastic diet without causing serious repercussions.

Zooarchaeological evidence suggests that, in the Mid-Saxon period, quantities of beef, pork and mutton were consumed within religious houses (Figure 83). This conforms to the historical evidence; according to Oswulf's charter, these meats could be eaten on a '*fugoldaeg*', that is, a day on which poultry could be eaten (cited in Magennis 1999, 60). If, as has been argued in the previous section, variety in assemblage composition can be linked to a sumptuary diet, it stands to reason that lack of variety may, in zooarchaeological terms, suggest frugality. With this in mind it is interesting to note that for the Mid-Saxon period, zooarchaeological assemblages from religious houses are the least varied: domestic mammal remains make up 97.3% with lower frequencies of birds, both domestic and wild, than are seen on either rural or urban sites. Where wild animals are represented it is noteworthy that they tend to be species with well-attested Christian connotations: for instance the roe deer (see Chapter 5) and the crane (see Sykes 2005, 98). By contrast animals such are wild boar and partridges, both of which were deemed to confer lascivious temperaments to the consumer, are notably absent from ecclesiastical assemblages (Sykes 2005, 95). To me, this indicates an adherence to concepts of dietary morality and confirms the idea that, in this period, fasting was largely restricted to monastics, with variation between the secular and ecclesiastic diet being greater than between high and low status consumption.

By the Late Saxon period the situation appears to have changed; Figure 84 shows that, in broad terms, assemblages from elite sites and religious houses are practically identical, both showing similar frequencies of domestic mammals, birds and game animals. Even at a more detailed level assemblages are difficult to separate: relative frequencies of the main domesticates (Figure 38), the ratios of different domestic birds (Figure 40), representation of deer species (Figures 69 and 70) and even the fish presence/diversity indices (Figures 58, 60 and 61) are parallel between the two site types. Again we are left to question what this means in terms of dietary piety – does it suggest that Late Saxon ecclesiastics had lapsed into gluttony or that the thegnly class were actively attempting to mimic the monastic diet? Unfortunately the answer is not clear cut and there are arguments for both scenarios.

The historical record demonstrates that Late Saxon ecclesiastics were frequently criticised for excessive consumption. In the late tenth/early eleventh-century poem *The Season for Fasting*, for instance, condemnation is lavished upon a lax priest who breaks the holy fast to consume wine, oysters and fish after mass (cited in Maggenis 1999, 90). However, as Magennis has argued, this poem is a product of the Benedictine reform which saw church leaders, such as Aethelwold and Dunstan, strive to re-establish Benedict's original standards of monastic living and impose a more rigorous regime of dietary restriction (Campbell 1991). Aethelwold drew up a supplement to the Rule of St Benedict (the *Regularis Concordia*) specifically for use by the English, and fasting became not only a religious requirement but a legal one too: the laws of Edgar (959-63), Aethelred (*c.* 1008) and Cnut (1020) all mention penalties for breaking a fast (Attenborough 1922). And it is here where the issue of dietary piety is complicated because it is possible that the diet of the secular and religious elite became similar due to this widespread religious reform, fasting being adopted by English kings and the *thegnly* class as a mark of moral and social superiority. A desire to manage fast-day resources may explain the tenth-century construction of what Rahtz (1979) labelled, a 'fowl house' in the Cheddar Palace complex. Interestingly, similar structures have been found on monastic sites, such as that of St Gall, Switzerland, perhaps reinforcing the idea of high-status mimicry of the ecclesiastic diet. If this hypothesis is correct, it would contradict the historically documented impression of the Saxons as irreligious gluttons.

On balance, the argument for Late Saxon monastic excess seems marginally more convincing, especially when the post-Conquest data are considered. It is evident from the writings of William of Malmesbury (*Gesta Regnum Anglorum*, cited in Handcock 1996, 315) that the memory of over-indulgences by Late Saxon ecclesiastics persisted even after 1066: 'some years before the coming of the Normans…the monks mocked the rule of their order…with the use of every kind of food'. But following the Conquest clerics seems to have been brought back in to line by their new Norman priors, some of whom insisted that their monks eat fish rather than meat (Serjeantson and Woolgar 2006, 104). The rationale for the promotion of fish consumption is easy to decipher. According to Galenic principles, the phlegmatic humour of fish would impart a suitably cool temperament to the consumer, and this combined with their Christian associations ('fishers of men') made them suitable for men of the cloth. Presence/diversity indices (Figure 58) suggest that monastic fish consumption did indeed increase during the Norman period. A return to frugality is also indicated by Figure 85 which shows that the composition of mid-eleventh to mid-twelfth-century assemblages from religious houses is closer to that seen in the Mid-Saxon period, domestic mammals accounting for 91% of the assemblage.

The Bayeux Tapestry's dining scenes could be taken as evidence that fish consumption was a trait of the Norman diet because although the image is copied from St Augustines Gospels, fish are not depicted in the original, thus the decision to include them in the Tapestry seems significant. However, as was seen in Chapter 5, presence/diversity indices (Figure 59) indicate that aristocratic fish consumption actually declined in late eleventh/early twelfth-century England. Combined with the increase in the consumption of game – sanguine and choleric meats thought to inspire lechery and carnal appetites – the diet of post-Conquest aristocracy seems far from abstemious. Of course, the hunting, distribution and consumption of game were important, in their own right, to the construction of Norman power and identity (Chapter 5 and Sykes 2006) but it is not immediately obvious how, with such conspicuous consumption of game meats, the Norman aristocracy can have simultaneously promoted their image of dietary piety.

The devout character attributed to the Normans must derive largely from the widespread church rebuilding and monastic foundation that followed the Conquest: before 1066 there were some 50 monastic houses but by the end of the eleventh century this number had increased to around 130 (Taylor 1996, 126). Ironically, it was actually the brutality of the Conquest itself that spurred these foundations, the Penitential Ordnance requiring all those who partook in, and benefited from, the Battle of Hastings to repent (Cowdery 1969). Whilst penitence was often demonstrated by monastic foundation – the new Norman lords now having the wealth to make these ostentatious acts of atonement – Bull (1998, 168-72) suggests that fasting was the staple of eleventh-century aristocratic penance. There are several reasons why this may not be apparent from the zooarchaeological record. First, avoidance of meat would, obviously, be undetectable in the animal bone record, and assemblages from late eleventh/early-twelfth century elite sites may simply reflect the infrequent consumption of refined, varied meals. More probable is that lords fasted vicariously through the monks of their religious houses: Bull (1998, 172-3) has shown that periods of abstinence were frequently commuted for alms or transferred to local monasteries in return for protection or property. The increased dietary piety apparent in assemblages from Norman period religious houses would certainly fit this view. Where fasting was undertaken by members of the secular aristocracy, periods of abstinence did not necessarily mean that they went without, they simply ate other things. Hieatt (2002, 25) has shown that Anglo-Norman recipe books are unique in their concern to create fast-day versions of popular dishes, perhaps reflecting the Norman elite's desire to be seen as pious without suffering any real hardship.

Concepts of excess and morality are culturally determined and it is, therefore, difficult to ascertain how 1066 affected levels of dietary piety. Whilst the post-Conquest aristocracy appear, on zooarchaeological grounds, to have been more over-indulgent than their pre-Conquest counterparts, the Norman aristocracy clearly managed to manipulate their consumption practices to control how they were perceived, both by themselves and by others.

8.4: Conclusion

Diet is a complex phenomenon representing much more than the fulfilment of biological needs. Social structure, religious beliefs and cultural choices all influence what and how food is eaten. Zooarchaeologically, it is often easy to highlight shifts in dietary composition, species representation providing a clear indication of change. Interpreting the meaning behind these patterns is, however, more difficult.

In this chapter a number of striking pre- to post-Conquest changes in diet have been identified. Those relating to aristocratic consumption are particularly evident, the overwhelming impression being that the diet of the Norman aristocracy became more 'luxurious' in every sense (van der Veen 2003). Assemblages from elite sites demonstrate a growth in the range and variety of meat animals (those obtained only for their primary products) consumed, with an increased representation of those species acquired through the 'unnecessary' activities of hunting and hawking. The hunting rights imposed on post-Conquest England can, in many respects, be seen as the precursors of the later medieval sumptuary laws. Similar to the legislation of 1363, Forest Law ensured that the lower classes were unable to access the dietary symbols (in this case game) most indicative of elite status. Instead, the peasant and urban population were required to consume animals from the domestic landscape that they worked and inhabited.

Whilst dietary hierarchy clearly became more pronounced as a result of the Conquest, Norman identity was constructed and reproduced by combining other elements with social status. Application of French terminology to particular foods and cuts of meat, served to link the Normans with elite identity. Connections with the Norman Empire, in particular Sicily, could then be embodied through the consumption of exotic animals: fallow deer, peafowl and perhaps pheasants and doves. The sumptuary cuisine that developed in post-Conquest England was enhanced by the introduction of new etiquettes, such as the dressing and presentation of peafowl and herons, both birds preferred by the French elite. Knowledge of these food rituals, or lack of the same, would have been a powerful mechanism for highlighting differences between the Saxon and Norman elite. But table manners also served another purpose; to reinforce feelings of Norman superiority. The zooarchaeological evidence suggests that the diet of the Norman elite was far more excessive than that of the Late Saxon aristocracy and, potentially, this sumptuary cuisine could have been perceived as gluttonous. However, by surrounding their diet with refinements in methods of preparation, presentation and consumption, the Normans were able to make over-indulgence appear abstemious. Perceptions of Norman piety were then seemingly reinforced by their associations with monastic foundation and their ability to fast vicariously through their religious leaders.

CHAPTER 9: THE INVISIBLE CONQUEST?

Debates concerning the Norman Conquest have centered on four main questions. Originally concerns focused on the issues of continuity versus change and whether 1066 impacted dramatically on the fabric of Anglo-Saxon society. Later, Anglo-Norman scholars went on to consider the degree to which influences represented the importation of a Norman Package. Unable to find a distinct material culture linking different areas of the Norman Empire, RCH Davies refuted the existence of the *Gens Normanorum*, suggesting it was a product of twelfth-century historians. Since then, a generation of historians have analysed the evidence for Norman identity. Archaeological research has contributed little to Anglo-Norman studies, with most examinations emphasizing the lack of evidence for 1066 and questioning whether changes were ephemeral or archaeologically tangible.

This book has sought to tackle these subjects from a novel perspective, that of zooarchaeology. Through a synthesis of animal bone data from fifth- to fourteenth-century sites on both sides of the Channel it has been possible to find new answers to these old questions.

9.1: Continuity versus Change

The Norman Conquest occurred at a time of socio-economic dynamism and, at the most basic level, there are no changes that can be attributed exclusively to 1066. Unlike the Roman Conquest, there is no clear evidence to suggest that domestic stock altered dramatically, either through the importation of new breeds or the deterioration of husbandry techniques. Pigs may prove the exception as metrical and ageing data suggest that new, possibly fast-maturing, types may have been introduced. Otherwise, there is little indication that the French economic set-up – a mixed regime geared towards local production – was imposed on post-Conquest England.

Some changes to the animal economy are apparent for the late eleventh and twelfth centuries: cattle and caprines were maintained to ages much older than in any preceding period and there is a significant decrease in cattle abundance with an associated increase in the frequency of sheep. When one looks more closely at the data, however, it becomes apparent that these pre- to post-Conquest shifts actually reflect long-term agricultural intensification. The increase in cattle and caprine age suggests a growing emphasis on secondary products, with cattle being managed for little other than traction, whilst sheep/goats were needed in increasing numbers to manure fields and provide wool for the growing market in grain and textiles.

Whilst these changes were clearly underway before the Conquest, having begun by at least the mid-ninth century, the mid-eleventh to mid-twelfth centuries do seem to mark a point of definition. Indeed, all areas of the animal economy appear to have witnessed intensification in this period. There is good evidence for the establishment and centralisation of professional tanners, horners and bone and antler workers – crafts which were undertaken on little more than a domestic scale before the mid-eleventh century. Butchery and skeletal representation data also suggest a commercialisation of aristocratic provisioning, whereby rather than obtaining whole animals through food rents, cuts of meat were increasingly purchased from urban markets. Indeed, it seems possible that the Norman re-labelling of choice joints of meat according to French titles may reflect this shift in the provisioning system. Again, the seeds of these changes were already sown in Late Saxon England – we know for instance that food rents were beginning to be commuted for cash payments during the reign of Edward the Confessor – but the foundation of towns by new Norman lords, together with the imposition of heavy taxes, must have been a strong impetus for the centralisation of artisans, for market exchange and the development of a coin-based economy.

One change cannot, however, be attributed to this new emphasis on agricultural production: the post-Conquest increase in the representation of pigs. Arable farming and pig husbandry are not always compatible and often an increase in grain production will be accompanied by a decline in the abundance of pigs, as for instance occurred between the mid-ninth and mid-eleventh centuries. The fact that pig frequencies increase at the point when arable production reached a peak suggests that this rise was dictated by cultural rather than economic factors. It seems possible that it reflects Norman dietary preferences.

Norman preferences may also explain pre- to post-Conquest shifts in wild animal exploitation, with frequencies of game mammals and birds increasing exponentially on elite sites. The Anglo-Saxon Chronicle refers to how William I enacted Forest Law to protect, above all, harts and hinds – red deer. The effectiveness of his ban is perhaps demonstrated by urban assemblages which show a dramatic pre- to post-Conquest decline in the representation of red deer. By contrast, assemblages from elite settlements show a contrary shift away from roe deer and towards red deer, a move that can probably be linked to changes in hunting strategies and, in particular, hunting grounds. In Saxon England aristocratic hunting appears to have been centered on enclosed wooded spaces. Roe deer are adapted for these environments and it is perhaps unsurprising that they were the prime targets of the pre-Conquest hunt. The Norman introduction of the 'Forest' concept changed perceptions of the hunting landscape away from these enclosed areas and towards the favoured habitat of British red deer: unbounded tracts of open moor, heath and agricultural land. The uptake of red deer hunting can thus be linked to the introduction of the Forest concept.

Changes in the ratio of red to roe deer are not the only pre- to post-Conquest shifts in deer representation; fallow deer also begin to appear regularly after 1066. The introduction date of this species has long been the subject of debate but, through detailed analysis of the zooarchaeological evidence, it is now clear that modern populations of fallow deer descend from animals that were brought to England within the first few decades after the Conquest. As exotic and highly-prized animals, fallow deer would not have been allowed to wander free; instead they would have been jealously guarded in enclosures: the twelfth-century increase in parks can be attributed directly to the growth in fallow deer populations.

Whilst fallow deer undoubtedly arrived in England whilst it was under Norman rule, the same cannot be said for rabbits. Both the historical and archaeological evidence suggests that they were not introduced before the end of the twelfth century: rabbits and, by association warrens, should not, therefore, be viewed as symbols of *Normanitus*.

The status of other managed species, notably domestic doves and pheasants, is more difficult to ascertain from the zooarchaeological evidence. In theory it is possible to differentiate the bones of these species from those belonging to other members of their family (pigeon and galliform respectively) but in practice few researchers attempt this. As a result it cannot be stated conclusively whether or not the Normans were responsible for their introduction. Other bird species can be more closely allied with the arrival of the Normans. Peafowl, for instance, are found much more regularly in post-Conquest assemblages, and have been recovered almost exclusively from high-status, in particular castle, sites. A striking pre- to post-Conquest change in bird representation is also seen for the heron, a species which is almost entirely absent from medieval assemblages pre-dating the mid-eleventh century. After the Conquest herons become well represented on both town and high-status sites. It seems improbable that this bird species was temporarily extinct in early medieval England. Instead, its absence in the archaeological record must simply be due to lack of exploitation. Why heron suddenly became the choice bird of the post-Conquest population is difficult to determine but it may be linked to a growth in the popularity of falconry. There is evidence that this hunting method was practiced, albeit sporadically, from the eighth century onwards but, in this earlier period, cranes were the species most commonly targeted. By the later medieval period heron had become a prized quarry and it seems that the origins of this tradition began shortly after the Conquest.

Pre- to post-Conquest changes in deer hunting are also apparent. On the basis of archaeological, historical and animal bone evidence it has been suggested that the most common pre-Conquest hunting technique was the 'drive', whereby groups of deer were chased into enclosures, known variously as *haga* or *haia*, and then killed with spears and knives. Once captured and killed, the deer carcasses appear to have been brought back to settlements whole, as assemblages from high-status sites (such the Cheddar Palaces, Goltho, Faccombe Netherton, Portchester Castle and Eynsham Abbey) contain deer bones from all parts of the body, forelimb elements being particularly well represented. Surprisingly, on all of these multi-period sites, this pattern of skeletal representation does not continue into the post-Conquest period. Where dating permits, it can be seen that almost at the point of 1066, deer body part patterns change dramatically. Far from being represented by large numbers of meat-bearing bones, elements from the upper forelimb (scapula and humerus) are scarce and the pelvis is almost entirely absent. Instead, bones from the lower hind limb, which carry negligible amounts of meat, are overwhelmingly the best represented parts of the skeleton. I have proposed that this shift reflects the origins of the 'unmaking' rituals, associated with *par force* hunting, that are well-documented in the later medieval period. By the thirteenth century, knowledge of these unmaking rituals and, in particular, the Gallic terminology surrounding them was deemed to be a mark of nobility. That the language of the hunt was French and that deer body part patterns suggestive of the unmaking appear shortly after 1066, strongly promotes the Normans as importers of the *chasse par force*. Indeed, the timing of many of the changes that I have outlined all suggest Norman influence. But do these changes represent the importation of a pre-existing Norman package?

9.2: A Norman Package?

Turning first to the issue of pork consumption. Without doubt pigs are the best-represented domestic animal on French elite sites, their bones vastly outnumbering those of cattle and sheep/goats, and it is tempting to conclude that these dietary preferences were brought to England by the Norman elite. Although pig frequencies in post-Conquest England never attained the level seen in France this probably reflects the agricultural set-up of England, the economy and landscape not being conducive to intensive pig husbandry, a regime far better suited to the wooded environment of northern France and Germany.

As well as providing pannage for pigs, woodland would have been the natural habitat of all manner of game animals. These wild resources appear to have been more regularly exploited in France than Anglo-Saxon England but even in France, where game was presumably more widely available, hunting never reached the level seen in post-Conquest England. From this it may be inferred that the well-cited Norman 'love of hunting' was not so much a Norman tradition but a passion that could be indulged only with occupation of England and the personal wealth that brought. Zooarchaeological evidence also confirms the idea that the Forest Law introduced on this side of the channel was more strictly applied than was the case in

Normandy, since wild animal assemblages from France demonstrate less status-based inequality than those from post-Conquest England.

Where direct Norman influence can be seen is in the pre- to post-Conquest changes in species representation, particularly on high-status sites. The sudden appearance of heron and peafowl, for instance, finds meaning when viewed against the French data set: both of these birds, although in particular heron, are found regularly in assemblages from elite sites in northern France. The clear pre- to post-Conquest shift from roe to red deer exploitation can also be linked to Norman influence: the red to roe deer ratio for post-Conquest England is almost identical to that for France, where red deer were also the preferred cervid species of the seigneurial hunt. The significance of this change in quarry should not be underestimated as it reflects a deliberate decision on the part of the hunters. As has been noted above, cultural perceptions of the hunting landscape – that the Saxon aristocracy hunted in parks whereas forest was a more common concept in Normandy – probably underlie this shift and it seems possible that these changes in quarry and hunting landscape were accompanied by the *chasse par force de chien* or at least the hunting rituals.

We have seen how this form of hunting, with its ritualised unmaking techniques, created distinctive skeletal patterning in deer bone assemblages. If the unmaking rituals were indeed a Norman tradition it should be abundantly clear from the zooarchaeological record. This is not the case. There is currently no evidence for skeletal patterning akin to that seen for post-Conquest England in eleventh or twelfth-century assemblages from Normandy, France or anywhere else in northern Europe. Patterns suggestive of the unmaking tradition do begin to appear in France from the fourteenth century but there is every indication that they reflect the uptake of English hunting traditions. Before it is concluded that the Anglo-Normans simply 'made up' a suite of new rituals, I would like to consider the evidence relating to the introduction of fallow deer.

Considering that the exploitation of red deer, heron and peafowl can be traced to practices in Normandy, one might anticipate that the same would be true for fallow deer. Again, this is not the case. Zooarchaeological evidence demonstrates that this deer species was absent from the rest of northern Europe until at least the late thirteenth century. In the twelfth century, England was the only area in northern Europe where fallow deer populations were established – elsewhere they were restricted to regions of the Mediterranean. With such a limited distribution it is not too difficult to predict where the British fallow deer were imported from: attention must surely be turned to the Norman kingdom Sicily. Sadly, there has been little archaeological investigation into medieval Sicily but evidence from one site, Brucato, does hint that both fallow deer and the unmaking rituals were of Sicilian origin. Certainly the strong relationship between the Normans of Sicily and England would have provided ample opportunity for the exchange of ideas. Concepts concerning hunting and the maintenance of deer no doubt formed part of this exchange. In addition to fallow deer and hunting rituals, there is a suite of elements (including recipes, a free-threshing variety tetraploid wheat and perhaps the pheasant) which are absent in Normandy and appear to have been introduced to England directly from Sicily. These findings add to Webber's (2005, 177) suggestion that the relationship between the two island kingdoms was stronger than that between England and Normandy.

9.3: The *Gens Normanorum*

As an ideological phenomenon evidence for ethnic identity will always be easier to determine from the historical than archaeological record. Guidance to the spheres in which the Normans may have chosen to objectify these assertions is provided by the numerous eleventh and twelfth-century histories that they wrote or commissioned. The self-image projected by these texts is one of ethnic and moral superiority founded, above all on the Norman's god-given military virtues. Whilst ethnic identity based on martial supremacy would appear to be inaccessible to zooarchaeological research, the evidence pertaining to hunting is actually very informative.

In many respects the pre-to post-Conquest changes in hunting are symptomatic of the time, being linked to widespread changes in social structure. The significant rise in aristocratic hunting can, for instance, be seen as reflecting the increasing social division that typified the eleventh and twelfth centuries. This was also a period of growing gender definition, with emphasis being placed increasingly on masculinity and military activity (Coss 1999), and hunting, with its references to warfare and male identity, must be symbolic of this. In England the origins of these social trends pre-date 1066 but the extent to which they were accelerated by the Conquest, is important. Differences in the social and cultural structures of inter-dependent groups are frequently the foundation for feelings and expressions of ethnicity (Jones 1997), thus any slight variation between Saxon and Norman social structure could have been magnified by the Conquest and employed as a signifier of Norman identity. There is some evidence to suggest that society in pre-Conquest Normandy was more patriarchal, at least in terms of inheritance, than was the case in Saxon England (Bates 1982; Coss 1999, 21). Combine this with the fact that the Conquest saw a male-dominated Norman elite dispossess a Saxon aristocracy that, as a result of the battle-depleted male line, was female-dominant, and it seems likely that masculine identity amongst the Normans of post-Conquest England would have been strong, being enmeshed with warfare, social status and authority. Anthropological studies have shown repeatedly that hunting is a mechanism for asserting precisely these traits (Kent 1989), and historical evidence demonstrates that this was certainly true in medieval England (Rooney 1993; Almond 2003). I suggest that the sudden and dramatic post-

Conquest increase in high-status wild resource exploitation reflects Anglo-Norman deployment of hunting as a device for demonstrating power over the subjugated population. This theory is born out by the imposition of Forest Law, which fundamentally altered perceptions of hunting space but also challenged ideology concerning the 'chain of being'; the relative position of animals to humans (see Figure 87).

In pre-Conquest England, hunting spaces were distinctly separate from the domestic landscape and it was only within aristocratic and royal reserves that hunting was restricted; elsewhere people were largely free to exploit wild resources. Under this system all humans, regardless of social status, were classified as higher beings than animals. With the introduction of Forest Law, geographic definition of wild and domestic space became blurred as the Normans superimposed their own hunting landscape over the existing domestic one. By restricting hunting rights to the elite, the wild became associated with high social status whereas domestic equated to the lower classes. Rather than all humans having domination over nature, Forest Law placed game animals in the same macro-category as the aristocracy, a category above the peasantry who were now essentially classified together with domestic animals.

These changes in hunting laws and landscape served to reinforce the divide between the conquerors and the native population, operating as they did on a basis of social exclusion. Imposition of Forest Law both restricted legitimate wild resource exploitation to the elite, in other words the Norman elite, but it also stamped Norman authority on the landscape. This is of particular significance as it is increasingly being recognised that landscape is a platform upon which group identity is inscribed and through which it is expressed (see for example Knapp and Ashmore 2000). As a reference point for the assertion of identity, the introduction of fallow deer was also significant in landscape terms. The menageries and parks built to retain them placed not only legal but also physical barriers on the hunting landscape, and the creation of fallow deer parks would have combined allusions to travel and the exotic with a strong statement of authority, rank and social exclusion (Herring 2003).

Although largely benefiting the Norman aristocracy, these post-Conquest changes to hunting did not exclude the wealthy Saxon elite who retained their lands and position after 1066 – these individuals were presumably still entitled to hunt. How the Normans set out to differentiate their hunting from that of the native aristocracy was, therefore, to introduce new and complicated hunting rituals, such as the unmaking procedure, that were steeped in Gallic terminology, itself a barrier to outside involvement.

The introduction of a new food terminology and novel dining rituals would have served a similar purpose, knowledge of etiquette separating the lords of Saxon and Norman descent. It is probably due to these new complexities surrounding food preparation, display and consumption that the Normans were able to cultivate an image of dietary piety. Despite the luxurious, possibly over-indulgent, nature of the post-Conquest cuisine, emphasis on table manners could be equated, as it is today, with breeding, morality and abstinence.

Ultimately it was the Conquest itself that enabled the development of a distinct Norman identity. Whilst 1066 conferred fantastic personal wealth upon the new Norman aristocracy, it also encouraged them to re-think how they wished to be perceived. Through the invention of new traditions and the manipulation of old ones, the Normans were able to project their image of martial and moral superiority and, by so doing, legitimise their rule over the defeated Saxon population.

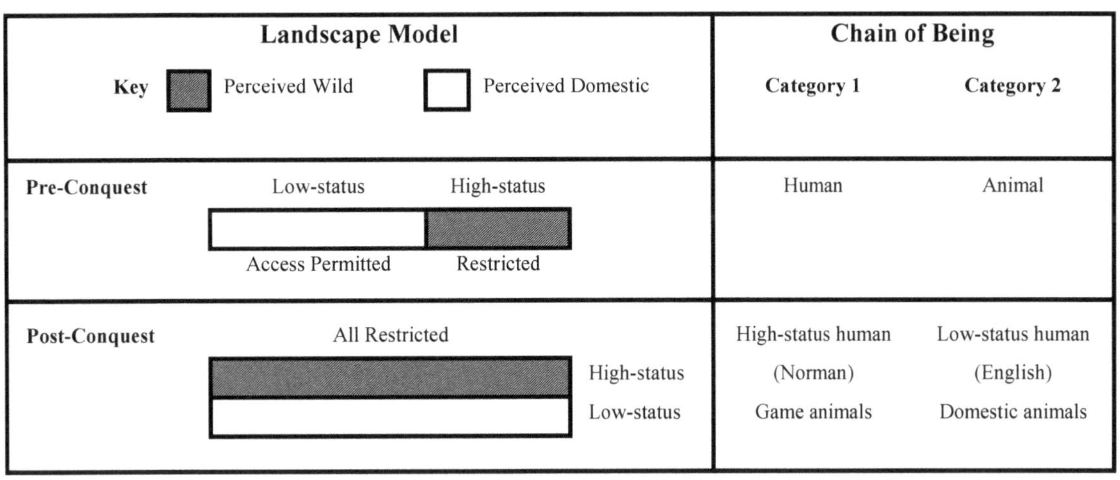

Figure 87: Pre- to post-Conquest changes in the perception of landscape and human-animal relationships

9.4: An Invisible Conquest?

When all these findings are taken into consideration, the Norman Conquest seems far from archaeologically invisible. Animal bone data provide evidence for Conquest-related change, the importation of a Norman package and the *Gens Normanorum*. Despite this, I agree that without historical evidence these changes would be difficult, if not impossible, to detect, let alone understand. However, there ***is*** historical evidence, and by using this in combination with other sources it has been possible to highlight areas of Norman influence that may not have been detectable through purely historical investigation. In this way I hope that this study serves as an example of how history and archaeology can be used to mutual advantage in studies of the past. More specifically, I hope it has highlighted how zooarchaeological research, so often seen as a peripheral specialism, can be employed profitably to address mainstream cultural questions.

BIBLIOGRAPHY

Albarella, U. 1997b. Shape variation of cattle metapodials; Age, Sex or Breed? Some examples from medieval and Post-medieval sites. *Anthopozoologica* **25-26**, 37-47.

Albarella U. 1997c. *The animal bones excavated in 1996 from Coslany Street, Norwich, Norfolk.* Ancient Monuments Laboratory Report 86/97, English Heritage.

Albarella, U. 1999. *The late Saxon and early medieval mammal and bird bones excavated in 1995 from Mill Lane, Thetford, Norfolk.* Ancient Monuments Laboratory Report 5/99, English Heritage.

Albarella U. 1999. "The mystery of husbandry": medieval animals and the problem of integrating historical and archaeological evidence. *Antiquity* **73**, 867-875.

Albarella U. 2006. Pig husbandry and pork consumption in medieval England, pp.72-87 in C. Woolgar, D. Serjeantson and T. Waldron (eds.) *Food in Medieval England: Diet and Nutrition.* Oxford University Press: Oxford.

Albarella, U., Beech, M., and Mulville, J. 1997. *The Saxon, medieval and Post-medieval mammal and bird bones excavated 1989-1991 from Castle Mall, Norwich (Norfolk),* Ancient Monuments Laboratory Report 00/97, English Heritage.

Albarella, U. and Davis, S. 1994. *The Saxon and medieval animal bones excavated 1985-1986 from West Cotton, Northamptonshire.* Ancient Monuments Laboratory Report 17/94, English Heritage.

Albarella, U. and Davis, S. 1996. Mammals and Birds from Launceston Castle, Cornwall: decline in status and rise in agriculture. *Circaea* **12**, 1-156.

Albarella, U., Dobney, K. and Rowley-Conwy, P. forthcoming. Size and shape of the Eurasian wild boar (*Sus scrofa*), with a view to the reconstruction of its Holocene history.

Allison, E. P. 1989 Bird bones, pp. 199-201 in R. Daniels (ed.) The Anglo-Saxon monastery at Church Close, Hartlepool, Cleveland. *Archaeological Journal* **145**, 158-201.

Almond, R. 2003. *Medieval Hunting.* Stroud: Sutton

Altuna, J. 1976. Los Mamiferos del yacimento prehistorico de Tito Bustillo (Asturias), pp. 151-99 in J. A. Moure Romanillo and M. Cano Herrera (eds.) *Excavaciones en la cuesa de "Tito Bustillo2: Trabajos de 1975 (Asturias).* Instituto de Estudios Asturianos: Oviedo.

Anderson, J. K. 1985. *Hunting in the Ancient World.* University of California Press: London.

Appadurai, A. 1986 (ed.) *The Social Life of Things: Commodities in Cultural Perspective.* Cambridge University Press: Cambridge.

Armitage, P. L. 1980. Mammalian Remains from the Roman and Saxo-Norman contexts, pp. 146-149 in D. M. Jones (ed.) *Excavations at Billingsgate Buildings 'Triangle', Lower Thames Street 1974* (London and Middlesex Archaeological Society Special Paper No 4) LAMAS: London.

Armitage, P. L. n.d. *The Mammalian Remains from the Roman, Medieval and Early Modern Levels at St Magnus, City of London.* Unpublished report.

Armitage, P.L. 1999. Animal bones, pp. 102-106 in I. Soden (ed.) A story of urban regeneration: excavation in advance of development off St Peter's Walk, Northampton 1994-7, *Northampton Archaeology* **28**, 61-128.

Armitage, P. L and Clutton-Brock, J. 1976. A system for classification and description of horn cores of cattle from archaeological sites. *Journal of Archaeological Science* **3**, 329-348.

Armitage, P. L. and Goodall, J. A. 1977. Medieval horned and polled sheep; the archaeological and iconographic evidence. *Antiquaries Journal* **57**, 73-89.

Ashby, S. P. 2002. The role of zooarchaeology in the interpretation of socioeconomic status: a discussion with reference to medieval Europe, *Archaeological Review from Cambridge*, **18**, 37-60.

Astill, G. 1991. Towns and Town hierarchies in Saxon England. *Oxford Journal of Archaeology* **10**, 95-114.

Astill, G. 1997. An archaeological approach to the development of agricultural technologies in Medieval England pp. 193-225 in G. Astill and J. Langdon (eds.) *Medieval Farming and Technology, the Impact of Agricultural Change in Northwest Europe.* Brill: New York.

Astill, G. and Grant, A. 1988. *The Countryside in Medieval England.* Blackwell: Oxford.

Astill, G. and Langdon, J., eds. 1997. *Medieval Farming and Technology, the Impact of Agricultural Change in Northwest Europe.* Brill: New York.

Aston, M. 1988. *Medieval Fish, Fisheries and Fishponds in England.* (BAR, British Series 182) British Archaeological Reports: Oxford.

Attenborough, F. L. 1922. *Laws of the Earliest English Kings*. Cambridge University Press: Cambridge.

Audoin, F. 1984. *Ossements animaux du Moyen Age au monastère de La Charité-sur-Loire*. Publications de la Sorbonne: Paris.

Audoin-Rouzeau, F. 1987. Medieval and early modern butchery: evidence from the monastery of La Charité-sur-Loire (Nievre). *Food and Foodways* **2**, 31-48.

Audoin-Rouzeau, F. 1989. *La faune de la rue de Lutèce à Paris*. Unpublished Manuscript.

Audoin-Rouzeau, F. 1991. *La Taille du Boeuf Domestique en Europe de l'Antiquité aux Temps Modernes*. Fiches d'Ostéologie Animale pour l'Archéologie. Série B: Mammifères **2**. Centre de Recherces Archéologiques du CNRS.

Audoin-Rouzeau, F. 1993. *Hommes et Animaux en Europe de l'époque antique aux Temps Modernes*. Corpus de données archéozoologiques et historiques (Dossiers de documentation archéologique 16). Paris: CNRS.

Audoin-Rouzeau, F. 1997. Les ossements du Cheptel médiéval. *Ethnozootechnie* **59**, 67-78.

Ayres, K. and Serjeantson, D. 2002. The animal bones, pp. 169-180 in T. Allen and J. Hiller (eds.) *The Excavation of a Medieval Manor House of the Bishops of Winchester a Mount House, Witney, Oxfordshire* (Thames Landscape Monograph no 13) Oxbow: Oxford.

Ayres, K.; Locker, A. and Serjeantson, D. 2003. Phases 2f-4a the medieval abbey: food consumption and production, pp. 360-406 in A. Hardy, A. Dodd and G. Keevil (eds) *Aelfric's Abbey: Excavations at Eynsham Abbey, Oxfordshire 1989-92* (Thames Valley Landscape Volume 16). Oxford: Oxford University School of Archaeology.

Baker, P. 1993. Le rôle de la chasse comme moyen de subsistance durant le haut Moyen Âge dans le nord d'Italie: une comparaison des sources zooarchéologiques et historiques. *Exploitation des Animaux Sauvage a Travers le Temps.* 505-516. XIIIe Rencontres Internationale d'Archéologie et d'Histoire d'Antibes, IVe Colloque international de l'homme et l'animal, Société de Recherche Interdisciplinaire.

Baker, P. and Clark, G. 1993. Archaeolozoological evidence for medieval Italy: a critical review of the present state of research, *Archeologia Medievale* **20**, 45-77.

Baker, P. 1998. *The vertebrate remains from Scole-Dickleburgh, excavated in 1993 (Norfolk and Suffolk) A140 and A143 road improvement project.* Ancient Monuments Laboratory Report 29/98, English Heritage.

Banham, D. 2004. *Food and Drink in Anglo-Saxon England.* Tempus: Stroud.

Barker, P. and Lawson, J. 1971. A pre-Norman field system at Hen Domen, Montgomery. *Medieval Archaeology* **15**, 58-72.

Barlow, F. 1966. The effects of the Norman Conquest, pp. 123-65 in D. Whitelock, D. C. Douglas, C. H. Lemmon and F. Barlow (eds) *The Norman Conquest: Its Settings and Impact.* Eyre & Spottiswoode: London.

Barrett, J., Locker, A. M. and Roberts, C. M. 2004. 'Dark Age Economics' revisited: the English fishbone evidence AD600 –1600 *Antiquity* **78**, 618-36.

Bartlett, R. 2000. *England Under the Norman and Angevin Kings 1075-1225*. Clarendon Press: Oxford.

Bates, D. 1982. *Normandy before 1066.* Longman: London.

Baxter, I. L. 2004. Animal, bird, reptile and amphibian bones, pp. 132-148 in N. Finn (ed.) *The Origins of a Leicester Suburb: Roman, Anglo-Saxon, Medieval and Post-Medieval Occupation on Bonners Lane.* (BAR British Series 172). BAR Publishing: Oxford.

Beech, M. 1986. The animal bones, pp. 45-7 in R. Holgate (ed.) Excavations at the Late Prehistoric and Romano-British Enclosure Complex at Carne's Seat, Goodwood, West Sussex. *Sussex Archaeological Collections* **124**, 35-51.

Bendrey, R. 2003. The identification of fallow deer (*Dama dama*) remains from Roman Monkton, the Isle of Thanet, Kent, pp. 15-18 in I.D. Riddler (ed.) *Materials of manufacture: the choice of materials in the working of bone and antler in northern and central Europe during the first millennium AD.* (BAR, International Series S1193) British Archaeological Reports: Oxford.

Bennett, H. S. 1971. *Life on the English Manor: A study of Peasant Conditions 1150-1400.* Cambridge University Press: Cambridge.

Bennett, K. 1983. Devensian late glacial and Flandrian vegetational history at Hockham Mere, Norfolk, England. *New Phytologist* **95**, 452-87.

Beresford, G. 1987. *Goltho: the development of an early medieval Manor c.850-1150.* English Heritage Archaeological Report: London.

Berg, D. 1993. Animal Bones from The Anglo-Saxon Deposits., pp. 30-34 in N. Field and K. Leahy (eds.) Prehistoric and Anglo-Saxon Remains at Nettleton Top, Nettleton. *Lincolnshire Archaeology* **28**, 9-38.

Bergman, C. 1847. Uber die Verhaltnisse der Warmekonomie der Thiere zu ihrer Grosse. *Gottingen Studien* **3**, 595-705.

Berke, H. 1997. Zur entwicklung der rinderhaltung und rinderzucht vom 1.-13. JHRH. in Köln. In *Anthropozoologica* **25-26**, 405-412.

Biddick, K. 1984a. Field Edge, Forest Edge: Early Medieval Social Change and Resource Allocation, pp. 105-118 in K. Biddick (ed.) *Archaeological Approaches to Medieval Europe*. Kalamazoo.

Biddick, K. 1984b. Pig Husbandry on the Peterborough Abbey estate from the twelfth to the fourteenth century AD, pp. 165-178 in C. Grigson and J. Clutton-Brock (eds.) *Animals in Archaeology* 4 *Husbandry in Europe*. (BAR, International Series 227) British Archaeological Reports: Oxford.

Birrell, J. 1982. Who poached the Kings deer? A study in 13th century crime. *Midland History*, 7, 9-25.

Birrell, J. 2006. procuring, preparing and serving venison in late medieval England pp. 176-88 in C. Woolgar, D. Serjeantson and T. Waldron (eds.) *Food in Medieval England: History and Archaeology* Oxford University Press: Oxford.

Blaauw, W. H. 1847. Remarks on the Nonae of 1340, as relating to Sussex. *Sussex Archaeological Collections* **1**, 58-64.

Bliese, J. R. E. 1991. The Courage of the Normans. A comparative study of Battle Rhetoric. In *Nottingham Medieval Studies* **35**. 1-27.

Bloch, M. 1966. *French Rural History: An essay on its basic characteristics.* Routledge and Kegan Paul: London.

Boessneck, J. 1973. *Die Tierknochenfunde aus dem Kabirenheiligtum bei Theben.* Munich: Institut für Palaeoanatomie, Domestikationsforschung und Geschicte der Tiermedizin.

Boisseau, S. and Yalden, D. W. 1998 The former status of the crane *Grus grus* in Britain, *Ibis*, **140**, 482-500.

Bökönyi, S. 1971. Angaben zum frühholozänen vorkommen des Damhirsches, Cervus (Dama) dama (Linné, 1758), in Europa. *Säugetierk. Mitt* **19**, 206-217.

Bökönyi, S. 1986. Animal remains from the Roman villa of San Potito-Ovindoli (L'Aquila) 1983-1984. A preliminary report. *Acta Archaeologica Acadmiae Scientarium Hungaricae* **38**, 89-91.

Bökönyi, S. 1989. Animal remains pp. 63-132 in M. Gimbutas and E. S. Elster (eds.) *Excavations at Sitagroi: a prehistoric village in north-east Greece*, Vol I. Los Angeles, California. Institute of Archaeology, University of California.

Bökönyi, S. 1995. The development of stockbreeding and herding in medieval Europe, pp. 62-75 in D. Sweeney (ed.) *Agriculture in the Middle Ages: Technology, Practice and Representation.* University of Pennsylvania Press: Philadelphia.

Bond, C. J. 1988. Monastic Fisheries, pp. 69-105 in M. Aston (ed.) *Medieval Fish, Fisheries and Fishponds in England I.* (BAR, British Series 182) British Archaeological Report Oxford.

Bond, C. J. and Chambers, R. A. 1988. Oxfordshire Fishponds, pp. 353-370 in M. Aston (ed.) *Medieval Fish, Fisheries and Fishponds in England I.* (BAR, British Series 182) British Archaeological Report Oxford.

Bond, J. 1988. Rabbits: The case for their Medieval introduction into Britain. *The Local Historian* **18**, 53-57.

Bond, J. 1994 The cremated animal bone, pp. 121-35 in J. McKinley (ed.) *The Anglo-Saxon cemetery at Spong Hill, North Elmham, Part VIII: the cremations* (East Anglian Archaeology Report 69) Norfolk Museum Service: Norwich.

Bond, J. M. 1995. Animal bone from the Early Saxon sunken-featured buildings and pits, pp. 142-6 in R. Rickett (ed.) *The Anglo-Saxon Cemetery at Spong Hill, North Elmham, Part VII: the Iron Age, Roman and Early Saxon Settlement.* (East Anglian Archaeology Report 73) Norfolk Museum Service: Dereham.

Bond, J. and O'Connor, T. 1999. Bones from Medieval Deposits at 16-22 Coppergate and Other Sites in York. *The Archaeology of York. The Animal Bones* 15/5. York Archaeological Trust.

Bossard, C. 1984a L'alimentation, pp. 749-993 in J. M. Pesez (ed.) *Brucato, histoire et archeologie d'un habitat medieval en Sicile*, Collection de l'Ecole Francais de Rome, **78** (2), Rome.

Bossard-Beck, C. 1984b Le mobilier ostéologique et botanique, pp. 615-671 in J. M. Pesez (ed.) *Brucato: histoire et archéologie d'un habitat médéival en Sicile*, Collection de l'Ecole Français de Rome **78** (2), Rome.

Bourdillon, J. 1979. The Animal Bones, pp. 207-212 in J. S. F. Walker (ed.) Excavations in Medieval Tenements on the Quilters Vault Site in Southampton. *Proceedings of the Hampshire Field Club Archaeological Society* **35**, 183-216.

Bourdillon, J. 1988. Countryside and town: the animal resources of Saxon Southampton, pp. 373-92 in D. Hooke (ed.) *Anglo-Saxon Settlements.* Blackwell: Oxford.

Bourdillon, J. 1992. *Animal bones from Late Saxon sites in Winchester, Hampshire.* Ancient Monuments Laboratory Report 42/92. English Heritage.

Bourdillon, J. 1993. Animal Bones, pp. 127-136 in Graham, A. H. and Davies, S. M. (eds.), *Excavations in Trowbridge, Wiltshire, 1977 and 1986-1988: The Prehistoric, Saxon and Saxo-Norman Settlements and the Anarchy Period Castle.* Wessex Archaeological Report **2**. Salisbury: Trust for Wessex.

Bourdillon, J. 1994 The animal provisioning of Saxon Southampton, pp. 120-125, in J. Rachkam (ed.) *Environment and Economy in Anglo-Saxon England,* (CBA Research Report 89), York: CBA.

Bourdillon, J. and Coy, J. 1980. The animal remains, pp. 79-120 in P. Holdsworth (ed.) *Excavations at Melbourne Street, Southampton 1971-76* (CBA Research Report 33) Council for British Archaeology: London.

Bradbury, J. 1997. *The Medieval Archer* Boydell: Woodbridge.

Brain, C. K. 1967. Hottentot food remains and their bearing on the interpretation of fossil bone assemblages. *Scientific Papers in the Namib Desert Research Station* **39**, 13-22.

Bramwell, D. 1975. Bird remains from Medieval London. *The London Naturalist* **54**, 15-20.

Bramwell, D. 1976. Report on the bird bones from Walton Aylesbury, pp. 287-288 In (Farley. M.) Saxon and Medieval Walton, Aylesbury Excavations 1973-74. *Records of Buckinghamshire* **20** (2). 153-290.

Bramwell, D. 1977. Bird bone, pp. 399-402 in H. Clarke, and A. Carter (eds) *Excavations in Kings Lynn 1963-70.*. Society for Medieval Archaeology, Monograph Series **7**.

Bramwell, 1979. The bird bones, pp. 333-4 in J.H. Williams (ed.) *St Peters Street Northampton: Excavations 1973-1976.* Northampton Development Corporation.

Bramwell, D. 1980. Identification and interpretation of bird bones, pp. 409-412 in P. Wade-Martins (ed.) *North Elmham Park. Vol II.* (East Anglian Archaeology, Report 9) Norfolk Museum Service: Dereham.

Brewer, E. 1992. *Sir Gawain and the Green Knight: sources and analogues.* Cambridge University Press: Cambridge.

Britnell, R. H. 1993 *The Commercialisation of English Society 1000-1500.* Cambridge University Press: Cambridge.

Brodribb, A. C., Hands, A. R. and Walker, D. R. 1972. *Excavations at Shakenoak III.* Privately published: Oxford.

Brown, D. 1997a. The distribution of North French Medieval Pottery in Southampton, pp. 161-169 in X. Delestre, X. and A. Woodcock (eds.) *Proximus.* Association for the Promotion of Archaeology in Haute-Normandie: Rouen.

Brown, D. 1997b. The social significance of imported medieval pottery, pp. 95-112 in C. Cumberpatch and P. W. Blinkhorn (eds.) *Not so Much A Pot, More A Way Of Life.* (Oxbow Monograph 83) Oxbow: Oxford.

Brown, D. 2002. *Pottery in Medieval Southampton c 1066-1510* (Southampton Archaeology Monograph 8, CBA Research Report 133). York: Council for British Archaeology.

Brown, M. A. 1998. The Feast Hall in Anglo-Saxon Society, pp. 1-15 in M. Carlin and J. T. Rosenthal (eds.) *Food and Feasting in Medieval Europe.* Hambledon Press: London.

Brown, R. A. 1969. An historian's approach to the origins of the castle in England. *Archaeological Journal* **126**, 131-48.

Brown, R.A. 1973. *Origins of English Feudalism: Historical Problems – Studies and Documents* **19**. Allen and Unwin: London.

Brown, R. A 1984. *The Norman Conquest* (Documents of Medieval History 5). Arnold: London.

Buckland-Wright, J.C. 1987. Animal bones in G. Sparey-Green, S. M. Davies and A. Ellison (eds.) *Excavations at Poundbury, Dorchester, Dorset 1966-182 Vol I: the Settlement.* (Dorset Natural History and Archaeology Society Monograph Series 7) Dorset Natural History and Archaeology Society: Dorchester.

Bull, M. 1998. *Knightly Piety and the Lay Response to the First Crusades: the Limousin and Gascony, c.970-c.1130.* Clarendon Press: Oxford.

Burnett, D. P. 1993. Animal bone from MK19, pp. 199-206, in R. J. Williams (ed.) *Pennyland and Hartigans: Two Iron Age and Saxon Sites in Milton Keynes* (Buckinghamshire Archaeological Society Monograph Series no 4) Buckinghamshire Archaeological Society: Ayelsbury.

Burnley, D. 1992. Lexis and semantics pp. 409-499 in N. Blake (ed.) *The Cambridge History of the English Language* Vol II 1066–1476. Cambridge University Press: Cambridge.

Cadman, G. 1983. Raunds 1977-1983: an excavation summary. *Medieval Archaeology* **27**, 107-22.

Callou, C. 2003. *De la Garenne au Clapier: Etude Archéozoologique du Lapin en Europe Occidental* (Mémoires du Muséum National d'Histoire Naturelle,

Tome 189) Publications Scientifiques du Muséum, Paris.

Campbell, J. 1991. *The Anglo-Saxons.* Penguin: London.

Caplan, P. 1997. Approaches to the study of food, health and identity, pp. 1-32 in P. Caplan (ed) *Food, Health and Identity.* Routledge: London.

Carandini, A. 1985. *Settefinestre. Una villa schiavistica nell'Etruria romana, III.* Modena: Panini Editore.

Carlin, M. 1998. Fast food and urban living standards in medieval England, pp. 27-53 in M. Carlin and J. T. Rosenthal (eds.) *Food and Feasting in Medieval Europe.* Hambledon Press: London.

Carpentier, V. 2002. Un habitat des XIe – XIIe siècles dans la campagne d'Argentan (Orne) *Archéologie Médiévale* **32**, 69-103.

Cartledge, J. 1979. Mammal bones, pp. 30-32 in B. Ayers and P. Murphy (eds.) *Waterfront Excavations at Whitefriars St Car Park, Norwich.* (East Anglian Archaeological Report 17) Norfolk Archaeological Unit: Norwich.

Cartledge, J. 1985. The Animal bones, pp. 220-222 in Atkin, M. (ed.), *Excavations in Norwich 1971-1978 Part II* (East Anglian Archaeology 26). Norwich Survey: Norwich.

Cartledge, J. 1988. Mammal Bones, pp. 111-115 in Ayers, B. (ed.) *Excavations at St-Martin-at-Palace Plain, Norwich 1981* (East Anglian Archaeological Report 37). Norfolk Archaeological Unit in conjunction with the Scole Archaeological Committee: Dereham.

Cartmill, M. 1993. *A View to a Death in the Morning: Hunting and Nature Through History.* Harvard University Press: Cambridge.

Cassady, R. F. 1986. *The Norman Achievement.* Guild Publishing: London

Casteel, R. W. 1977. Characterization of faunal assemblages and the minimum number of individuals determined from paired elements: continuing problems in archaeology. *Journal of Archaeological Science* **4**, 125-134.

Chaplin, R. E. 1971 *The Study of Bones from Archaeological Sites.* Seminar Press: London.

Chapman, N. and Chapman, D. 1975. *Fallow Deer: their history, distribution and biology.* Terence Dalton: Lavenham.

Charbonnier, P. 2002. The economy and society of France in the later Middle Ages: on the eve of crisis, pp. 47-60 in D. Potter (ed.) *France in the Middle Ages.* Oxford University Press: Oxford.

Charles, B. and Ingrem, C. 2001. Animal bones, pp.28-33 in A. Hardy and C. Bell (eds.) *The Excavations of a Medieval Rural Settlement at the Pepperhill Lane Electricity Substation, Northfleet, Kent.* (Oxford Archaeology Occassional Paper No 10) Oxford Archaeology: Oxford.

Cherryson, A. 2002 The identification of archaeological evidence for hawking in Medieval England, *Acta Zoologica Cracoviensia*, **45**, 307-14.

Chibnall, M. 1986. *Anglo-Norman England 1066-1166.* Basil Blackwell: Oxford.

Chibnall, M. 1999. *The Debate on the Norman Conquest.* Manchester University Press: Manchester.

Churchill, D. M. 1965. The kitchen midden site at Westward Ho! Devon, England: ecology, age and relation to changes in land and sea level. *Proceedings of the Prehistoric Society* **31**, 74-84.

Clapham, J. 1966. *A Concise Economic History of Britain: from the earliest times to 1750.* Cambridge University Press: Cambridge.

Clark, W. B. 2006. *A Medieval Book of Beasts. The Second-Family Bestiary: Commentary, Art, Text and Translation.* Boydell: Woodbridge.

Clason, A.T. 1967. *Animal and Man in Holland's Past: an investigation of the animal world surrounding Man in prehistoric and early historical times in the provinces of north and south Holland.* Volume A. Groningen.

Clavel, B. 2001. *L'animal dans l'alimentation Medievale et Moderne en France du Nord (XIIe - XVIIe siècles).* Revue Archeolgique de Picardie No Spécial 19.

Clutton-Brock, J. 1976. The animal resources, pp. 373-92 in D, M. Wilson (ed.) *The Archaeology of Anglo-Saxon England.* Methuen: London.

Clutton-Brock, J. 1977. The Animal bones, pp. 86 in R. Bruce-Mitford (ed.) *Mawgan Porth a Settlement of the Late Saxon period on the North Cornish Coast: Excavations 1949-52, 1954 and 1974.* English Heritage: London.

Clutton-Brock, J. 1984. The Master of the Game: the animals and rituals of medieval venery. *Biologist* **31** (3), 167-71.

Clutton-Brock, T. H., Albon, S. D. and Guiness, F. E. 1982. *Red deer: behaviour and ecology of two sexes.* University of Chicago Press: Chicago.

Cockayne, O. 1851. *Leechdoms, Wortcunning and Starcraft of Early England.* I-III Roll Series. Reprint 1961. Holland Press: Oxford.

Cocks, A.H. 1921, A Romano-British Homestead in the Hambledon Valley, Berkshire. *Archaeologia* **71**, 163-6.

Columella, L.J.M. (ed. & trans. by E.S. Forester & E.H. Heffner). 1955. *On Agriculture* (Loeb Classical Library 361) London: Heinemann.

Comet, G. 1997. Technology and Agricultural Expansion in the Middle Ages: the example of France north of the Loire, pp. 11-41 in G. Astill and J. Langdon (eds.) *Medieval Farming and Technology, the Impact of Agricultural Change in Northwest Europe*. Brill: New York.

Connell, B., Davis, S. J. M. and Locker, A. 1997. Animal bones from Camber Castle, East Sussex, 1963-1983. Ancient Monuments Laboratory Report 107/97. English Heritage.

Coss, P. 1999. *The Lady in Medieval England 1000-1500*. Sutton: Stroud.

Cotter. J. 1997. *A twelfth-century pottery kiln at Pound Lane, Canterbury: Evidence for an Immigrant Potter in the late Norman period.* (Canterbury Archaeological Trust, Occasional Paper 1). Canterbury Archaeological Trust: Canterbury.

Counihan, C. and Van Esterik, P. 1997. *Food and Culture: A Reader* Routledge: New York.

Cowan, D. P. 1991. Rabbit *Oryctolagus cuniculus* pp. 508-518 in G. B. Corbet and S. Harris (eds.) *The Handbook of British Mammals*. Blackwell: Oxford.

Cowdrey, H. E. J. 1969. Bishop Ermenfrid of Sion and the Penitential Ordinance following the Battle of Hastings. *Journal of Ecclesiastical History* **20**, 225-42.

Coy, J 1977. *Bird Bones from Wirral Park Farm (the Mound) Glastonbury 11th/12th century.* Unpublished Manuscript.

Coy, J. 1980. The animal bones, pp. 41-51 in J. Haslam (ed.) A Middle Saxon iron smelting site at Ramsbury, Wiltshire. *Medieval Archaeology* **24**, 1-68.

Coy, J. 1981a. The animal bones, pp. 106-110 in R. J. Silvester (ed.) An excavation on the Post-Roman site of Bantham, South Devon. *Devon Archaeological Society Proceedings* **39**, 89-118.

Coy, J. 1981b. The bird bones. microfiche 200-201 in J. H. Williams and M. Shaw (eds.) Excavations in Chalk Lane, Northampton, 1975-1978. *Northamptonshire Archaeology* **16**, 87-136.

Coy, J. 1982. The role of wild vertebrate fauna in urban economies in Wessex, pp. 107-116 in A. R. Hall and H. K. Kenward (ed.) *Environmental Archaeology in the Urban Context* (CBA Research Report 43) Council for British Archaeology: London.

Coy, J. 1983. The animal bone, pp. 91-96 in K. S. Jarvis (ed.) *Excavations on Christchurch 1969-1980* (Dorset Natural History and Archaeological Society Monograph Series No 5). Friary Press: Dorchester.

Coy, J. 1985. Animal bones from site 5, pp. 109-111 in E. Lewis (ed.) Excavations in Bishop's Waltham, 1967-78 *Proceedings of the Hampshire Field Club* **16**, 81-126.

Coy, J. 1986. *Animal Bones from Newbury Berkshire: Excavations in Cheap Street 1981-2*. Ancient Monuments Laboratory Report 56/86. English Heritage.

Coy, J. 1987. *Animal Bones from Abbots Worthy (Itchen Abbas Road), Hampshire*. Ancient Monuments Laboratory Report 156/87. English Heritage.

Coy, J. 1989a. Animal bones, pp. 111-134 in Astill, G. G and Lobb, S. J. Excavations of Prehistoric, Roman and Saxon deposits at Wraysbury, Berkshire. *Archaeological Journal* **146**, 68-134.

Coy, J. 1989b. The provisioning of fowls and fish for towns, pp. 25-41 in D. Serjeantson and T. Waldron (eds.) *Diet and Crafts in Towns: The evidence of animal remains the Roman to the Post-Medieval periods* (BAR, British Series 199) British Archaeological Reports: Oxford.

Coy, J 1995. Animal bones, pp. 132-9 in (Fasham, P. J. and Keevill, G. with Coe, D.) *Brighton Hill South (Hatch Warren): an Iron Age Farmstead and Deserted Medieval Village in Hampshire.* (Wessex Archaeological Report 7). Trust for Wessex Archaeology: Salisbury.

Coy, J. 1997a. Animal bones, pp. 75-79 and 134-141 in A. G. Vince, S. J. Lobb, J. C. Richards and L. Mepham (eds.) *Excavations in Newbury, Berkshire 1979-1990.* (Wessex Archaeological Report 13). Trust for Wessex Archaeology: Salisbury.

Coy, J. 1997b Comparing bird bones from Saxon sites: problems of interpretation, *International Journal of Osteoarchaeology*, 7, 415-21.

Coy, J., Bramwell, D. and Jewell, J. 1977. Animal Bones, pp. 119-122 in R. Cramp, P. Everson, and D. N. Hall (ed.) Brixworth Archaeological Research Committee. Excavations at Brixworth 1971 and 1972. *Journal of the British Archaeological Association* **130**, 52-122.

Crabtree, P. 1983. *Report on the Animal Bones from the Chapter House at St Albans Abbey (Hertfordshire).* Unpublished manuscript.

Crabtree, P. 1990. *West Stow: Early Anglo-Saxon Animal Husbandry.* (East Anglian Archaeology 47) Suffolk

County Planning Department in conjunction with The Scole Archaeological Committee Ltd: Bury St. Edmunds.

Crabtree, P. 1991. Zooarchaeology and complex societies: some uses of faunal remains for the study of trade, social status and ethnicity *Archaeological Method and Theory* **2**, 155-205.

Crabtree, P. 1994. Animal exploitation in East Anglian villages, pp. 40-54 in J. Rackham (ed.) *Environment and Early Economy in Anglo-Saxon England.* (CBA Report 89) Council for British Archaeology: London.

Crabtree, P. J. 1995. The symbolic role of animals in Anglo-Saxon England: evidence from burials and cremations, pp. 20-26 in K. Ryan and P. J. Crabtree (eds) *The Symbolic Role of Animals In Archaeology* (MASCA Research Papers in Science and Archaeology 12). University of Pennsylania: Philadelphia.

Crabtree, P. 1996. Production and consumption in an early complex society: animal use in Middle Saxon East Anglia. *World Archaeology* **28** (1), 58-75.

Crossely-Holland, K. 1993. *The Exeter Book of Riddles.* Penguin: London.

Cummins, J. 1988. *The Hound and the Hawk: the Art of Medieval Hunting.* Weidenfield and Nicholson: London.

Cunliffe, B. 1978. *Iron Age Communities in Britain: an account of England, Scotland and Wales from the seventh century BC until the Roman Conquest.* Routledge and Kegan Paul: London.

Currie, T. C. 1993. The animal bones, pp. 72-74 in C. K. Currie (ed.) Excavations at the Postern Mill site, Malmesbury, 1986-87. *The Wiltshire Archaeological and Natural History Magazine* **86**. 58-74.

Cutting, D. L. 1955. *Fish Saving: a history of fish processing from ancient to modern times.* Leonard Hill: London.

Darby, H. C. 1973. *A New Historical Geography of England.* Cambridge University Press: Cambridge.

Darby, H. C. 1977 *Domesday England.* Cambridge University Press: Cambridge.

Darling, F. F. 1937. *A Herd of Red Deer.* Oxford University Press: Oxford.

Darvill, T. and Coy, J. 1977. *Report on the Faunal Remains from Wirral Park Farm, (The Mound), Glastonbury.* Ancient Monuments Laboratory Report 2431, English Heritage.

Darvill, T. C. and Coy, J. 1985. Report on the faunal remains from the Mound, Glastonury, Somerset Archaeology and Natural History **129**, 56-60.

Davidson, A. 1999. *Oxford Companion to Food.* Oxford University Press: Oxford.

Davis, R. H. C. 1976. *The Normans and their Myth.* Thames and Hudson: London.

Davis, S. J. M. 1987. *Prudhoe Castle, a report on the animal remains.* Ancient Monuments Laboratory Report 168/87, English Heritage.

Davis, S. J. M. 1991a. *Late Saxon to Medieval Farmstead at Eckweek, Avon.* Ancient Monuments Laboratory Report 35/1991, English Heritage.

Davis, S. J. M. 1991b. Faunal remains from Closegate I and II, Newcastle, Tyne and Wear, 1988 and 1990 excavations. Ancient Monuments Laboratory Report 81/91, English Heritage.

Davis, S. J. M. 1997b. *Animal bones from the Roman site Redlands Farm, Stanwick, Northamptonshire, 1990 excavations.* London, HBMC Ancient Monuments Laboratory Report 106/97. English Heritage.

Davis, S. J. M. 2005. *Animal bones from Roman São Pedro, Fronteira, Alentejo.* Trabalhos do CIPA 88, Lisbon, Instituto Português de Arqueologia.

Davis, S. J. M. 2006. *Faunal remains from Alcáçova de Santarém, Portugal* (Trabalhos de Arcueologia 43) Instituto Português de Arqueologia: Lisbon.

Davison, B. K. 1969. Early Earthwork Castles: A New Model. *Château-Gaillard* **3**, 37-47.

Davison, B. K. 1977. Excavations at Sulgrave, Northamptonshire, *Archaeological Journal* **134**, 105-14.

Devailly, G. 1973. *Le Berry du Xe siècle au milieu du XIIIe* Mouton: Paris.

Dobney, K. 2001 A place at the table: the role of vertebrate zooarchaeology within a Roman research agenda for Britain, pp. 36-45 in S. James and Millet, M. *Britons and Romans: Advancing the Research Agenda.* (CBA Research Report 125) Council for British Archaeology: York.

Dobney, K. and Jaques, D. 2002 Avian signatures for identity and status in Anglo-Saxon England, *Acta Zoologica Cracoviensia,* **45**, 7-21.

Dobney, K. M., Jaques, S. D., and Irving, B. G. 1995. Of Butchers and Breeds: Report on vertebrate remains from the city of Lincoln. *Lincoln Archaeological Studies* **5**.

Done, G. 1983. The animal bones, pp. 87 in D. W. Williams 16 Bell Street, Reigate: Excavations of a medieval and post-medieval site 1974-6. *Surrey Archaeological Collections* **74**, 47-89.

Done, G. n.d. *The Animal Bones from 45 Saxon huts at Mucking, Essex.* Unpublished Manuscript.

Douglas, D. C. and Greenaway, G. W., eds. 1953. *English Historical Documents 1042-1189.* Eyre and Spottiswode: London.

Douglas, M. 1975. *Implicit Meanings: Essays in Anthropology.* Routledge and Kegan Paul: London.

Drew, K. F. 1972. *The Burgundian Code.* University of Pennsylvania Press: Philadelphia.

Driver, J. C. 1984. Zooarchaeological Analysis of Raw-Material Selection by a Saxon Artisan. *Journal of Field Archaeology* **11**, 397-403.

Driver, J. C. 1990. Animal bones in J. C. Driver, J. Rady and M. Sparks (eds.) Excavations in the Cathedral Precinct, 2 Linacre Garden, 'Master Omers' and St Gabriels Chapel. Canterbury Archaeological Trust by the Kent Archaeological Society: Maidstone.

Dulley, A. J. F. 1969. The early history of the Rye fishing industry. *Sussex Archaeological Collections* **107**, 36-64.

Dyer, C. C. 1988. Change in diet in the Late Middle Ages: the case of harvest workers. *Agricultural History Review* **36**, 21-37

Dyer, C. C. 1994. *Everyday Life in Medieval England.* Hambledon Press: London.

Eastham, A. 1976. The bird bones, pp.287-295 in Cunliffe, B. W. (ed.), *Excavations at Portchester Castle Vol. II: Saxon.* (Reports of the Research Committee, Society of Antiquaries of London 33) Society of Antiquaries: London

Eastham, A. 1977. The bird bones, pp. 233-238 in Cunliffe, B. W. (ed.), *Excavations at Portchester Castle III: Medieval.* (Reports of the Research Committee, Society of Antiquaries of London 34) Society of Antiquaries: London.

Effros, B. 2002. *Creating Community with Food and Drink in Merovingian Gaul.* New York: Palgrave Macmillan.

Eisenstadt, S. N. 1969. Some observation on the Dynamics of Traditions. *Comparative Studies in Society and History* **11**, 451-475.

Ellison, A. 1975. Animal skeletal material, pp. 146-151 in P. L. Drewett (ed.) Excavations at Hadleigh Castle Essex, 1971-72. *Journal of the British Archaeological Association*, **38**, 90-154.

English, B. 1979. *The Lords of Holderness 1086-1260.* Oxford University Press: Oxford.

Ervynck, A., Cooremans, B., and Van Neer, W. 1994. De voedselvoorziening in de Sint-Salvatorsabdij te Ename (Stad Oudenaarde, prov. Oost-Vlaanderen) 3. Een latrine bij de abtswoning (12 de-begin 13de eeuw), *Archeologie in Vlaanderen* **4**, 311-322.

Ervynck, A., Van Neer, W., Lentacker, A. 1999. Introduction and extinction of wild animal species in historic times: the evidence from Belgium, pp. 399-407 in Benecke, N. (ed.), *The Holocene history of the European vertebrate fauna: modern aspects of research workshop 6th-9th April 1998.* Verlag Marie Leidorf GmbH: Rahden/Westf.

Ervynck, A., Van Neer, V., Hüster-Plogmann, H. and Schibler, J. 2003 Beyond affluence: the zooarchaeology of luxury, *World Archaeology* **34** (3), 428-41.

Everton, R. F. 1982. The animal bone, pp. 144 in R. Leech (ed.) *Excavations at Catsgore, 1970-1973.* (Western Archaeological Trust Monograph 2) Western Archaeological Trust: Bristol.

Fernie, E. 1982. *The Architecture of the Anglo-Saxons.* Batsford: London.

Finberg, H. P. R. 1947. Some Early Tavistock Charters. *English History Review* **62**, 365.

Fisher, C. T. 1986. Bird bones from the excavation at Crown car park, Nantwich, Cheshire. *Circaea* **4** (1) 55-64.

Flambard-Hériche, A-M. 1996a. Premiers résultats des fouilles du château de Vatteville-la-Rue (Seine-Maritime). *Actes du XVII colloque international Château Gaillard tenu a Abergavenny (G.B)* en Septembre 1994: Caen.

Flambard-Hériche, A-M. 1996b *Vatteville-la-Rue (76) le <Vieux Château>, Rapport de fouille 1994-1996.* Service Régional de l'Archéologie du haut Normandie, pp 145.

Flambard-Hériche, A-M. 1998a Le shell keep de Vatteville-la-Rue (Seine-Maritime) in *Proximus*, 169-179. Publication of the Archaeological Round Table, Dieppe 17th and 18th September.

Flambard-Hériche, A-M. 1998b. La cuisine du logis seigneurial de Vieux Château de Vatteville-la-Rie (Seine-Maritime). *Mémoires et Documents de la Société Historique et Archéologique de l'Orne* **1**, 131-145.

Fleming, R. 2001 The new wealth, the new rich and the new political style in Late Anglo-Saxon England, *Anglo-Norman Studies*, **23**, 1-22.

Foch, J. 1966. *Metrische Untersuchungen an Metapodien eniger Europäischer Rindererassen.* Dissertation, University of München.

Forde, S., Johnson, L., and Murray, A. 1995. *Concepts of National Identity in the Middle Ages* (Leeds Texts and Monographs New Series 14) School of English, University of Leeds: Leeds.

Fourquin, G. 1975. Le premier moyen âge pp. 286-371 in G. Duby and A. Wallon (eds.) *Histoire de la France Rurale* Seuil: Tours.

Fowler, R. W. 1972. *Wulfstan's Canons of Edgar* (Early English Texts Series) Oxford University Press: London.

Freeman, E. A. 1867-79. *The History of the Norman Conquest*. Clarendon Press: Oxford.

Furnivall, F. J. 1868. Early English Meals and Manners. Early English Text Society/Oxford University Press: London.

Gamble, C. S. 1983. The Animal Bones, 114-117 in P. J. Fasham (ed.) Excavations in Banbury, 1972: second and final report. *Oxoniensia* **48**, 71-118.

Gardiner, M. 1997 The exploitation of sea-mammals in medieval England: bones and their social context, *Archaeological Journal* **154**, 173-95.

Garmonsway, G. N. 1967. *The Anglo-Saxon Chronicle*. Dent and Dutton: London.

Gautier, A. 1972. La faune d'un puits de l'abbaye de Saint Avit-Senieur (XI à XIII siècle, Dordogne, France) *Archéologie Médiévale* **2**, 355-379.

Gautier, A 1984a. How do I count you, let me count the ways? Problems of archaeozoological quantification, pp. 237-52 in C. Grigson and J. Clutton-Brock (eds.) *Animals and Archaeology: 4 Husbandry in Europe*. (BAR, International Series 227) British Archaeological Reports: Oxford.

Gautier, A. 1984b Enkele dierresten uit de abdji Ter Duinen te Koksijde (ca. 1175-ca. 1250, ca. 1250-1318, 1593-1601 A.D), *De Duinen* **13-14**, 61-63.

Gautier, A. 2006. *Le Festin dans l'Angleterre Anglo-Saxonne (Ve-XIe siècle)* Presses Universitaires de Rennes: Renne.

Gautier, A. in press. Deer parks in Sussex and the Godwinesons *Anglo-Norman Studies* **29**.

Gautier, A. n.d. Saxons et Normands à table (fin XIème - début XIIème siècle): frontières et identités dans l'œuvre des historiens anglo-normands.

Gebbles, A. 1977. The animal bones, pp. 277-284 in M. Bell (ed.) Excavations at Bishopstone. *Sussex Archaeological Collections* **115**, 1-299.

Getty, R. 1975. *Sisson and Grossman's the Anatomy of the Domestic Animals*. W. B. Saunders and Co.: Philadelphia.

Gidney, L. J. 1990. *The animal bones from the 1987 excavations of the Saxon contents at Morrison Hall, Hartlepool, Cleaveland*. Ancient Monuments Laboratory Report 111/90. English Heritage.

Gidney, L. J. 1991 *The animal bones from the 1987 Excavations: the animal bones from the Medieval deposits at St Peters Lane*. Ancient Monuments Laboratory Report 116/91. English Heritage.

Gidney, L. 1991. *Little Lane medieval contexts* Ancient Monuments Laboratory Report 57/91, English Heritage.

Gilbert, J. 1979. *Hunting and Hunting Reserves in Medieval Scotland*. John Donald: Edinburgh.

Giles, R. 1991. *L'avifaune d'Andone aux Xe et Xièmes siècles*, Mémoire de DEA Archéologie et Environment, MNHN, Université de Paris I et V.

Golding, B. 1994. *Conquest and Colonisation: the Normans in Britain, 1066-1100*. Macmillan: London.

Goody, J. 1982. *Cooking, Cuisine and Class: A study in comparative sociology* Cambridge University Press: Cambridge.

Graham-Campbell, J. 1990. *The Viking World* (second edition) Frances Lincoln, Weidenfeld and Nicolson: London.

Grant, A. 1971. The animal bones, 90–93 in G. F. Bryant and J. M. Steene (ed.) Excavations at the deserted medieval settlement at Lyveden. A third interim report *Journal of the Northampton County Borough Museum and Art Gallery* 9.

Grant, A. 1974. Excavations at St John's, Bedford. Unpublished manuscript.

Grant, A. 1975. The faunal remains, pp. 378-408 in B. Cunliffe (ed.) *Excavations at Portchester Castle 1: Roman* (Reports of the Research Committee, Society of Antiquaries of London 32) Society of Antiquaries: London.

Grant, A. 1976. The Faunal Remains, pp. 262-287 in Cunliffe, B. W. (ed.), *Excavations at Portchester Castle Vol. II: Saxon*. (Reports of the Research Committee, Society of Antiquaries of London 33) Society of Antiquaries: London.

Grant, A. 1977. The Animal Bones, mammals, pp. 213-239 in Cunliffe, B. W. (ed.), *Excavations at Portchester Castle III: Medieval*. (Reports of the Research Committee, Society of Antiquaries of London 34) Society of Antiquaries: London.

Grant, A. 1979. The Animal Bones (Citizen's House), pp. 60-70 in Cunliffe, B. W. (ed.), *Excavations in Bath, 1950-1975.* Bristol: Committee for Rescue Archaeology in Avon Gloucestershire and Somerset.

Grant, A. 1981. The significance of deer remains at occupation sites of the Iron Age to the Anglo-Saxon period, pp. 53-64 in M. Jones and G. Dimbleby (eds.) *The Environment of Man: The Iron Age to the Anglo-Saxon period* (BAR, British Series 87) British Archaeological Reports: Oxford.

Grant, A. 1982. The use of tooth wear as a guide to the age of domestic ungulates, pp. 91-108 in B. Wilson, C. Grigson and S. Payne (eds.) *Ageing and Sexing Animals from Archaeological sites* (BAR, British Series 109) British Archaeological Reports: Oxford.

Grant, A. 1987. Some observations on butchery in England from the Iron Age to the medieval period. *Anthropozoologica* **Premier Numéro Spécial**, 53-57.

Grant, A. 1988a. The animal resources, pp. 149-197 in G. Astill and A. Grant (eds.) *The Countryside of Medieval England.* Blackwell: Oxford.

Grant, A. 1988b. Food, status and religion in the Middle Ages: an archaeozoological perspective. *Anthropo-zoologica* **Second Numéro Special**, 139-146.

Grant, A. 2002 Food, status and social hierarchy, pp. 17-23 in P. Miracle and N. Milner (eds.) *Consuming passions and patterns of consumption.* McDonald Institute for Archaeological Research: Cambridge.

Grant, M. 2000 *Galen on Food and Diet*, London: Routledge.

Grayson, D. K. 1984. *Quantitative Zooarchaeology.* Academic Press: London.

Griffith, N. J. L., Halstead, P. L. J., MacLean, A., and Rowley-Conwy, P. A. 1983. Faunal remains and economy, pp. 341-8 in Mayes, P. and Butler, L. A. S. (eds.), *Sandal Castle Excavations 1964-1973.* Wakefield Historical Publications: Wakefield.

Grigson, C. 1982. Sex and age determination of some bones and teeth of cattle; a review of the literature, pp. 7-24 in B. Wilson, C. Grigson and S. Payne (eds.) *Ageing and sexing animal bones from archaeological sites*, (BAR, British Series 109) British Archaeological Reports: Oxford.

Groundes-Peace, Z. 1971. *Mrs Groundes-Peace's Old Cookery Notebook.* David and Charles: Exeter.

Grundy, G. B. 1932. The Saxon charters of Somerset Part IV, *Proceedings of the Somerset Archaeological and Natural History Society* **78**, 186.

Hagen, A. 1998. *A Handbook of Anglo-Saxon Food: processing and consumption.* Anglo-Saxon Books: Norfolk.

Hagen, A. 2002 *A Second Handbook of Anglo-Saxon Food and Drink: Production and Distribution*, Frithgarth: Anglo-Saxon Books.

Hall, A. 1988. York 700-1050, pp. 125-132 in R. Hodges and B. Hobley (eds.) *The Rebirth of Towns in the West AD 700-1050.* (CBA Research report 68) Council for British Archaeology: London.

Halstead, P. 1985. A study of the mandibular teeth from Romano-British contexts at Maxey, pp. 219-24 in F. Pryor (ed.) *Archaeology and Environment of the Lower Welland Valley Vol. 1.* (East Anglian Archaeology Report 27) Cambridgeshire Archaeological Committee in conjunction with the Fenland Project Committee and the Scole Archaeological Committee: Cambridge.

Hambleton, E. 1999. *Animal Husbandry regimes in Iron Age Britain* (BAR, British Series 282) British Archaeological Reports: Oxford.

Hamilakis, Y. 2000. The Anthropology of food and drink consumption and the Agean Archaeology, pp. 55-64 in W. Carlson and S. Vaugham (eds.) *Aspects of Paleodiet in the Aegean.* Oxford: Oxbow.

Hamilakis, Y. 2003. The sacred geography of hunting: wild animals, social power and gender in early farming societies, pp. 239-247 in E. Kotjabopoulou, Y. Hamilakis, P. Halstead, C. Gamble and V. Elafanti (eds). *Zooarchaeology in Greece: Recent Advances.* British School at Athens: London.

Hamilton, R. 1971. Animal remains pp. 165 in K. Branigan (ed.) *Latimer: Belgic, Roman, Dark Age and Early Modern Farm.* Chess Valley Archaeological Society: Bristol.

Hamilton-Dyer, S. 2001. Animal bone, pp.102-109 in J. Pine (ed.) The excavation of a Saxon settlement at Cadley Road, Collingbourne Ducis, Wiltshire. *Wiltshire Archaeological and natural History Magazine* **94**, 88-117.

Handcock, W. D. 1996. *English Historical Documents Volume 9.* Routledge: London.

Hammond, P. W. 1995. *Food and Feast in Medieval England.* Alan Sutton: Stroud.

Harman, M. 1979a. Mammalian bone, 77-78 in J. H. Williams (ed.) Excavations on Marefair, Northampton, 1977. *Northamptonshire Archaeology* **14,** 38-79.

Harman, M. 1979b. The mammalian bones, pp. 328-332 in J.H. Williams (ed.) *St Peters Street Northampton: Excavations 1973-1976.* Northampton Development Corporation.

Harman, M. 1985a. The animal and bird bones, pp. 190-2 in G. Lambrick (ed.) Further Excavations on the Second Site of the Dominican Priory, Oxford. *Oxoniensia* **50**, 131-208.

Harman, M. 1985b. The other mammalian bones, pp. 328-331 in Williams, J. H., Shaw, M., and Denham, V. (eds.), *Middle Saxon Palaces at Northampton*. Northampton: Northampton Development Corporation.

Harman, M. 1993. The animal bones, pp. 223-239 in M. J. Darling and D. Gurney (eds.) Caister-on-Sea Excavations by Charles Green, 1951-55 (East Anglian Archaeology Report No 60) Field Archaeology Division, Norfolk Museum Service: Gressenhall.

Harper-Bill, C. 1979. The Piety of the Anglo-Norman Knightly Class. In *Proceedings of the Battle Conference* **2**. 63-77.

Harvey, B. 1993. *Living and Dying in England 1100-1540. The Monastic Experience*. Clarendon Press: Oxford.

Harvey, B. 2006. Monastic pittances in the Middle Ages, pp. 215-227 in C. Woolgar, D. Serjeantson and T. Waldron (eds.) *Food in Medieval England: History and Archaeology*, Oxford: Oxford University Press.

Harvey, J. 1982. *Medieval Gardens*. Batsford: London.

Hawes, R. O. 1984. Pigeons, pp. 351-56 in I. L. Mason (ed.) *Evolution of Domesticated Animals*. Longman: London.

Hawkes, A. J. 1997 Symbolic lives: the visual evidence, pp. 311-44 in J. Hines (ed.) *The Anglo-Saxons from the Migration Period to the Eighth Century: An Ethnographic Perspective*. Boydell: Woodbridge.

Heighway, C. M., Garrod, A. P., and Vince, A. G. 1979. Excavations at 1 Westgate Street, Gloucester 1975. *Medieval Archaeology* **23**, 182-5.

Henderson, A 1997. From Coney to Rabbit: the story of a managed coloniser, *The Naturalist* **122**, 101–21.

Herring, P. 2003. Cornish medieval deer parks, pp. 34-50 in R. Wilson-North (ed.) *The Lie of the Land: Aspects of the Archaeology and History of the Designed Landscape in the South West of England*. The Mint Press: Exeter.

Hesse, B. and Wapnish, P. 1985. *Animal Bones in Archaeology: from objectives to analysis*. Taraxacum: Washington.

Hieatt, C. B. 2002 Medieval Britain, pp. 19-45 in M. Weiss Adamson (ed.) *Regional Cuisines of Medieval Europe: A Book of Essays*. Routledge: London.

Hieatt, C. B. and Jones, R. F. 1986. Two Anglo-Norman Culinary Collections ed. from BL MSS Addit. 32085 and Royal 12.C.xii. *Speculum: A Journal of Medieval Studies* **61** (4), 859-882.

Higgs, E. and Greenwood, W. 1979 Faunal report, pp. 354-362 in Rahtz, P.(ed.), *The Saxon and Medieval Palaces at Cheddar Excavations 1960-2* (BAR British Series 65) British Archaeological Reports: Oxford.

Higgs, E. S. and Jarman, M. 1977. Yeavering: faunal report, pp. 325-332 in B. Hope-Taylor, (ed.) *Yeavering: an Anglo-British Centre of early Northumbria*. (Dept. of the Environment Archaeological Reports 7) H.M.S.O: London.

Higham, C. 1969. The metrical attributes of two samples of Bovine limb bones. *Journal of Zoology* **157**, 63-74.

Hilton, R. 1992. *English and French Towns in Feudal Society*. Cambridge University Press: Cambridge.

Hinton, D. A. 1990. *Archaeology, Economy and Society: England from the fifth to fifteenth century*. Seaby: London.

Hinton, M. A. 1979. On the remains of vertebrate animals found in the middens of Rayleigh Castle. *Essex Naturalist*, **17**, 16-21.

Hodges, R. 1989. *The Anglo-Saxon Achievement*. Duckworth: London.

Hodges, R. 2000. *Towns and Trade in the Age of Charlemagne*. Duckworth: London.

Hoffmann, R. C. 1996. Economic Development and Aquatic Ecosystems in Medieval Europe, *The American Historical Review* **101** (3), 631-669.

Holmes, J. M. 1993. The animal bones, pp. 133-158 in R. J. Williams (ed.) *Pennyland and Hartigans: Two Iron Age and Saxon Sites in Milton Keynes* (Buckinghamshire Archaeological Society Monograph Series no 4) Buckinghamshire Archaeological Society: Ayelsbury.

Holmes, U. T. Jr. 1952. *Daily living in the twelfth century*. University of Wisconsin Press: Wisconsin.

Hooke, D. 1989. Pre-conquest woodland: its distribution and usage, *Agricultural History Review* **37** (2), 113-129.

Hooke, D. 1998. *The Landscape of Anglo-Saxon England*. Leicester University Press: Leicester.

Hooper, W. D. and Ash, H. B. (translated). 1967. *Marcus Porcius Cato on agriculture. Marcus Terentius varro on agriculture*. Loeb Classical Library No 283. Heinemani, London.

Hough, C. 2001. Place-name evidence for an Anglo-Saxon animal name: OE **pohha/*pocca* fallow deer, *Anglo-Saxon England* **30**, 1-14.

Huntley, J. P. and Stalibrass, S. 1995. *Plant and vertebrate remains from archaeological sites in northern England: data reviews and future directions*. Research Report 4. Architectural and Archaeological Society of Durham, Northumberland: Durham.

Hurst, J. G. 1976. The pottery, pp. 283-384 in D. M. Wilson (ed.) *The Archaeology of Anglo-Saxon England.* Methuen: London.

Hutton-MacDonald, R. 1999. Animal bones from St Nicholas' Street and Guildhall Street, pp. 75-82 in P. Andrews and K. Penn (eds.) *Excavations in Thetford, North of the River, 1989-90.* (East Anglian Archaeology Report 87) Field Archaeology Division, Norfolk Museums Service: Dereham.

Ijzereef, G. F. 1989. Social Differentiation from Animal Bone Studies, pp. 41-55 in D. Serjeantson and T. Waldron (eds.) *Diet and Crafts in Towns: The Evidence of Animal Remains the Roman to the Post-Medieval Periods.* (BAR, British Series 199) British Archaeological Reports: Oxford.

Ingrem, C. n.d. *Assessment of the animal remains from Bishopstone, Sussex* unpublished report to the Sussex Archaeological Society, Centre for Applied Archaeological Analyses (University of Southampton) 2005.

Jameson, E. 1938. The Sicilian Norman kingdom in the mind of Anglo-Norman contemporaries, *Proceedings of the British Academy* **24**, 237-85.

Jessop, O. 1996. A new artifact typology for the study of medieval arrowheads. *Medieval Archaeology* **40**, 192-205.

Jewel, P. A 1968. In (Brodribb, A. C., Hands, A. R., and Walker, D. R.) *Excavations at Shakenoak I.* 112-3. Privately published: Oxford.

Jones, A. K. G. 1983 The fish remains, pp. 44 in D. Allen and C. H. Dalwood (ed.) Iron Age Occupation, a Middle Saxon Cemetery and 12th-19th century Urban Occupation: Excavations in George Street, Aylesbury. *Records of Buckinghamshire* **25**, 1-60.

Jones, A. K. G. 1984. Fish bones, pp. 192-194 in A. Rogerson and C. Dallas (eds.) *Excavations in Thetford 1948-59 and 1973-80.* (East Anglian Archaeology Report 22). Norfolk Archaeological Unit in conjunction with the Scole Archaeological Committee: Dereham.

Jones, A. K. G. 1993. Fish remains, Pp. 191 in C. Dallas (ed.) *Excavations in Thetford by B. K. Davidson between 1964 and* 1970. (East Anglian Archaeology 62) Norfolk Archaeological Unit in conjunction with the Scole Archaeological Committee: Dereham.

Jones, A. K. G. 1994. Fish remains, pp.396-7 in T. Allen (ed.) A medieval grange of Abingdon Abbey at Dean Court Farm, Cumnor. *Oxoniensia* **59**, 219-447.

Jones, A. K. G. 2002. Fish bones pp.158-9 in N. Baker (ed.) Shrewsbury Abbey: Studies in the Archaeology and History of an Urban Abbey (Shropshire Archaeological and Historical Society Monograph Series no 2).

Jones, A. K. G. and Scott, S. A. 1985. The fish bones, pp. 223-228 in M. Atkin, A. Carter and D. H. Evans (ed.) *Excavations in Norwich 1971-1978 Part II.* (East Anglian Archaeology Report 26) Norwich Survey in conjunction with Scole Archaeological Committee: Norwich.

Jones, G. 1980. The Animal Bones. In (Farley, M.) Mid Saxon Occupation at Chichley, Buckingham. *Records of Buckinghamshire* **22**, 90-102.

Jones, G. 1983 The medieval animal bones, pp. 31-44 in D. Allen and C. H. Dalwood (ed.) Iron Age Occupation, a Middle Saxon Cemetery and 12th-19th century Urban Occupation: Excavations in George Street, Aylesbury. *Records of Buckinghamshire* **25**, 1-60.

Jones, G. 1984. Animal bones, pp. 187-192 in A. Rogerson and C. Dallas (eds.) *Excavations in Thetford 1948-59 and 1973-80.* (East Anglian Archaeology Report 22) Norfolk Archaeological Unit in conjunction with the Scole Archaeological Committee: Dereham.

Jones, G. 1993. Animal and bird bones, pp. 176-192 in C. Dallas (ed.) *Excavations in Thetford by B. K. Davidson between 1964 and 1970.* (East Anglian Archaeology 62) Norfolk Archaeological Unit in conjunction with the Scole Archaeological Committee: Dereham.

Jones, G. 1994a. Mammal and bird bones, pp. 37-43 in B. S. Ayers (ed.) *Excavations at Fishergate, Norwich, 1985.* (East Anglian Archaeology 68) Norfolk Archaeological Unit in conjunction with the Scole Archaeological Committee: Dereham.

Jones, G. 1994b. Animal bones, pp. 386-396 in T. Allen (ed.) A medieval grange of Abingdon Abbey at Dean Court Farm, Cumnor. *Oxoniensia* **59**, 219-447.

Jones, G. G. 2002. The excavated faunal remains, pp. 145-159 in N. Baker (ed.) Shrewsbury Abbey: Studies in the Archaeology and History of an Urban Abbey (Shropshire Archaeological and Historical Society Monograph Series no 2).

Jones, K. T. 1993. The Archaeological Structure of a Short-term Camp, pp. 101-114 in J. Hudson (ed.) *From*

Bones to Behaviour: Ethnoarchaeological and Experimental Contributions to the Interpretation of Faunal Remains. (Occasional Paper 21) Southern Illinois University, Center for Archaeological Investigations.: Carbondale.

Jones, R.T. and Serjeantson, D. 1983. *The Animal Bones from Five Sites at Ipswich*. Ancient Monuments Laboratory Report 13/83. English Heritage.

Jones, R. T., Sly, J., Simpson, D., Rackham, J., and Locker, A. 1985a. *The terrestrial vertebrate remains from the Castle, Barnard Castle*. Ancient Monuments Laboratory Report 7/85. English Heritage.

Jones , R. T., Levitan, B., Stevens, P. And Hocking, L. 1985b. Castle Lane, Brackley, Northamptonshire. The Vertebrate Remains. Ancient Monuments Laboratory Report 4812. English Heritage.

Jones, R. T. and Ruben, I. 1987. Animal bones, with some notes on the effects of differential sampling, pp. 197-206 in G. Beresford (ed.) *Goltho: the development of an early medieval manor c.850-1150*. English Heritage Report: London.

Jones, R. T., Reilley, K., and Pipe, A. R. 1997. The animal bones, pp. 123-131 in B. Morley and D. Gurney (eds.) *Castle Rising Castle, Norfolk*. (East Anglian Archaeology 81) Norfolk Museum Service: Norwich.

Jones, S. 1997. *The Archaeology of Ethnicity: Constructing Social Identities in the Past and Present*. Routledge: London.

Jope, M. 1961–2. The animals bones, 197–201 in M. Biddle, The medieval village of Seacourt, Berkshire. *Oxoniensia* **26–27**, 70-210 .

Jourdan, L. 1977. La faune de la Bourse (II au V siècles) Marseille: aspects de la faune de Rougiers (Moyen Age), pp, 95-112 in *L'élevage en Méditerranée occidentale, Actes du Colloque international d l'Institut de Recherche méditerranée*, Mai 1976. Ed du CNRS: Paris.

Keene, D. 1985. *Survey of Medieval Winchester* Volume 1, Part I. Clarendon Press: Oxford.

Kelly, F. 1997 *Early Irish farming: the evidence of the law-texts*. Dublin Institute for Advanced Studies: Dublin.

Kenyon, D. M. E. 1997. Patterning in butchery as cultural residue. *Anthropozoologica* **25-26**, 131-136.

Kent, S. 1993. Variability in faunal assemblages: the influence of hunting skill, sharing, dogs and mode of cooking on faunal remains at the sedentary Kalahari community. *Journal of Anthropological Archaeology* **12**, 323-385.

Kent, S. 1989. *Farmers as Hunters: The Implications of Sedentism*. Cambridge: Cambridge University Press.
Kerridge, E. 1967. *The Agricultural Revolution*. Allen and Unwin: London.

Kessinger, K. M. 1989. Hunting and male domination in Cashinahua society, pp. 18-26 in S. Kent (ed.) *Farmers as Hunters: The Implications of Sedentism*. Cambridge University Press: Cambridge.

King, A. 1978. A comparative survey of bone assemblages from Roman Britain. *Bulletin of the Institute of Archaeology* **15**, 207-232.

King, A. 1982. The animal bones, pp. 193-205 in P. Bennett, S. Frere and S. Stow (ed.) *The Excavations at Canterbury Castle*. (The Archaeology of Canterbury Vol. I) Kent Archaeological Society: Maidstone.

King, A.C. 1985. I resti animali: I mammiferi, I rettili e glianfibi, in A. Carandini (ed.) *Settefinestre. Una villa schiavistica nell'Etruria romana, III*: 278-300. Modena: Panini Editore.

King, A. 1991. Food production and consumption – meat, pp. 15-20 in R. Jones (ed.) *Britain in the Roman Period: Recent Trends*. J.R. Collis: Sheffield.

King, A. C. 2002. Mammals: evidence from wall paintings, sculpture, mosaics, faunal remains and ancient literary sources, pp. 401-50 in W. Feemster Jashemski and F.G. Meyer (ed.) *The Natural History of Pompeii*. Cambridge University Press: Cambridge.

King, J.E. 1962. Report on animal bones pp. 355-361 in J.J. Wymer (ed.) Excavations at the Maglemosian sites at Thatcham, Berkshire, England. *Proceedings of the Prehistoric Society* **28**, 329-361.

Klein, R. G. and Cruz-Uribe, K. 1984. *The Analysis of Animal Bones from Archaeological Sites*. University of Chicago Press.

Knapp, A. B. and Ashmore, W. 2000. Archaeological landscapes: constructed, conceptualized, ideational, pp. 1-30 in W. Ashmore and A. B. Knapp (eds.) *Archaeologies of Landscape: Contemporary Perspectives*. Blackwell: Oxford.

Knowles, D. 1940. *The Monastic Order in England: A History of its Development from the Times of St Dunstann to the Fourth Lateran Council 940-1216*. Cambridge University Press: Cambridge.

Kühnhold, B. 1971. *Die Tierknochenfunde aus Unterregenbach, einer mittelalterlichen Siedlung Württembergs*. Inaugural Dissertation zur Erlangung der veterinärmedizinscchen Doktorwürde de Tierärztlichen Kakultät der Ludwig-Maximilians-Universität. München.

Küpper, W. 1972. *Die Tierknochenfunde von der Burg Schiedberg beir Sagogn in Grabünden. II die Kleinen Wiederkaüer, die Wildtiere und das gefügel.* Inaugural Dissertation zur Erlangung der veterinärmedizinscchen Doktorwürde de Tierärztlichen Kakultät der Ludwig-Maximilians-Universität. München.

Kylie, E. 1911. *The English Correspondance of St Boniface.* Chatto and Windus.

Lamb, H. H. 1981. Climate from 1000BC to 1000AD, pp. 53-65 in M. Jones and G. Dimbleby (eds.) *The Environment of Man from the Iron Age to Anglo-Saxon Period* (BAR, British Series 87) British Archaeological Reports: Oxford.

Langbein, J. and Chapman, N. *Fallow Deer* Mammal Society: London.

Langdon, J. 1986. *Horses, Oxen and Technological Innovation: the Use of Draught Animals in English Farming from 1066 to 1500.* Cambridge University Press: Cambridge.

Langdon, J. 1988. Agricultural equipment, pp. 86-107 in G. Astill and A. Grant (eds.) *The Countryside in Medieval England.* 86-107. Blackwell: Oxford.

Langenwalter, P. E. II. 1980. The archaeology of 19th century Chinese subsistence at the lower China store, Madera County, California, pp.102-12 in R. L. Schuyler (ed.) *Archaeological Perspectives of Ethnicity in America.* Baywood: New York.

Lasdun, S. 1991 *The English Park: Royal, Private and Public.* Andre Deutsch Ltd: London.

Lasota-Mokalewska, A. 1984. Anatomical distribution of bone remains from the medieval castle at Ciechanow, Poland, pp. 155-159 in C. Grigson and J. Clutton-Brock (eds.) *Animals and Archaeology Volume 4, Husbandry in Europe.* (BAR, International Series 227) British Archaeological Reports: Oxford.

Laurioux, B. 1988. Le lièvre lubrique et la bête sanglante. Réflexions sur quelques interdits alimentaires du Haut Moyen Âge, *Anthropozoologica,* **special number 2**, 127-32.

Lauwerier, R. C. G. M. 1988. *Animals in Roman Times in the Dutch Eastern River Area.* (Neaderrlandse Oudheden 12) ROB: Amersfoort.

Lauwerier, R. C. G. M. and Zeiler, J. T. 2001. Wishful thinking and the introduction of the rabbit to the Low Countries, *Environmental Archaeology* **6**, 87-90.

Lawrance, P. 1982. Animal bones, pp. 275-294 in J. G. Code and A. D. S. Streeten (eds.) Excavations at Castle Acre, Norfolk 1972-77: county house and castle of the Norman earls of Surrey. *Archaeological Journal* **139**, 138-301.

Le Blay, J.-C. 1993. Hasard, taphonomie ou redevance? Le cas du porc mediéval. *Anthropozoologica* **17**, 7-10.

Le Blay, J.-C., Lepetz, S., and Yvinec, J.-H. 1997. L'èlevage dans l'Antiquité tardive en Ile-de-France, pp. 50-67 in *Les campagnes de l'Ile-de-France de Constantin à Clovis.* Actes de la 2nd journèe. Documentation de travail No 3 Décembre 1997.

Lemanski, S. J. 2005. *The Rectitudines Singularum Personarum: Anglo-Saxon Landscapes in Transition.* MA thesis, University of Akron.

Lennard, R. 1959. *Rural England 1086-1135. A Study of Social and Agrarian Conditions.* Clarendon Press: Oxford.

Le Patorel J. 1976. *The Norman Empire.* Oxford University Press: Oxford.

Lepetz, S., Meniel, P., and Yvinec, J.-H. 1995. Archéozoologie des installations rurales de la fin de l'Âge du Fer au début du Moyen Âge, pp. 169-82 in *L'Histoire rurale en France.* Actes du colloque de Rennes (Oct 94), Histoire et Sociétés Rurales no 3 Ier semestre 1995.

Lepetz, S. and Yvinec, J.-H. 1998. L'élevage à la période Gallo-Romaine et au Haut Moyen âge en Normandie l'apport de l'archéozoologie, pp. 169-182 in *Le Monde Rural en Normandie..* Annales de Normandie, Séries des Congrès des Sociétés Historiques et Archéologique de Normandie, Volume 3. Caen, Musée de Normandie.

Lepetz S. and Yvinec J.-H. 2002. Présence d'espèces animales d'origine méditerranéennes en France du nord aux périodes romaine et médiévale: actions anthropiques et mouvements naturels, pp. 33-42 in Gardeisen A. (ed.), *Mouvements ou déplacements de populations animales en Méditerranée au cours de l'Holocène* (BAR, International Series 1017). Oxford: John and Erica Hedges.

Lepiksaar, J. 1966-1978. *Restes animaux provenant du Grand Besle.* Meddelan den Fran Lunds Universitets Historika Museum, 85-116.

Lepiskaar, J. 1980. The bird bones, 158-160 in P. Wade Martin (ed.) Village sites in the Launditch Hundred (East Anglian Arch Report).

Leslie, R. F. 1966. *The Wanderer.* Manchester University Press: Manchester.

Lever, C. 1979. *The Naturalised Animals of the British Isles.* Granada Publishing: St Albans.

Levi-Strauss, C. 1966. The Culinary Triangle. *Partizan Review* **3**, 586-595.

Levi-Strauss, C. 1969. *The Raw and the Cooked.* Harper and Row: New York.

Levitan, B. 1982. The faunal remains, pp. 269-283 in P. Leach (ed.) *Ilchester Vol 1: Excavations 1974-5* (Western Archaeological Trust Monograph 3 Western Archaeological Trust: Bristol.

Levitan, B. 1984. The Vertebrate Remains, pp. 108-148 in Rahtz, P. and Rowley, T. (eds.), *Middleton Stoney: Excavations and Survey in a North Oxfordshire Parish 1970-1982*. Oxford University Dept of External Studies: Oxford.

Levitan, B. 1984b. Faunal remains from Priory Barn and Benham's Garage, pp. 167-194 in P. Leach (ed.) *The Archaeology of Taunton.* (Western Archaeological Trust Monograph 8) Western Archaeological Trust: Bristol.

Levitan, B. 1985. The animal bones, pp. 130-5 in A. Saville (ed.) Salvage recording of Romano-British, Saxon, Medieval and Post-Medieval remains at North Street, Winchcombe, Gloucestershire. *Transactions of the Bristol and Gloucestershire Archaeological Society* **103**, 101-139.

Levitan, B. 1987a. *Medieval Animal Husbandry in South West England. A Selective Review and Suggested Approach.* Ancient Monuments Laboratory Report, 223/87. English Heritage.

Levitan, B. 1987b. *Bone Analysis and urban Economy: examples of selectivity and a case for comparison.* Ancient Monuments Laboratory Report 224/87. English Heritage.

Levitan, B. 1990. The vertebrate remains, pp. 220-239 in M. Bell (ed.) *Brean Down Excavations 1983-1987* (Historic Buildings and Monuments Commission Report 15) English Heritage: London.

Levitan, B. 1992. The animal bone, pp. 74-5 in G. D. Keevill (ed.) An Anglo-Saxon Site at Audlett Drive Abingdon, Oxford. *Oxoniensia* 57, 56-79.

Levitan, B. and Locker, A. 1987. The vertebrate remains from Westward Ho!, Devon, pp. 213-222 in Balaam, N. D., Levitan, B., and Straker, V. (eds.), *Studies in Palaeoeconomy and Environment in South West England* (BAR British Series 181) Oxford: British Archaeological Reports.

Lewis, C. Mitchell-Fox, P., and Dyer, C. 1997. *Village, Hamlet and Field: changing medieval settlements in central England.* Manchester University Press: Manchester.

Lewis, M. 2007 (in press). Identity and status in the Bayeux Tapestry: the iconographic and artefactual evidence. *Anglo-Norman Studies* **29**.

Liddiard, R. 2000. *Landscapes of Lordship: The Castle and the Countryside in Medieval Norfolk, 1066-1200.* (BAR, British Series 309) British Archaeological Reports: Oxford.

Liddiard, R. 2003. The deer parks of Domesday Book *Landscapes* **4** (1), 4-23.

Liddiard, R. 2005. *Castles in Context: Power, Symbolism and Landscape, 1066 to 1500*, Windgather Press, Macclesfield.

Lister, A. M. 1984. Evolutionary and ecological origins of British deer, *Proceedings of the Royal Society of Edinburgh. Section B.* **82** (part 4), 205-229.

Lister, A. M. 1996. The Morphological distinction between bones and teeth of fallow deer (*Dama dama*) and red deer (*Cervus elaphus*). *International Journal of Osteoarchaeology*, **6**, 119-143.

Lloyd, T. H. 1977. *The English Wool Trade in the Middle Ages.* Cambridge University Press: Cambridge.

Locker, A. 1985. Animal and plant remains, pp. 183-89 in J. N. Hare (ed.) *Battle Abbey. The eastern range and the excavations 1978-80.* (Historic Buildings and Monuments Commission for England Archaeological Report 2) Historic Buildings and Monuments Commission for England: London.

Locker, A. 1987. The fish remains, pp. 114-117 in B. Ayers (ed.) *Excavations at St-Martin-at-Palace Plain, Norwich 1981.* (East Anglian Archaeological Report 37) Norfolk Archaeological Unit: Dereham.

Locker, A. 1989. *The fish bones from a 13th/14th century deposit from Cleeve Abbey, Somerset.* Ancient Monuments Laboratory Report 2/89, English Heritage.

Locker, A. 1990. The animal bones, pp. 205-12 in D. S. Neal, A. Wardle, and J. Hunn (ed..) *Excavations at Gorhambury, St Albans.* Historic Monuments and Buildings Commission for England: London.

Locker, A. 1992. Animal bone, pp. 172-181 in S. Woodiwiss (ed.) *Iron Age and Roman Salt Production and the Medieval Town of Droitwich: Excavations at the Old Bowling Green and Friars Street.* (CBA Research Report 81) Council for British Archaeology: London.

Locker, A. 1993a. The fish remains, pp. 149-150 in R. Cowie and R. Whitehead (eds.) Two Middle Saxon Occupation Sites: Excavations at Jubilee Hall and 21-22 Maiden Lane. *London and Middlesex Archaeological Society Transactions* **39**, 47-163.

Locker, A. 1993b. The fish remains, pp. 148-150 in R. Cowie (ed.) Excavations at the Peabody Site, Chandos

Place and the National Gallery. *London and Middlesex Archaeological Society Transactions* **40**, 35-176.

Locker, A. 1994. Fish bones, pp. 42-45 in B. Ayers, B. (ed.) Excavations at Fishergate, Norwich, 1985. (East Anglian Archaeology 68) Field Archaeology Division, Norfolk Museums Service in conjunction with the Scole Archaeological Committee: Dereham.

Locker, A. 1996. *The Fish Remains from St Marys, Clerkenwell.* Unpublished report for Museum of London Archaeological Services.

Locker, A. 1997. *The Saxon, Medieval and Post-Medieval Fish Bones from Excavations at Castle Mall, Norwich, Norfolk.* Ancient Monuments Laboratory Report 85/97, English Heritage.

Locker, A. 2001. *The Role of Stored Fish in England 900-1750 AD: The Evidence from Historical and Archaeological Data.* Sofia: Publishing Group Limited.

Losco-Bradley, P. M. and Salisbury, C. R. 1988 A Saxon and Norman fish weir at Colwick, Nottinghamshire, pp. 39-68 in M. Aston (ed.) *Medieval Fish, Fisheries and Fishponds in England* I (BAR, British Series 182) British Archaeological Reports: Oxford.

Loud, G.A. 1981. The '*Gens Normannorum*': myth or reality? *Anglo-Norman Studies* **4**, 104-16.

Loud, G. A. 2003. The Kingdom of Sicily and the Kingdom of England 1066–1266, *History* **88**, 540–67.

Loveluck, C. 2001 Wealth, waste and conspicuous consumption. Flixborough and its importance for mid and late Saxon settlement studies, pp. 78-130 in H. Hamerow and A. MacGregor (eds) *Image and Power in the Archaeology of Early Medieval Britain: Essays in honour of Rosemary Cramp.* Oxbow: Oxford.

Loyn, H.R. 1970. *Anglo-Saxon England and the Norman Conquest.* Longman: London.

Lucy, S. 2000. *The Anglo-Saxon Way of Death: Burial Rites in Early England.* Sutton: Gloucester.

Luff, R. 1982. *A zooarchaeological study of the Roman north-western provinces.* (BAR, International Series 137) British Archaeological Reports: Oxford.

Luff, R. 1993. The animal bone in W. Rodwell and K. Rodwell (eds.) Rivenhall: investigations of a Roman Villa, Church and Village, 1950-77 (CBA Report 80, Chelmsford Archaeological Trust report 4.2) Chelmsford Museums Service and the Council for British Archaeology: London.

Luff R. 1993. Animal bones from excavations in Colchester, 1971-85. (Colchester Archaeological Report, 12). Colchester Archaeological Trust: Colchester.

Luff, R. 1994. Butchery at the workmen's village (WV), Tell-el-Amarna, Egypt, pp. 158-170 in R. Luff and P. Rowley-Conwy (eds.) *Whither Environmental Archaeology?* (Oxbow Monograph 38) Oxbow: Oxford.

Lyman, R. L. 1994. *Vertebrate Taphonomy.* Cambridge University Press; Cambridge.

McCormick, F. 1991. The effect of the Anglo-Norman settlement on Ireland's wild and domesticated fauna, pp. 41-52 in P. Crabtree and K. Rym (eds.) *Animal Use and Cultural Change.* 41-52. MASCA Research Papers in Science and Archaeology Supplement to Volume 8. MASCA, the University Museum of Archaeology and Anthropology, University of Pennsylvania, Philadelphia.

McEneaney-Schneider, E. 1979. *Tierknochenfunde von der Heunebeurg einem frühkeltischen Herrensitz bei Hundersingen an der Donau (Grabungen 1966 bis 1979). Wiederkäuer ohne die Bovini.* Inaugural Dissertation zur Erlangung der veterinärmedizinscchen Doktorwürde de Tierärztlichen Kakultät der Ludwig-Maximilians-Universität. München.

McPeake, J. C., Bulmer, M. and Rutter, J. A. 1980. Excavations in the Garden of No.1 Abbey Green, Chester, 1975-77. *Journal of Chester Archaeological Society* **63**, 15-37.

MacGregor, A. 1985. *Bone, Antler, Ivory and Horn.* Croom Helm: London.

MacGregor, A. 1991. Antler, Bone and Horn, pp. 355–78 in J. Blair and N. Ramsay (eds.) *English Medieval Industries: Craftsmen, Techniques, Products.* Hambledon Press: London.

MacGregor, A. 1996. Swan rolls and beak markings. Husbandry exploitation and regulation of *Cygnus olor* in England, c.1100-1900. *Anthropozoologica* **22**, 39-69.

MacGregor, A. Mainman, A. J. and Rogers, N. S. H. 1999. *Craft, Industry and Everyday Life: Bone, Antler, Ivory and Horn from Anglo-Scandinavian and Medieval York.* The Archaeology of York. The Small Finds 17/12. Council for British Archaeology. York Archaeological Trust.

Mackinnon, M. 2004. *Production and consumption of animals in Roman Italy: Integrating the Zooarchaeological and Textual Evidence* (Journal of Roman Archaeology. Supplementary series no. 54). Journal of Roman Archaeology: Portsmouth.

Magennis, H. 1999. *Anglo-Saxon Appetites: Food and Drink and Their Consumption in Old English and Related Literature.* Four Court Press: Bodmin.

Maltby, M. 1979a. *Faunal Studies on Urban Sites: the animal bones from Exeter 1971-1975.* (Exeter

Archaeological Reports 2) Department of Prehistory and Archaeology, University of Sheffield: Sheffield.

Maltby, M. 1979b. The animal bones, pp. 182-5 in C. M. Heighway, A. P. Garrod and A. G. Vince (eds.) Excavations at 1 Westgate Street Gloucester 1975. *Medieval Archaeology* **23**, 159-213.

Maltby, M. 1981. Iron Age, Romano-British and Anglo-Saxon animal husbandry: a review of the faunal evidence, pp. 185-90 in M. Jones and G. Dimbleby (eds.) *The Environment of Man: The Iron Age to the Anglo-Saxon period.* (BAR, British Series 87) British Archaeological Reports: Oxford.

Maltby, M. 1982 Animal and bird bones., pp. 114-135 in R. A. Higham (ed.) Excavations at Oakhampton Castle, Devon. Part 2 - The Bailey. *Devon Archaeological Society* **40**, 19-151.

Maltby, M 1983. The animal bone in M Millett and S. James (eds.) Excavations at Cowdery's Down, Basingstoke, Hampshire 1978-81. *Archaeological Journal* **140**, 151-280.

Maltby, M. 1989. Urban rural variations in the butchering of cattle in Romano-British Hampshire, pp. 75-106 in D. Serjeantson and T. Waldron (eds.) *Diet and Crafts in Towns.* (BAR, British Series 199) British Archaeological Reports: Oxford.

Maltby, M. 1996. Animal bone, pp. 155-163 in N. Holbrook and A. Thomas (eds.) The Roman and Early Anglo-Saxon settlement at Wantage, Oxfordshire: excavations at Mill Street, 1993-4 Oxoniensia **61,** 109-79.

Maltby, M. 2003. Animal Bone, pp. 71-76 in C. Bateman, D. Enright and N. Oakey (eds.) Prehistoric and Anglo-Saxon settlements to the rear of Sherbourne House, Lechdale: excavations in 1997, *Transactions of the Bristol and Gloucestershire Archaeological Society* **121**. 23-96

Mariezkurrena, K. 1986. La cabona ganadera del castillar de Mendavia (Navarra). *Separat Munibe* **86**. Sociedad de Ciencias Aranzadi San Sebastian.

Marples, B. J. 1976. The animal bones, 302-4 in T. G. Hassal (ed.) Excavations at Oxford Castle, 1965-1973. *Oxoniensia* **41**, 232-308.

Marples, B. J. 1977. Animal bones, pp. 166-7 in B. Durham (ed.) Archaeological investigations in St Aldates, Oxford. *Oxoniensia* **42**, 82-203.

Marvin, G. 2001. The problem of foxes: legitimate and illegitimate killing in the English countryside pp. 189-211 in J. Knight (ed.) *Natural Enemies: People-Wildlife Conflict in Anthropological Perspective.* London: Routledge.

Marvin, W. P. 2006. *Hunting Law and Ritual in Medieval English Literature*. Brewer: Cambridge.

Mason, D. 1985. *Excavations at Chester: 26-42 Lower Bridge Street, 1974-6, The Dark Age and Saxon Periods.* (Archaeological excavation and survey report 3) Chester City Council, Grovesnor Museum Archaeological: Chester.

Masseti, M. 1996. The postglacial diffusion of the genus *Dama* Frisch, 1775, in the Mediterranean region. *Supplemento alle Ricerche di Biologia della Selvaggina* **25,** 7-29.

Mayes, P. and Butler, L. A. S. 1983. *Sandal Castle Excavations 1964-1973.* Wakefield Historical Publications: Wakefield.

Mayr-Harting, H. M. and Moore, R. I. 1985. *Studies in Medieval History presented to R.C. H. Davis.* Hambledon: London.

Meddens, B. 1990. *Animal Bones from Catterick Bridge (CEU 240) a Roman town (North Yorkshire) excavated in 1983.* Ancient Monuments Laboratory Report 98/90, English Heritage.

Meddens, B. 1997. The animal bones, pp. 100-106 in J. D. Hurst (ed.) *A Multi-Period Salt Production Site at Droitwich. Excavations at Upwich.* (CBA Research Report 107) Council for British Archaeology: London.

Meddens, B. 2000. The animal bone, pp.315-355 in Ellis, P (ed.), *The Roman Baths at MACELLUM at Wroxeter: Excavations by Graham Webster 1955-85* (English Heritage Archaeological Report 9). London: English Heritage.

Mennell, S. 1985. *All Manners of food: eating and taste in England and France from the Middle Ages to the present.* Blackwell: Oxford.

Miles, D. 1984. *Archaeology at Barton Court Farm, Abingdon, Oxon.* (Oxford Archaeological Unit Report 3, CBA Research Report 50) Oxford Archaeological Unit: Oxford.

Millais, J. G. 1906. *Mammals of Great Britain and Ireland.* Longman: London.

Miller, E. and Hatcher, J. 1995 *Medieval England: Towns Commerce and Crafts 1068-1348.* Longman: London.

Miller, T. 1890-1. *OE version Bede's Ecclesiastical History* Volumes I and II. Early English Texts Series.

Millet, M. 1990. *The Romanization of Britain: An Essay in Archaeological Interpretation.* Cambridge University Press: Cambridge.

Moffett, L. 1989. The Archeobotanical evidence for free-threshing tetraploid wheat in Britain, pp. 233-43 in S. Vytlacock (ed.) *Palaeoethnobotany and Archaeology (International Work-Group for Palaeoethnobotany 8th Symposium Nitra-Nové Vozokany 1989)*. Nitra: Slovak Academy of Sciences (Acta Interdisciplinaria Archaeologica 7).

Montón Subias, S. 2002. Cooking in Zooarchaeology: is the issue still raw? pp. 7-15 in P. Miracle and N. Milner (eds) *Consuming Passions and Patterns of Consumption* Cambridge: McDonald Institute for Archaeological Research.

Morel, C. 1985. *Inventaire des restes osseux provenant du site archéologique moyenâgeux de Saint-Denis - Contribution à l'étude de l'alimentation de l'homme.* Thèse pour le Doctorat vétérinaire, Ecole nationale vétérinaire d'Alfort.

Morris, G. 1990. Animal bone and shell in S. W. Ward (ed.) *Excavations at Chester: the lesser medieval religious houses, sites investigated 1964-1983.* Chester City Council Department of Leisure Services: Chester.

Morris, A. J. n.d. An analysis of the past status of non-domestic avian species from bones in the archaeological record. Unpublished manuscript, University of Southampton.

Müldner, G and Richards, M. 2006. Diet in medieval England: the evidence from stable isotopes, pp. 228-238 in C. Woolgar, D. Serjeantson and T. Waldron (eds.) *Food in Medieval England: History and Archaeology*, Oxford: Oxford University Press.

Mulville, J. 2003. Phases 2a-2e: Anglo-Saxon Occupation, 343-60 in A. Hardy, A. Dodd and G. Keevil (eds) *Aelfric's Abbey: Excavations at Eynsham Abbey, Oxfordshire 1989-92* (Thames Valley Landscape Volume 16). Oxford University School of Archaeology: Oxford.

Mulville, J., Ayres, K. and Clark, K. 2004. Animal bone, pp. 325-350 in G. Hey (ed.) *Yarnton: Saxon and Medieval Settlement and Landscape* (Thames Valley Monograph No 20) Oxford Archaeological Unit: Oxford.

Murphy, P. 1987. Avian eggshell, pp. 113-4 in B. Ayers, (ed.) *Excavations at St-Martin-at-Palace Plain, Norwich 1981.* (East Anglian Archaeological Report 37) Norfolk Archaeological Unit in conjunction with the Scole Archaeological Committee: Dereham.

Murray, E. and Hamilton-Dyer, S. 2001 Bird and fish remains, 255-261 in M. Gardiner, R. Cross, N. Macpherson-Grant and I. Riddler, Continental Trade and Non-Urban Ports in Mid-Saxon England: Excavations at Sandtun, West Hythe, Kent, *Archaeological Journal* **158**, 161-290.

Myres, J. N. L. 1970. The Angles, Saxon and Jutes. *Proceedings of the British Academy* **56**, 145-174.

Nachtsheim, H. 1949. *Vom Wildtier zum Haustier*. Paul Parey: Hamburg and Berlin.

Napier, A. 1999. *The Old English Versions of the enlarged Rule of Chrodegang* Early English Text Series.

Nicholson, R. 1995. Fish bones, pp. 128-31 in P. Andrews (ed.) *Excavations at Redcastle Furze, Thetford 1988-9.* (East Anglian Archaeology Report 72) Field Archaeology Division Norfolk Museums Service: Dereham.

Noddle, B. 1969 The animal bones, pp. 124-126 in (P. Rahtz (ed.) Upton Gloucestershire, 1964-68 second report. *Transactions of the Bristol and Gloucestershire Archaeological Society* **88**, 74-126.

Noddle, B. A. 1970. The animal bones, pp. 37-40 in P. J. Fowler, K. S. Gardner, and P. A. Rahtz (eds.) *Cadbury Congresbury, Somerset, 1968*: an introductory report. Department of Extramural Studies, University of Bristol: Bristol.

Noddle, B. 1975. A comparison of the animal bones from 8 Medieval sites in southern Britain, pp. 248-260 in (Clason, A. T. ed.) *Archaeozoological Studies. Paper of the Archaeozoological Conference 1974, held at the Biologisch archaeologisch Institut of the State University of Groningen.* North Holland Publishing: Amsterdam.

Noddle, B. 1976. Report on the animal bones from Walton, Aylesbury, pp. 269-287 in M. Farley (ed.) Saxon and Medieval Walton, Aylesbury Excavations 1973-74. *Records of Buckinghamshire* **20** (2). 153-290.

Noddle, B. 1977. Mammal bone, pp 378-399 in H. Clarke and A. Carter (eds.) *Excavations in Kings Lynn 1963-70.*. Society for Medieval Archaeology, Monograph Series 7.

Noddle, B. 1978. Some minor skeletal differences in sheep, pp. 133-142 in D. R. Brothwell, K. D. Thomas and J. Clutton-Brock (eds.) *Research problems in Zooarchaeology*. (Institute of Archaeology Occasional Paper 3) Institute of Archaeology: London.

Noddle, B. 1980. Identification and interpretation of animal bones., pp. 375-408 in P. Wade-Martins (ed.) *North Elmham Park. Vol II.* (East Anglian Archaeology, Report 9) Norfolk Archaeological Unit: Gressenhall.

Noddle, B. 1985a. Animal bones, pp. 84-94 in R. Shoesmith (ed.) *Hereford City Excavations, Vol 3. The Finds.* (CBA Research Report 56) Council for British Archaeology: London.

Noddle, B. 1985b. Animal and bird bones, pp. 176-180 in L. Watts and P. Rahtz (ed.) *Mary-Le-Port, Bristol Excavations 1962/3.* (Bristol Museum and Art Gallery Monograph 7) Bristol Museum and Art Gallery: Bristol.

Noddle, B. A. 1985c. The animal bones, pp. 82-92 in G. Webster, P. Fowler, B. Noddle, and L. Smith (ed.) The Excavation of a Romano-British Rural Establishment at Barnsley Park, Gloucestershire 1961-1979: Part III. *Transactions of Bristol and Gloucester Archaeological Society* **103**, 73-100.

Noddle, B. A. 1987. *Animal Bones from Jarrow, 3rd Report*. Ancient Monument Laboratory Report 80/87. English Heritage.

Noddle, B. and Bramwell, 1975. The animal bones, pp. 332-339 in C. Platt and R. Coleman-Smith (eds.) *Excavations in Medieval Southampton 1953-1969 Vol. I: the excavation reports*. 332-339. Leicester University Press: Leicester.

Noddle, B. and Bramwell, D. 1983. Animal bones from Area A (East), pp. 34-37 in M. C. H. Carver (ed.) Two town houses in medieval Shrewsbury, *Transactions of the Shropshire Archaeological Society* **61**.

O'Connor, T. P.1979. The animal bones, pp. 147 in D. J. Freke (ed.) Excavations in Tanyard Lane, Steyning 1977. *Sussex Archaeological Collections* **117**, 135-51.

O'Connor, T. P. 1982a. The animal bones for Flaxengate, Lincoln c. 870-1500. *The Archaeology of Lincoln 18 (1)*. Council for British Archaeology: London.

O'Connor, T. P. 1982b. *The Archaeozoological Interpretation of Morphometric Variability British Sheep Limb Bones*. Unpublished PhD. University College London.

O'Connor, T. P. 1984. *Selected Groups of Bones from Skeldergate and Walmgate*. (The Archaeology of York: the Animal Bones 15/1) Council for British Archaeology for the York Archaeological Trust: London.

O'Connor, T. P. 1987. Why bother looking at archaeological wild mammal assemblages? *Circaea* **4** (2), 107-14.

O'Connor, T. P. 1988. *Bones from the General Accident Site, Tanner Row*. (The Archaeology of York: the Animal Bones 15/2). Council for British Archaeology for the York Archaeological Trust: London.

O'Connor, T. P. 1989a. *Bones from the Anglo-Scandinavian levels at 16-22 Coppergate*. (The Archaeology of York: The Animal Bones 15/3) Council for British Archaeology for the York Archaeological Trust: London.

O'Connor, T. P. 1989b. What shall we have for dinner? Food remains from urban sites, pp. 13-25 in D. Serjeantson and T. Waldron (eds.) *Diet and Crafts in Towns: The Evidence of Animal Remains the Roman to the Post-Medieval Periods*. (BAR, British Series 199) British Archaeological Reports: Oxford.

O'Connor, T. P. 1991. *Bones from 46-54 Fishergate*. (The Archaeology of York: the Animal Bones 15/4). Council for British Archaeology for the York Archaeological Trust: London.

O'Connor, T. P. 1992. Provisioning urban communities: a topic in search of a model. *Anthropozoologica* **16**, 101-106.

O'Connor, T. P. 1994. 8th-11th century economy and environment in York, pp. 136-147 in J. Rackham (ed.) *Environment and Economy in Anglo-Saxon England*. (CBA Research Report 89) Council for British Archaeology: London.

O'Connor, T. P. 2001 On the interpretation of animal bone assemblages from *Wics*, pp. 54-60 in D. Hill and R. Cowie (eds.) *Wics: The Early Medieval Trading Centres of Northern Europe*. Sheffield Academic Press: Sheffield.

O'Connor, T. P. 2007. Thinking about beastly bodies, pp. 1-10 in A. Pluskowski, (ed.) *Breaking and Shaping Beastly Bodies: Animals as Material Culture in the Middle Ages* Oxbow, Oxford.

O'Neil, H. E. 1945. The Roman Villa at Park Street near St Albans, Herts. Report on the Excavations of 1943-45. *Archaeological Journal* **102**, 21-110.

Orwin, C. S. 1949. *A History of English Farming*. Nelson: London.

O'Shea, R. 1992. The animal remains, pp. 98-100 in P. L. Drewett (ed.) Excavations at Lewes Castle, East Sussex. *Sussex Archaeological Collections* **130**, 69-106.

O'Shea, R. 1993. The animal remains, pp. 55-57 in M. Gardiner (ed.) The excavation of a Late Anglo-Saxon Settlement at Market Field, Steyning, 1988-9. *Sussex Archaeological Collections* **131**, 21-67.

O'Shea, R. 1996. The animal remains, pp. 115 in M. Gardiner, M. Russel and D. Gregory (eds.) Excavations at Lewes Friary 1985-6 and 1988-9. *Sussex Archaeological Collections* **134**, 71-124.

O'Sullivan, A. 2003. Place, memory and identity among estuarine fishing communities: interpreting the archaeology of early medieval fish weirs. *World Archaeology* **35** (3), 449-468.

Payne, S. 1973. Kill-off patterns in sheep and goats: the mandibles from Asvan Kale. *Anatolian Studies* **23**, 139-47.

Payne, S. and Bull, G. 1988. Components of variation in measurements of pig bones and teeth, and the use of measurements to distinguish wild boar form domestic pig remains. *Archaeozoologia* **2**, 27-66.

Peck, R. W. 1986. *Applying contemporary analogy to the understanding of animal behaviour on Roman villa sites*. Unpublished doctoral thesis, University of Southampton.

Pemberton, J. M. and Smith, R. H. 1985. Lack of biochemical polymorphism in British fallow deer. *Heredity* **55**, 199-207.

Pernetta, J. 1974. The animal bones, pp. 112-115 in M. Robinson (ed.) Excavations at Copt Hay, Tetsworth, Oxon. *Oxoniensia* **38**, 41-115.

Petit-Dutaillis, C. 1914. *Studies and Notes Supplementary to Stubb's Constitutional History II.* Translated by Waugh, W. T. Manchester University Press: Manchester.

Pierce, D. n.d. *What Became of the Wild Boar?* Unpublished undergraduate dissertation, University of Southampton.

Pinter-Belows, S. 1992. The vertebrate remains from sites 94 and 95 pp. 69-79 in G. Milne, G. and J. D. Richards (eds.) *Two Anglo-Saxon Buildings and Associated Finds*. York University Archaeology 9: York.

Pinter-Belows, S. 1997. Animal bone, pp. 65-73 in C. Jones, G. Eyre-Morgan, S. Palmer and N. Palmer (eds.) Excavations in the outer enclosure of Boteler's Castle, Oversley, Alcester, 1992–93. *Transactions of the Birmingham and Warwickshire Archaeological Society* **101**.

Pinter-Bellows, S. 2000. The Animal Remains, pp. 167-184 in P. A. Stamper and R. A. Croft (eds.) *The South Manor Area. Wharram: A Study of Settlement on the Yorkshire Wolds,* 8. University of York: York.

Platt, C. 1976. *The English Medieval Town*. Secker and Warburg: London.

Pollington, S. 2003. *The Mead-Hall: Feasting in Anglo-Saxon England.* Anglo-Saxon Books: Frithgard.

Poole, K. n.d. *Status, economy and diet at 12th-century Ludgershall Castle: a zooarchaeological study.* Unpublished undergraduate dissertation, University of Southampton.

Pooley, E. 1945. *The Guilds of the City of London.* Williams: London.

Poplin, F. 1984. Contribution osteo-archéologique à la connaissance des astragales de l'Antre corycien. *Bulletin de Correspondences Helléniques Supplément* **9**, 381-393.

Poplin, F. 1987. La Decoupe et le partage du cerf en venerie. *Anthropozoologica* **premier numéro spéciale**. 19-22.

Postan, M. M. 1966. *Cambridge Economic History of Europe*. Volume 1. Cambridge University Press: Cambridge.

Postles, D. 1984. The Osney Abbey Flock. *Oxoniensia* **49**, 141-152.

Potts, C. 1995. *Atque unum ex diversis gentibus populum effecit*: historical tradition and the Norman Identity, *Anglo-Norman Studies* **17** 139-175.

Powell, A. 1998. *Château de Mayenne: Assessment f the Animal Bones*. Unpublished assessment for Centre for Human Ecology and Environment, University of Southampton.

Powell, A. 2002. Animal bone, pp. 44-49 in S. Foreman, J. Hiller, and D. Petts (eds) *Gathering the people, settling the lands. The archaeology of a Middle Thames landscape. Vol 3: Anglo-Saxon to Post-Medieval* (Thames Valley Landscape Monograph 14) Oxford.

Powell, A. and Clark, K. 2002. Animal bones, pp. 101-108 in A. Mudd (ed.) Excavations at Melford Meadows, Brettenham, 1994: Romano-British and Early Anglo-Saxon Occupations (East Anglian Archaeology No 99) Oxford Archaeological Unit: Oxford.

Powell, A., Serjeantson, D. and Smith, P. forthcoming. food consumption and disposal: the animals remains. In (Hicks, M. and Hicks, A. eds.) *Excavations at St Gregory's Priory, Canterbury*. (The Archaeology of Canterbury, New Series Volume 2) Canterbury Archaeological Trust: Canterbury.

Powell, A. and Serjeantson, D. forthcoming. *Animal Bones from Pevensey Castle.*

Power, E. 1941. *The Wool Trade in English Medieval History*. Oxford University Press: Oxford.

Price, T. D. and Getsdóttir, H. 2006. The first settlers of Iceland: an isotopic approach to colonisation. *Antiquity* **80**, 130-144.

Probert, W. 1823. *The Ancient Laws of Cumbria.* London.

Prummel, W. 1975. Some reflections on the faunal remains of the Roman castellum Valkenburg, excavations 1962, pp. 225-229 in A. T. Clason (ed.) *Archaeozoological Studies: Papers of the Archaeozoological Conference 1974, held at the Biologisch archaeologisch Institut of the State University of Groningen*. Amsterdam: North Holland Publishing.

Prummel, W. 1997. Evidence of Hawking (Falconry) from bird and mammal bones. *International Journal of Osteoarchaeology* **7**, 333-338.

Rackham, D. J. 1989. Animal Remains pp. 146-158 in D. Austin (ed.) *The Deserted Medieval Village of Thrislington, County Durham: Excavations 1973-4.* (Society for Medieval Archaeology Monograph 12) Society for Medieval Archaeology: Lincoln.

Rackham, D. J. n.d. *The fauna from a 9th-century farmstead at Ribblehead, West Yorkshire.* Ancient Monuments Laboratory Report No. 2970, English Heritage.

Rackham, J. 1994a. Economy and Environment in Saxon London, pp. 126-136 in J. Rackham (ed.) *Environment and Economy in Anglo-Saxon England.* (CBA Research Report 89) Council for British Archaeology: London.

Rackham, J. 1994b. *Animal Bones, Interpreting the Past.* British Museum: London.

Rackham, J. 2003 Animal bones, pp. 258-273 in G. Taylor (ed.) An early to middle Saxon settlement at Quarrington, Lincolnshire, *Antiquaries Journal* **83**, 231-80.

Rackham, O. 1997. *The History of the Countryside: The Classic history of Britain's Landscape, Flora and Fauna.* Phoenix: London.

Rahtz, P. 1979 *The Saxon and Medieval Palaces at Cheddar Excavations 1960-2.* (BAR, British Series 65) British Archaeological Report: Oxford.

Recarte, J. M., Vincent, J. P. and Hewison, A. J. M. 1998. 'Flight responses of park fallow deer to the human observer' *Behavioural Processes* **44**, 65-72.

Reinken, G. 1997. *Wieder-Verbreinung. Verwendung und Namesgebungung des Damhirsches Cervus dama I. In Europa.* Z. Jagdwiss **43**, 197-206.

Reynolds, A. 1999. *Later Anglo-Saxon England: Life and Landscape.* Tempus. Stroud.

Richards, J. 1991. *The Viking Age.* English Heritage: London.

Richards, M. 2003. Explaining the dietary isotope evidence for rapid adoption of the Neolithic in Britain, pp. 31-36 in M. Parker-Pearson (ed.) *Food, Culture and Identity in the Neolithic and Early Bronze Age.* (BAR, international series 1117) BAR Publishing: Oxford.

Ritvo, H. 1987. *The Animal Estate: The English and other creatures in the Victorian age.* Harvard University Press.

Roberts, B. and Wrathmell, S. 2002. *Region and Place: A study of English Rural Settlement.* English Heritage: London.

Robertson, A. J. 1939 *Anglo-Saxon Charters.* Cambridge University Press: Cambridge.

Robinson, R. 1984. Rabbit, pp. 239-246 in I. L. Mason (ed.) *Evolution of Domesticated Animals.* Longman: London.

Rooney, A. 1993. *Hunting in Middle English Literature.* Boydell Press: Cambridge.

Round, J. H. 1895. *Feudal England: Historical Studies on the XIth and XIIth centuries.* Sonnenschein and Co: London.

Round, J. H. 1899. Some Early Sussex Charters. *Sussex Archaeological Collections* **42**, 73-86.

Round, J. H. 1903. *Victoria County History Essex I:* London.

Rowley, T. 1981. *The Origins of Open Field Agriculture.* Croom Helm: London.

Rowley, T. 1984. *The Norman Heritage 1066-1200.* Granada: London.

Rowley, T. 1997. *Norman England. An Archaeological Perspective on the Norman Conquest.* Batsford/ English Heritage: London.

Rowley, T. 1999. *The Normans.* Tempus: Stroud.

Rutter, J. D. and Duke, G. 1958. *Excavations at Crossgates near Scarborough, 1947-56.* Scarborough.

Ryder, M. L. 1964. The history of sheep breeds in England. *Agricultural History Review* **9**, 1-12.

Ryder, M. L. 1975. Some Phoenician animal remains from Sicily, pp. 213-218 in Clason, A. T. (ed.), *Archaeozoological studies: Papers of the Archaeozoological Conference 1974, held at the Biologisch archaeologisch Institut of the State University of Groningen.* Amsterdam: North Holland Publishing.

Ryder, M. L. 1983. *Sheep and Man.* Duckworth: London.

Ryder, M. L. 1984. Medieval sheep and wool types, *Agricultural History Review* **32**, 14-28

Sabine, E. L. 1933. Butchering in medieval London. *Speculum: A Journal of Medieval Studies* **8,** 335-353.

Sackett, J. R. 1986. Style, function and assemblage variability: a reply to Binford. *American Antiquity* **51** (3) 628-634.

Sadler, P. 1989. The animal bones, pp. 209-17 in A. Hawkins (ed.) Medieval and post-medieval occupation at the teachers centre, Walton Road. *Records of Buckinghamshire* **31**, 191-217.

Sadler, P. 1990. The Faunal remains, pp. 462-508 in J. R. Fairbrother (ed.) *Faccombe Netherton: Excavtions of a*

Saxon and Medieval Manorial Complex II. (British Museum Occasional Paper 74) British Museum: London.

Sadler, P. 1991. The use of tarsometatarsi in sexing and ageing domestic fowl (Gallus gallus L.), and recognising five toed breeds in archaeological material. *Circaea* **8**(1), 41-48.

Salisbury, C. R.1981. An Anglo-Saxon weir at Colwick, Nottinghamshire. *Transactions of the Thoroton Society* **87**, 54-59.

Salisbury, John of, 1993. *Policraticus I–IV* (ed. K. S. B. Keats-Rohan), Turnholt: Brepols.

Salisbury, J. E. 1994. *The Beast Within: Animals in the Middle Ages*. Routledge: London.

Sawyer, P. H. 1965. The wealth of Britain in the Eleventh century. *Transactions of the Royal Historical Society 5th series* **14**, 155-64.

Sawyer, P. H. 1968. *Anglo-Saxon Charters: An Annotated List and Biography*. Royal Historical Society: London.

Sayers, W. 2002. Some fishy etymologies: Eng. *Cod*, Norse *ˎorskr*, Du. *Kabeljauw*, Sp. *Bacalao*. *NOWELE: North-Western European Language Evolution* **41**:17-30.

Schmid, E. 1973. Ziegenhörner als Gerberei-Abfall, *Scheiz Volkskunde* **63**, 65-6.

Schmidt-Pauly, I. 1980. *Römerzeitliche und Mitelalterliche tierknochenfunde aus Breisach im Breisgau.* Inaugural Dissertation zur Erlangung der veterinärmedizinscchen Doktorwürde de Tierärztlichen Kakultät der Ludwig-Maximilians-Universität. München.

Schibler, J. and Furger, A. R. 1993. *Die tierknochenfunde aus Augusta Raurica (Grabungen 195501974).* Forschungen in Augst 9.

Schibler, J. 1988. *Die Tierknochenfunde aus Augusta Raurica (Grabungen 1955-1974).* Augst: Romermuseum.

Scott, E. M. 1996. Who ate what? Archaeological food remains and cultural diversity pp. 339-356 in E. J. Reitz, L.A. Newsom and S. J.Scudder (eds.) *Case Studies in Environmental Archaeology*. Plenum Press: London.

Scott, S. 1991. The animal bones, pp. 216-233 in P. Armstrong, D. Tomlinson and D. H. Evans (eds.) *Excavations at Lurk Lane, Beverley 1979-82*. (Sheffield Archaeological Reports 1) Collis: Sheffield.

Scott, S. 1992. The animal bones, pp. 236-251 in D. H. Evans and D. G. Tomlinson (eds.) *Excavations at 33-35 Eastgate, Beverley 1983-86.* (Sheffield Excavation Reports 3) Archaeology Unit, Property Services, Humberside County Council: Sheffield.

Scott, S. 1994. The Animal Bone, pp. 284-293 in K. Steedmand (ed.) Excavation of a Saxon Site at Riby Cross Roads, Lincolnshire. *Archaeological Journal* **151**, 212-306.

Scully, T. 1995. *The Art of Cookery in the Middle Ages*. Boydell: Suffolk.

Scully, T. 2002 Medieval France: A. the north, pp. 19-45 in M. Weiss Adamson (ed.) *Regional Cuisines of Medieval Europe: A Book of Essays*. Routledge: London.

Seddon, D., Calvocoressi, D., and Cooper, C., with Higgs, E. S. 1964. Fauna, pp. 69-71 in P. V. Addyman, (ed.) A Dark Age Settlement at Maxey, Northants. *Medieval Archaeology* **8,** 20-73.

Senecal, C. 2001 Keeping up with the Godwinesons: in pursuit of aristocratic status in Late Anglo-Saxon England, *Anglo-Norman Studies* **23**, 251-66.

Serjeantson, D. 1989 Animal remains and the tanning trade, pp. 129-147 in D. Serjeantson and T. Waldron (eds.) *Diet and Crafts in Towns: The Evidence of Animal Remains the Roman to the Post-Medieval Periods*. (BAR, British Series 199) British Archaeological Reports: Oxford.

Serjeantson, D. 2000a. Good to eat and good to think with: classifying animals from complex sites, pp. 179-89 in P. Rowley-Conwy (ed.) *Animal Bones, Human Societies*. Oxbow: Oxford.

Serjeantson, D. 2000b. Bird bones, pp. 182-185 in C. J. Young (ed.) *Excavations at Carisbrooke Castle, Isle of Wight, 1921-1996*. (Wessex Archaeological Report 18) Trust for Wessex: Salisbury.

Serjeantson, D. 2006. Birds as food and markers of status, in C. Woolgar, D. Serjeantson and T. Waldron (eds.) *Food in Medieval England: History and Archaeology*, Oxford University Press: Oxford.

Serjeantson, D. and Locker, A. 1997. *Fish consumption in medieval southern England: the archaeological evidence.* Paper to ICAZ fish conference.

Serjeantson, D. and Woolgar, C. 2006. Fish consumption in medieval England, pp. 102-130 in C. Woolgar, D. Serjeantson and T. Waldron (eds.) *Food in Medieval England: History and Archaeology*. Oxford University Press: Oxford.

Shaw, M. 1993. The excavation of a late 15th- to 17th-cetury tanning complex at The Green, Northampton. *Post-Medieval Archaeology* **30**, 63-127.

Sheail, J. 1971. *Rabbits and their History*. Davis and Charles: Newton Abbot.

Sheail, J. 1984. The Rabbit. *Biologist* **31** (3), 135-140.

Shopkow, L. 1997. *History and Community: Norman Historical Writing in the 11th and 12th centuries*. Catholic University of America Press.

Silver, I. A. 1969. The ageing of domestic animals, pp. 283-302 in D. Brothwell and E. S. Higgs (eds.) *Science in Archaeology* (revised edition), Thames and Hudson: London.

Simoons, F. J. 1994. *Eat Not This Flesh* (2nd edition). University of Wisconsin Press: Wisconsin.

Smith, A.H. 1964. *The Placenames of Gloucestershire Part I.* (English Placenames Society Vol 38) English Placenames Society.

Smith, R. M. 1988. Human resources, pp. 188-212 in G. Astill and A. Grant (eds.) *The Countryside in Medieval England.* Blackwell: Oxford.

Smith, P. 1994. *The Early Norman Animal Bone from Carisbrooke Castle, the Isle of Wight.* Ancient Monuments Laboratory Report 49/94, English Heritage.

Smith, P. 1995. *The Fish Bones from Launceston Castle, Cornwall.* Ancient Monuments Laboratory Report 56/95, English Heritage.

Spiegeleire, M. A. de 1982. Tourinnes-Saint-Lambert. Les restes d'animaux du site gallo-romain. *Wavriensia* **31**, 100-108.

Stafford, P. 1980 The "Farm of One Night" and the organisation of King Edwards's Estates in Domesday. *Economic History Review 2nd series* **33**, 491-502.

Starr, R.J. 1992. Silvia's deer (Vergil, Aeneid 7.479-502): game parks and Roman law. *The American Journal of Philology* **113** (3), 435-9.

Steane, J. M 1988. The royal fishponds of medieval England, pp. 39-68 in M. Aston (ed.) *Medieval Fish, Fisheries and Fishponds in England* I. (BAR, British Series 182) British Archaeological Reports: Oxford.

Stenton, F. 1908. *William the Conqueror and the Rule of the Normans*. London.

Stenton, F, 1943. *Anglo-Saxon England.* Oxford University Press: Oxford.

Stevens, P. 1990. The animal bones, pp. 258-259 in M. Gardiner (ed.) An Anglo-Saxon and Medieval settlement at Boltophs, Bramber, West Sussex. *Archaeological Journal* **147**, 216-75.

Stevens, P. 1992. The vertebrate remains from site 39, pp. 67-69 in G. Milne, G. and J. D. Richards (eds.) *Two Anglo-Saxon Buildings and Associated Finds*. York University Archaeology 9: York.

Stevens, P. 1997. The animal bones, pp. in150-162 in M. Lyne (ed.) *Lewes Priory: excavations by Richard Lewis 1969-82*. Lewes Priory Trust: Lewes.

Stevens, P. 2000. Animal bones, pp. 42 in C. Butler (ed.) *Saxon Settlement and Earlier Remains at Friars Oak, Hassocks, West Sussex*. (BAR British Series, 295) British Archaeological Reports: Oxford.

Sutermeister, H. 1976. Burpham: A settlement site within the Saxon defences. *Sussex Archaeological Collections* **114**, 194 207.

Swanton, M. 1975. *Anglo-Saxon Prose*. Dent: London.

Sykes, N. J. 1997. *Animal bones from the Saxo-Norman site in North Street, Lewes: a re-analysis and re-interpretation*. Unpublished MSc dissertation, Institute of Archaeology, University College London.

Sykes, N. J. 2001. The Norman Conquest: A Zooarchaeological Perspective. Unpublished Ph.D. thesis, University of Southampton.

Sykes, N. J. 2004a Neolithic and Saxon animal bone, pp. 87-91 in C. J. Ellis (ed.) *A Prehistoric Ritual Complex at Eynesbury, Cambridgeshire* (East Anglian Archaeology Report 17). Salisbury: Trust for Wessex.

Sykes, N.J. 2004b. The introduction of fallow deer (*Dama dama*): a zooarchaeological perspective. *Environmental Archaeology* **9**, 75-83.

Sykes, N. J. 2005a. The zooarchaeology of the Norman Conquest, *Anglo-Norman Studies* **27**, 185–197.

Sykes, N. J. 2005b. Hunting for the Normans: zooarchaeological evidence for medieval identity, pp. 73–80 in A. Pluskowski (ed.) *Just Skin and Bones? New Perspectives on Human-Animal Relations in the Historical Past.* (BAR, British Series s1410) BAR Publishing: Oxford.

Sykes, N. J. 2006. Tracking animals using strontium isotopes in teeth: the role of fallow deer (*Dama dama*) in Roman Britain, *Antiquity* **80**, 948-959.

Sykes, N. J. 2007. Taking sides: the social life of venison in medieval England, pp. 150-61 in A. Pluskowski, (ed.) *Breaking and Shaping Beastly Bodies: Animals as Material Culture in the Middle Ages* Oxbow, Oxford.

Sykes, N. J., 2007. Animal bones and animal parks. *In:* R. Liddiard (ed.) *The Medieval Deer Park: new perspectives*. Windgather Press: Macclesfield.

Sykes, N. J. in prep. Sheep horn cores: new methods for determining sex and level of improvement.

Sykes, N. J. n.d. The animal bones from Clay Hill, Sussex.

Sykes, N. J., Ingrem, C. and Ayres K. 2005. The bones and shell, pp. 116-132 in R. Poulton (ed.) *A Medieval Royal Complex at Guildford: Excavations at the Castle and Palace*. Surrey Archaeological Society: Guildford.

Taylor, M. 1996. The impact of the Norman ecclesiastical revival in Britain after 1066 - a summary, pp. 125-129 in in X. Delestre, X. and A. Woodcock (eds.) *Proximus*. Association for the Promotion of Archaeology in Haute-Normandie: Rouen.

Teichert, M. 1984. Size variation in cattle from Germani Romana and Germania libera, pp. 93-103 in C. Grigson and J. Clutton-Brock (eds.) *Animals in Archaeology 4: Husbandry in Europe*. (BAR, International Series 227) British Archaeological Reports: Oxford.

Thawley, C. 1981. The Mammal, Bird and Fish Bones, pp. 173-175 in J. E. Mellor and T. Pearce (eds.) *The Austin Friars, Leicester*. (CBA Research Report 35) Leicester County Council: Leicester.

Thibébaux, M. 1967. The Medieval Chase. *Speculum: A Journal of Medieval Studies* **47**, 260-274.

Thomas, H. M. 2003. *The English and the Normans: Ethnic Hostility, Assimilation and Identity 1066–c. 1220*. Oxford University Press: Oxford.

Thomas, R. 1988. A statistical evaluation of criteria used in sexing cattle metapodials. *Archaeozoologica* **2** (1.2), 83-92.

Thomas, R. 1989. *Cattle and the Romano-British Economy: A metrical analysis of size variation*. Unpublished PhD thesis, University of Southampton.

Thomas, R. 2005. *Animals, economy and status : integrating zooarchaeological and historical data in the study of Dudley Castle, West Midlands (c.1100-1750)* (BAR, British Series 392) BAR Publishing: Oxford.

Thompson, M. 1991. *The Rise of the Castle*. Cambridge University Press.

Thompson, M. 1995. *The Medieval Hall: the basis of secular domestic life 600-1600AD*. Scholar Press: Aldershot.

Thorn, F. and Thorn, C. 1986. *Domesday Book, 25, Shropshire*, Phillimore, Chichester.

Thorpe, L. 1973. *The Bayeux Tapestry and the Norman Invasion*. The Folio Society: London.

Tittensor, A. M. and Tittensor, R. M. 1985. The Rabbit Warren at West Dean near Chichester. *Sussex Archaeological Collections* **123**, 151-87.

Toynbee, J. M. C. 1973. *Animals in Roman Life and Art*. Thames and Hudson: London.

Tranier, M. 1973. Antre Corycien, Note sur la faune des vertèbrès associée au gisement préhistorique. *Bulletin de correspondences Helléniques* **97**, 528-538.

Trevelyan, G. M. 1957. *Illustrated English social history. Vol 3. The eighteenth century*. Longman: London.

Trow-Smith, R. 1957. *A History of British Livestock Husbandry to 1700*. Routledge and Kegan Paul: London.

Uerpman. H.-P. 1970. *Studien überfrühe Tierknockenfunde von der Iberischen Halbinsel*. Inaugural dissertation zur Erlangung der tiermedizinschen Doktorwürde der Tierärztlichen Fakultät der Ludwig-Maximilaus-Universität München. Institut für Palaeoanatomie, Domestikationsforschung un Geschicte der Tiermedizin der universität München.

Uerpman, H.-P. 1987. *The Ancient Distribution of Ungulate mammals in the Middle East* Beihefte zum Tübunger Atlas des Vorderen Orients, Reihe A Naturwissenschaften 27. Dr. Ludwig Reichert Verlag: Wiesbaden.

Vadet, A. and Villette, P. 1986. *Les ossements animaux du puits S.14 à Douai. Mémoires de la Société Académique du Boulonnais*, Tome I Fascicule 3. Société d'Histoire Naturelle du Boulonnais.

Vanderford, S. 1997. Biometrical analysis of medieval and post-medieval rabbits Unpublished MSc dissertation. Institute of Archaeology, London.

Van der Veen, M. 2003. When is food a luxury? *World Archaeology* **34**(3), 405-427.

Van Damme, D. and Ervynck, A. 1988. Medieval ferrets and rabbits in the castle of Laarne (East Flanders Belgium): A contribution to the history of a predator and its prey. *Helinium* **28** (2), 278-284.

Varro, M.T. 1934 (trans. by William Davis Hooper). *On Agriculture* (Loeb Classical Library 233) London: Heinemann.

Veale, E. M. 1957. The Rabbit in England. *English History Review* **5**, 85-90.

Vince, A. 1989. *Finds and Environmental Evidence* (Aspects of Saxo-Norman London 2, London and Middlesex Archaeological Society Special Paper 12) London and Middlesex Archaeological Society: London.

Vince, A. 1994. Saxon Urban economies: an archaeological perspective, pp. 109-119 in J. Rackham (ed.) *Environment and Economy in Anglo-Saxon England: A Review of Recent Work on the Environmental Archaeology of Rural and Urban Anglo-Saxon Settlements in England.* (CBA Research Report 89) Council for British Archaeology: London.

von den Dreisch, A. and Boessneck, J. 1974. Kritische Anmerkungen zur Widerristhohen-Berechnung aus langmassen vor- und fruhgeschichtlicher Tierknochen. *Saugetierkundliche Mitteilungen,* **22** (4), 325-48.

von den Dreisch, A. 1976. *A Guide to the Measurement of Animal Bones from Archaeological Sites.* (Peabody Museum Bulletin 1) Harvard University Press.

von Strassburg, G. 1960. *Tristan* (translated by Hatto, A. T.) London: Penguin.

Wade, K. 1988. Ipswich, pp. 93-100 in R. Hodges and B. Hobley (eds.) *The Rebirth of Towns in the West AD 700-1050.* (CBA Research report 68) Council for British Archaeology: London.

Wapnish, P. and Hesse, B. 1988. Urbanisation and the organisation of animal production at Tell Jemmeh in the Middle Bronze Age Levant. *Journal of Near Eastern Studies* **47** (2), 81-89.

Warman, S. 2000. Morphometric investigation of dental variation to examine genetic relationships between pig populations. Unpublished PhD. Thesis, Institute of Archaeology, UCL.

Warry, J. 1988. The ancient history of rabbits. *The Local Historian* **18**, 13-15.

Webb, C. C. J. 1932. *John of Salisbury.* London.

Webber, N. 2005. *The Evolution of Norman Identity 911-1154.* Boydell: Woodbridge.

Webster, L. 1986. Anglo-Saxon England AD 400-1100, pp. 119-160 in I. Longworth and J. Cherry (eds.) *Archaeology in Britain since 1945.* British Museum Press: London.

Weinstock, J. 2002. The medieval and post-medieval bone remains from Heigham Street, Norwich. *English Heritage Centre for Archaeology publication 33/2002.*

Welch, M. 1992. *Anglo-Saxon England.* English Heritage/Batsford: London.

West, B. and Rackham, J. 1993a. Birds and mammals, pp. 150-154 in R. Cowie and R. Whytehead (ed.) Two Middle Saxon occupation sites: excavations at Jubille Hall and Maiden Lane. *Transactions of the London and Middlesex Archaeological Society* **39** (1988), 47-164.

West, B. and Rackham, J.1993b. Animal remains, pp. 148-168 in R. Cowie (ed.) Excavations at the Peabody Site, Chandos Place and the National Gallery. *Transactions of the London and Middlesex Archaeological Society* **40** (1989), 35-176.

Westly, B. 1974. The animal bones in W. H. Manning (ed.) Excavations on Late Iron Age, Roman and Saxon sites from Ufton Nervet, Berkshire. *Berkshire Archaeological Journal* **67**, 1-61.

Westly, B. 1976. Animal bones, pp. 319-320 in E. W. Holden (ed.) Excavations at Old Erringham, Shoreham, West Sussex Part I: a Saxon weaving hut. *Sussex Archaeological Collections* **114**, 306-22.

Westly, B. 1977. The animal bones, pp. 67-8 in K. J. Barton and E. W. Holden (eds.) Excavations at Bramber Castle, Sussex, 1966-7. *Archaeological Journal* **134**, 11-79.

Westly, B. 1980. Animal bones, pp. 294 in E. W. Holden (ed.) Excavations at Old Erringham, Shoreham, West Sussex Part II: The 'Chapel' and Ringwork. *Sussex Archaeological Collections* **118**, 257-97.

Wheeler, A. 1979. Fish Remains, pp. 21-23 in M. Parrington (ed.) Excavations at Stert Street, Abingdon, Oxon. *Oxoniensia* **44**, 1-25.

Wheeler, R. E. M. and Wheeler, T. V. 1932. *Report on the Excavation of the Prehistoric, Roman and Post-Roman site in Lydney Oak, Gloucester.* (Report of the Research Committee of the Society of Antiquaries, London 9) Society of Antiquaries: London.

White D. A. 1964. Excavations at War Ditches, Cherry Hinton. *Proceedings of the Cambridge Antiquity Society* **56**, 9-29.

White, J. n.d. The Animal bones from Wells Museum unpublished undergraduate dissertation, University of Southampton.

White, K. D. 1970. *Roman Farming.* Thames and Hudson: London.

Whitehead, K. G. 1972. *Deer of the World.* Constable: London.

Whitehouse, R. and Whitehouse, D. 1974. pp. 215 in C. Stanford, (ed.) *Croft Ambrey: Excavations Carried Out for the Woolhope Naturalists' Field club.* Adams and Sons: Hereford.

Whitlock, D. 1955. *English Historical Documents c.500-1042.* Eyre and Spottiswoode: London.

Wickham, C. 1994. *Land and Power; Studies in Italian and European Social History, 400-1200.* British School at Rome: London.

Williams, A. 1997. *The English and the Norman Conquest*. Boydell: Woodbridge.

Williams, J. H. 1979. *St Peters Street Northampton: Excavations 1973-1976*. (Archaeological Monograph 5, Northampton Development Corporation 2). Northampton Development Corporation: Northampton.

Williams, R. 1993 *Pennyland and Hartigans: Two Iron Age and Saxon Sites in Milton Keynes* (Buckinghamshire Archaeological Society Monograph Series no 4) Buckinghamshire Archaeological Society: Ayelsbury.

Wilson, B. 1962. Anglo-Saxon rural economy: a survey of the archaeological evidence and a suggestion, Agricultural History Review **10** (2), 65-79.

Wilson, B. 1975. Bone report, pp. 98-101in D. Miles (ed.) Excavations at West St, Helen Street, Abingdon, 1972. *Oxoniensia* **40**, 79-104.

Wilson, B. 1976. The animal bones, pp. 144-147 in K. A. Rodwell (ed.) Excavations on the Site of Banbury Castle. *Oxoniensia* **41**, 90-147.

Wilson, B. 1979. The mammal bones and other environmental records, pp. 16-21 in M. Parrington (ed.) Excavations at Stert Street, Abingdon, Oxon. *Oxoniensia* **44**, 1-25.

Wilson, B. and Allison, E. 1985. Faunal remains, pp. 120-121, in J. Sharpe Osney Abbey, Oxford: Archaeological Investigations, 1975-8 *Oxoniensia* **50**, 94-130.

Wilson, B., Bramwell, D., and Wheeler, A. 1975. The animal bones from the Broad Street and Old Gaol sites. *Oxoniensia* **40.** 105-121.

Wilson, B. and Bramwell, D. 1980. Animal bone and shell, pp. 198-9 in N. Palmer (ed.) A Beaker burial and medieval tenement in The Hamel, Oxford. *Oxoniensia* **45**, 124-225.

Wilson, B., Locker, A. and Marples, B. 1989. Medieval animal bone and marine shells from Church Street and other sites in St Ebbes, Oxford. *Oxoniensia* **54**, 258-268.

Wilson, B. and Locker, A. 1995. Animal bone, pp. 241-3 in M. Roberts (ed.) Excavations at Jowett Walk, Oxford. *Oxoniensia* **60**, 225-246.

Wilson, B. 1996. *Spatial Patterning among Animal Bones in Settlement Archaeology: An English Regional Exploration*. (BAR, British Series 251) British Archaeological Reports: Oxford.

Wilson, C. A. 2003 *Food and Drink in Britain from the Stone Age to the 19th Century*, Chicago: Academy Chicago.

Wilson, D. M. 1976. *The Archaeology of Anglo-Saxon* England. Methuen: London.

Wilson, J. 1909. *Evolution of British Cattle and the fashion of breeds* Vinton & Co., Ltd: London.

Wilson, T. 1995. Animal bones, pp. 120-128 in P. Andrews (ed.) *Excavations at Redcastle Furze, Thetford 1988-9*. (East Anglian Archaeology Report 72) Field Archaeology Division Norfolk Museums Service: Dereham.

Wiseman, J. 1986. *A history of the British Pig*. Duckworth: London.

Woolgar, C. M. 1999. *The Great Household in Late Medieval England*. Yale University Press.

Woolgar, C. M. 2000. "Take this penance now and afterwards the fare will Improve": Seafood and late medieval diet, pp. 7-25 in D. J. Starkey, C. Reid and N. Ashcroft (eds.) *England's Sea Fisheries: the commercial sea fisheries of England and Wales since 1300*. Chatam: London.

Woolgar, C. M. 2006. Group diets late medieval England, pp. 191-200 in C. Woolgar, D. Serjeantson and T. Waldron (eds.) *Food in Medieval England: History and Archaeology*, Oxford: Oxford University Press.

Yalden, D. 1999. *A History of British Mammals*. T. and A. Poyser: London.

Yannouli, E and Trantalidou, K. 1999. The fallow deer (*Dama dama* Linnaeus, 1758): Archaeological presence and representation in Greece, pp. 247-282 in Benecke, N. (ed.), *The Holocene history of the European vertebrate fauna: modern aspects of research workshop 6th-9th April 1998*. Verlag Marie Leidorf GmbH: Rahden/Westf.

Yapp, B. 1981 *Birds in Medieval Manuscripts*, London: the British Library.

Yapp, W. B. 1983. Game-birds in medieval England. *Ibis* **125**, 218-221.

Yealland, S. and Higgs, E. S. 1966. The economy, pp. 139-40 in R. A. Hilton and P. A. Rahtz (ed.) Upton Gloucesteshire, 1959-1964. *Transactions of the Bristol and Gloucester Archaeological Society* **85**, 70-140.

Yvinec, J.-H. 1988. Alimentation Carnée au début du Moyen Âge. *Anthropozoologica* **Second Numero Spécial**. 122-128.

Yvinec, J.-H. 1993. La part du gibier dans l'alimentation du haut Moyen Âge, pp. 491-504 in *Exploitation des Animaux Sauvage a Travers le Temps*. Association pour la Promotion et la Diffusion des Conaissances Archéologiques: Juan-les-Pins (Société de recherche interdisciplinaire, XIIIe Rencontres Internationale d'Archéologie et d'Histoire d'Antibes IVe Colloque international de l'homme et l'animal).

Yvinec, J.-H. 1997. Fouilles de sauvetage sous la place du marché à Compiègne (Oise) 1991/1993:L'évolution urbaine de l'aire palatiale du haut moyen âge aux marché Médiéval et moderne. *Revue archéologique de Picardie.* No Spécial **13**.

Yvinec, J-H. 1999. Étude Archéologique du site de Dury 'le Moulin' (Somme), *Revue Archéologique de Picardie* **1**, 247-56.

Zeuner, F. E. 1963. A *History of Domesticated Animals*. Hutchinson: London.

Appendix Ia: Taxa representation (NISP) by site type and date. See Table 2 for references.

Site	Domestic Mammals			Wild Mammals							Domestic Birds				Wild	Total
	cattle	sheep	pig	red	roe	fallow	deer'	boar	rabbi	hare	fowl	goose	duck	bird	Bird	NISP
5th - mid-9th Rural																
Wolverton Turn, Buckinghamshire	211	335	117	1			1	1			33	6			2	707
Pennylands, Buckinghamshire	1170	894	330				14			2	68	28	1		2	2509
Hartigans, Buckinghamshire	167	28														195
Chichley, Buckinghamshire	10	106	13								1	1				131
Eynesbury, Cambridgeshire	188	218	125	1	2						9	9			2	554
Lot's Hole, Essex	196	111	78					1			10	4	1			401
Lechdale, Gloucestershire	266	154	22								16	11	1		9	479
Northbrook, Hampshire	43	67	4								7					121
Old Down Farm, Hampshire	125	140	25								14					304
Cadley Road, Wiltshire	377	573	66	18	29						74	21			26	1184
Bonners Lane, Leicestershire	35	53	66								8	2	2		1	167
Quarrington, Lincolnshire	640	253	111	4							19	8		10		1045
Quarrington, Lincolnshire	589	345	62							1	14	6	2	11		1030
Spong Hill, Norfolk	634	75	10	5												724
Maxey, Northamptonshire	194	191	59											45		489
Barton Court Farm, Oxfordshire	377	573	354	10	1					1					15	1331
Audlett Drive, Oxfordshire	91	67	15								2	1	1			177
Wantage, Oxfordshire	94	70	10	1							2	1			1	179
Eynsham, Oxfordshire	230	291	121	5	7					2	16	18				690
New Wintles, Oxfordshire	435	281	104				5				4	3			5	837
Yarnton, Oxfordshire	366	203	90								8	14				681
Yarton (Cresswell Field), Oxfordshire	199	107	48							2	4	11				371
Yarton (Worton), Oxfordshire	101	42	25								4	29				201
Melford Meadows, Suffolk	286	130	46	3							2	6				473
West Stow, Suffolk	7873	11148	3903	58	33					10	521	285			89	23920
Bishopstone, Sussex	35	55	24	1	1						6	3				125
Friars Oak, West Sussex	65	23	11	1	1											101
Riby Cross, Yorkshire	902	546	70	1							14	30				1563
Total NISP	**15899**	**17079**	**5909**	**109**	**74**		**20**	**2**		**18**	**856**	**497**	**8**	**66**	**152**	**40689**
Total NISP by Group		**38887**					**223**					**1427**			**152**	**40689**
Average Percentage NISP by Group		**95.6**					**0.5**					**3.5**			**0.4**	**100**
5th - mid-9th Urban																
Abbots Worthy, Hampshire	502	443	108	19							70	17			9	1168
Hamwic, Hampshire	23896	14606	6953	12	8						800	353			47	46675
Sandtun, Kent	91	109	51								31	13			23	318
Maiden Lane, London	2898	861	1547	28							80	61	4		5	5484
Jubilee Hall, London	843	336	365	6	1						24	14				1589
Peabody Buildings, London	63	76	71		1					1	33	15	1		47	308
Marefair, Northamptonshire	193	176	130								12	21	5		1	538
National Gallery, London	475	661	470	1	2						45	30	2			1686
Redcastle Furze, Norfolk	407	301	105								19	5			2	839
St Aldates, Oxfordshire	69	27	8								6	1				111
Ipswich, Suffolk	4282	2206	3130	27	12					1	513	129	9		12	10321
Ipswich, Suffolk	3408	1809	1973	4	7						397	176	11		24	7809
Fishergate, Yorkshire	8296	3421	1295	2	5			3			603	222			18	13865
Total NISP	**20934**	**9874**	**9094**	**68**	**28**			**3**		**2**	**1732**	**674**	**32**		**109**	**42550**
Total NISP by Group		**86661**					**140**					**3722**			**188**	**90711**
Average Percentage NISP by Group		**95.5**					**0.2**					**4.1**			**0.2**	**100**
5th - mid-9th Elite																
Wicken Bonhunt, Essex	5138	3858	20954	16	143					4	3082	2038	147		107	35487
Lake End Road, Essex	2108	704	1288	9	18			3		2	198	64	67		34	4495
Portchester Castle, Hampshire	5627	2892	1857	318	206					29	31	33	2		6	11001
Saxon Palaces, Northamptonshire	194	476	151		1					4	30	29	11		4	900
Caister-on-Sea, Norfolk	305	108	77	13	1					1	52	16	18		8	599
Eynsham 2b, Oxfordshire	350	1011	407	7	8					2	271	118	37		24	2235
Eynsham 2c, Oxfordshire	33	43	32		4						9	4	1			126
Wroxeter, Shropshire	10106	2212	3793	302	44			59		51	391	6	23		212	17199

Appendix Ia: Taxa representation (NISP) by site type and date. See Table 2 for references.

Site	Domestic Mammals			Wild Mammals							Domestic Birds				Wild	Total
	cattle	sheep	pig	red	roe	fallow	deer'	boar	rabbi	hare	fowl	goose	duck	bird	Bird	NISP
Boltophs, Bramber, Sussex	113	55	98	10							2	1				279
Brandon, Suffolk	13441	24652	9121	50	108					28	1306	964	755		143	50568
Ramsbury, Wiltshire	339	247	249	20	42									52		949
Total NISP	**37754**	**36258**	**38027**	**745**	**575**			**62**		**121**	**5372**	**3273**	**1061**	**52**	**538**	**123838**
Total NISP by Group		**112039**					**1503**					**9758**			**538**	**123838**
Average Percentage NISP by Group		**90.5**					**1.2**					**7.8**			**0.5**	**100**
5th - mid-9th Religious																
Hartlepool, Cleveland	36	146	16								9	15	1		2	225
St Albans, Hertfordshire	136	84	510	2	41						23	11	3		3	813
North Elmham, Norfolk	2424	2808	2182	3	43					3	37	18	5		17	7540
Total NISP	**2596**	**3038**	**2708**	**5**	**84**					**3**	**69**	**44**	**9**		**22**	**8578**
Total NISP by Group		**8342**					**92**					**122**			**22**	**8578**
Average Percentage NISP by Group		**97.3**					**1**					**1.4**			**0.3**	**100**
mid-9th - mid-11th Rural																
Walton Vicarage, Buckinghamshire	603	497	331	12							20	44			3	1510
Goltho, Lincolnshire	103	32	10	1	2					1	2	3	1			155
West Cotton, Northampton	46	35	28								5	4	1		4	123
Steyning, Sussex	624	1768	260							1	62	19	23		11	2768
Wharram Percy, Yorkshire	299	121	69								96	8				593
Total NISP	**1675**	**2453**	**698**	**13**	**2**					**2**	**185**	**78**	**25**		**18**	**5149**
Total NISP by Group		**4826**					**17**					**288**			**18**	**5149**
Average Percentage NISP by Group		**93.7**					**0.3**					**5.6**			**0.4**	**100**
mid-9th - mid-11th Urban																
Newbury, Berkshire	485	623	202		1					4	40	14	1		1	1371
Winchcombe, Gloucestershire	343	512	76		2						36	5			1	975
West Gate, Gloucestershire	62	68	49	1							3	3	1			187
Winchester, Hampshire	1411	1176	670	5	4					7	161	6	14		2	3456
Victoria Road, Winchester, Hampshire	150	297	58								11		1			517
Hereford City, Herefordshire	1648	433	492	84	9	3					58	4	1			2732
Lincoln city, Lincolnshire	1037	449	203	5							63	13			5	1775
Fishergate, Norfolk	893	222	216	3						1	13	10	1			1359
Brandon Road, Norfolk	1425	1050	483		1	1			1		100	35	15			3111
St Nicholas' Street, Norfolk	112	58	21							1		1				193
Thetford 1092, Norfolk	919	650	394	4							29	13	3		7	2019
Mill Lane, Thetford, Norfolk	548	336	318		3					1	69	11	11		2	1299
Marefair, Northamptonshire	140	122	55								7				1	325
St Peters Street, Northamptonshire	162	228	88							1	2	4			1	486
Codrington Library, Oxford	27	33	25								11	3				99
Shrewsbury, Shropshire	35	15	26	5	2						6				2	91
Ipswich, Suffolk	398	114	119								24	5	3		1	664
Coppergate, Yorkshire	17832	7024	4588	91	5			6		9	1217	348	1		126	31247
Skeldergate, Yorkshire	354	129	53	1												537
Lurk Lane, Yorkshire	4083	1533	1263	1	63			5		6	88	109	3		16	7170
Total NISP	**32064**	**15072**	**9399**	**200**	**90**	**4**		**11**	**1**	**30**	**1938**	**584**	**55**		**165**	**59613**
Total NISP by Group		**56535**					**336**					**2577**			**165**	**59613**
Average Percentage NISP by Group		**95**					**0.6**					**4**			**0.3**	**99.9**
mid-9th - mid-11th Elite																
Portchester Castle, Hampshire	1029	626	385	74	27					26	511	147	62		225	3112
Faccombe Netherton, Hampshire	284	449	405	28	56					18	61	15			49	1365
Goltho, Lincolnshire	234	152	177	8	47	25			2	1	36	15	5			702
Castle Mall, Norfolk	541	236	277		2	1			5		191	22	9		3	1287
Saxon Palaces, Northamptonshire	733	785	349	2	1					3	113	68	19		7	2080
Bishopstone, Sussex	511	1057	748		16					20	388	76	1		1	2818
Cheddar Palaces, Somerset	1806	1064	1288	52	64			2						11		4287
Trowbridge Castle, Wiltshire	718	415	342	10	19						20	7			5	1536
Ramsbury, Wiltshire	447	307	353	16	31									99	1	1254
Total NISP	**6303**	**5091**	**4324**	**190**	**263**	**26**		**2**	**7**	**68**	**1320**	**350**	**96**	**110**	**291**	**18441**

Appendix Ia: Taxa representation (NISP) by site type and date. See Table 2 for references.

Site	Domestic Mammals			Wild Mammals							Domestic Birds				Wild	Total
	cattle	sheep	pig	red	roe	fallow	deer'	boar	rabbi	hare	fowl	goose	duck	bird	Bird	NISP
Total NISP by Group		**15718**					**556**					**1876**			**291**	**18441**
Average Percentage NISP by Group		**85.2**					**3**					**10.1**			**1.6**	**99.9**
mid-9th - mid-11th Religious																
St Albans Abbey, Hertfordshire	104	112	506		35					2	22	8	4		2	795
Westminster Abbey, London	769	535	615	4	18				2	7	205	12	17		36	2220
Jarrow, Northumberland	319	106	63	1	2						94	56	1		2	644
Eynsham 2d, Oxfordshire	294	444	226	1	22					8	45	16	9		3	1068
Eynsham 2e, Oxfordshire	118	166	133	4	15					1	9	7	1		5	459
Eynsham 2f, Oxfordshire	260	322	261	4	47					8	29	11	9		17	968
Lewes Priory, Sussex		5	2								9				1	17
Total NISP	**1864**	**1690**	**1806**	**14**	**139**				**2**	**26**	**413**	**110**	**41**		**66**	**6171**
Total NISP by Group		**5360**					**181**					**564**			**66**	**6171**
Average Percentage NISP by Group		**86.9**					**3**					**9.1**			**1**	**100**
mid-11th - mid-12th Rural																
Walton Vicarage, Buckinghamshire	726	871	396	20							56	19	2		3	2093
Brighton Hill South, Hampshire	198	140	98								6				1	443
Pepperhill Lane, Kent	91	24	27								1					143
Total NISP	**1015**	**1035**	**521**	**20**							**63**	**19**	**2**		**4**	**2679**
Total NISP by Group		**2571**					**20**					**84**			**4**	**2679**
Average Percentage NISP by Group		**96**					**0.7**					**3.1**			**0.2**	**100**
mid-11th - mid-12th Urban																
Newbury, Berkshire	82	39	25			1			12		37	8	2		1	207
Bartholomew Street, Newbury, Berkshire	59	69	42								24	11			1	206
Exeter, Devon	3191	893	445	3	2	2			1	5	204	42	1		26	4815
Lincoln, Lincolnshire	1110	899	114	2	1						44	19			4	2193
Lincoln city, Lincolnshire	88	110	26	1							8	13			5	251
Kings Lynn, Norfolk	603	715	350			1			1		77	42	5		13	1807
Brandon Road, Norfolk	1757	1577	687	1					1	1	166	69	8		6	4273
St Martins-at-palace plain, Norfolk	953	702	660	3	5	7	4			5						2339
St Nicholas Street, Norfolk	838	937	382	1	1				8	7	111	49	7		6	2347
Fishergate Norfolk	952	309	289	1	1					1	113	33	4		2	1705
Redcastle Furze, Norfolk	313	386	149		1				2							851
St Aldates, Oxfordshire	482	313	64								66	15				940
Codrington Library, Oxfordshire	186	278	102								75	29	2			672
Benhams Garage, Somerset	374	242	27								1	1				645
Ipswich, Suffolk	295	121	132								32	13	3			596
North Street, Lewes, Sussex	614	1699	540	6	16	5					200	13			23	3116
Coppergate, Yorkshire	1096	384	255	7							50	15			11	1818
General accident, Yorkshire	139	41	33	2				1			11	2			8	237
Fishergate, Yorkshire	1025	660	237		2			1	1	1	159	40			11	2137
Eastgate, Yorkshire	2734	4632	810		10	3		7		23	855	662	131		34	9901
Total NISP	**16891**	**15006**	**5369**	**27**	**39**	**19**	**4**	**9**	**26**	**43**	**2233**	**1076**	**163**		**151**	**41056**
Total NISP by Group		**37266**					**167**					**3472**			**151**	**41056**
Average Percentage NISP by Group		**90.8**					**0.4**					**8.4**			**0.4**	**100**
mid-11th - mid-12th Elite																
Bristol Castle, Avon	1237	972	959			93					252	555			202	4270
Launceston Castle, Cornwall	54	59	62	3	1					1	5	1			4	190
Portchester Castle, Hampshire	136	78	77	1		1					10				2	305
Faccombe Netherton, Hampshire	333	372	471	611	246	91			30	29	629	164			182	3158
Castle Mall, Norfolk	446	264	216			3				2	99	30	9		4	1073
Castle Rising Castle, Norfolk	68	66	59		3	7			10	1	41	7	1		11	274
Wirral Park, Somerset	175	236	180				29				14	10	1		4	649
Oxford Castle, Oxfordshire	165	328	108								29	2	1		5	638
Middleton Stoney, Oxfordshire	78	129	123	192							1055	98			208	1883
Guildford Castle, Surrey	79	123	62	1	1	3					11				4	284
Pevensey Castle, Sussex	291	366	341	3	5	2		1		6	49	28	10		26	1128
Bramber Castle, Sussex	274	182	254			10			6		87	7				820

Appendix Ia: Taxa representation (NISP) by site type and date. See Table 2 for references.

Site	Domestic Mammals			Wild Mammals							Domestic Birds				Wild	Total
	cattle	sheep	pig	red	roe	fallow	deer'	boar	rabbi	hare	fowl	goose	duck	bird	Bird	NISP
Dudley Castle, West Midlands	108	67	78	17	10	8			3	3	20	6	3		3	326
Trowbridge, Wiltshire	619	771	350	12	14	4				8	96	33			13	1920
Sandal Castle, Yorkshire	31	16	64	1		5										117
Total NISP	**4094**	**4029**	**3404**	**841**	**280**	**227**	**29**	**1**	**49**	**50**	**2397**	**941**	**25**		**668**	**17035**
Total NISP by Group		**11527**					**1477**					**3363**			**668**	**17035**
Average Percentage NISP by Group		**67.6**					**8.7**					**19.7**			**4**	**100**
mid-11th - mid-12th Religious																
Jarrow, Northumberland	218	167	159	2							129	27	1		4	707
Eynsham Abbey, Oxfordshire	1432	908	1325	31	84	6			1	33	217	47	11		34	4129
Osney Abbey, Oxfordshire		1	1												1	3
Lurk Lane, Yorkshire	2535	1887	1061		42	1		7		9	198	61	2		23	5826
Total NISP	**4185**	**2963**	**2546**	**33**	**126**	**7**		**7**	**1**	**42**	**544**	**135**	**14**		**62**	**10665**
Total NISP by Group		**9694**					**216**					**693**			**62**	**10665**
Average Percentage NISP by Group		**91**					**2**					**6.4**			**0.6**	**100**
mid-12th - mid-14th Rural																
Walton Vicarage, Buckinghamshire	645	847	292	22	3						31	18	2		100	1960
Thrislington Tofts, County Durham	163	132	51	1		1			6	1	117	6	2		5	485
Rivenhall, Chelmsford, Essex	110	117	48	10	6					1	12	5	1		3	313
Upton, Gloucestershire	106	452	23						1		8					590
Boteler's Castle, Alcester, Leicestershire	122	167	41	4	1				1	2	8	1			1	348
Lyveden, Northamptonshire	122	161	82	17	1	79					13					475
West Cotton, Northamptonshire	760	658	318	1	1					4	68	40	21		27	1898
Seacourt, Oxfordshire	87	76	24	2							10	1				200
Eckweek, Somerset	113	333	54						2		30	11	1		1	545
Total NISP	**2228**	**2943**	**933**	**57**	**12**	**80**			**10**	**8**	**297**	**82**	**27**		**137**	**6814**
Total NISP by Group		**6104**					**167**					**406**			**137**	**6814**
Average Percentage NISP by Group		**89.6**					**2.4**					**6**			**2**	**100**
mid-12th - mid-14th Urban																
Newbury, Berkshire	229	168	125						13	1	80	37	3		8	664
Bartholomew Street, Berkshire	299	368	106				6				30	17			3	829
George Streeet, Buckinghamshire	509	425	194			2			2		168	39	2		1	1342
The Lanes, Cumbria	337	66	60	9	1						17	5			2	497
Old Grapes Lane, Cumbria	57	19	17	1							13	35				142
Exeter, Devon	2108	2320	763	9	3	5			1	30	352	135	3		41	5770
West Gate Gloucestershire	162	91	84								29	9			4	379
Southampton, Hampshire	145	73	104	3						2	6	2	2		1	338
Quilter's vault, Hampshire	500	501	150	2		1			8	1	53	9	2		1	1228
Victoria Road 700, Hampshire	89	109	97		4	9			3	5	57	7	1		1	382
Victoria Road 975, Hampshire	116	104	102						2	1	39	37	2		3	406
St Johns Street, Hampshire	95	111	72			1			2	11	61	14	1		2	370
Eastern Suburbs 53, Hampshire	335	293	214	6	2	14			4	7	73	26	2		2	978
Hereford city, Herefordshire	1006	310	185	20	4	2					13	7	1		1	1549
Little Lane, Leicestershire	211	223	77						2		56	25				594
Little Lane, Leicestershire	70	48	24						1		10	5			1	159
Lincoln city, Lincolnshire	218	143	42	1							9	31			3	447
Closegate I and II, Newcastle	39	71	13	1	1				2	1	8	1	2		2	141
Coslany Street, Norfolk	27	25	19								4	4			2	81
Heigham Street, Norfolk	63	24	29										1			117
Alms Lane, Norfolk	532	559	184			1			1	1	93	86	5		2	1464
Castle Mall, Norfolk	171	144	62			1			5	2	83	18	3		2	491
Marefair, Northamptonshire	193	484	80							1	77	63	1		5	904
Banhams Garage, Somerset	1500	1436	131				1				1	11	1		4	3085
Brandon Road, Suffolk	1051	1066	411	1						4	98	33	5		2	2671
Marlipins, Sussex	10	18	19	1					1		27	9	1		15	101
General Accident, Yorkshire	4640	1338	732	20	7			2			192	52	4		43	7030
Total NISP	**14712**	**10537**	**4096**	**74**	**22**	**36**	**7**	**2**	**47**	**67**	**1649**	**717**	**42**		**151**	**32159**
Total NISP by Group		**29345**					**255**					**2408**			**151**	**32159**
Average Percentage NISP by Group		**91.2**					**0.8**					**7.5**			**0.5**	100

Appendix Ia: Taxa representation (NISP) by site type and date. See Table 2 for references.

Site	Domestic Mammals			Wild Mammals							Domestic Birds				Wild	Total
	cattle	sheep	pig	red	roe	fallow	deer'	boar	rabbit	hare	fowl	goose	duck	bird	Bird	
mid-12th - mid-14th Elite																
Launceston Castle, Cornwall	503	553	490	25	18	67			20	27	504	60	5		80	2352
Oakhampton Castle, Devon	382	422	538	112	36	48				24	178	15	4		26	1785
Thrislington, County Durham	76	63	48								110	9			41	347
Portchester, Hampshire	192	159	88	9		7					139	12	4		19	629
Portchester, Hampshire	193	189	174	12		8					91	48	14		46	775
Castle Rising, Norfolk	402	297	521	2	9	152			333	96	952	291	5		265	3325
Banbury Castle, Oxfordshire	48	67	42		2				9	4	41	6	1		2	222
Guildford Castle, Surrey	586	654	578	50	16	124			24	16	600	30	12		122	2812
Dudley Castle, West Midlands	1213	404	861	138	103	386			189	60	367	62	33		92	3908
Trowbridge, Wiltshire	305	339	213	13	3	5				4	33	9			7	931
Total NISP	**3900**	**3147**	**3553**	**361**	**187**	**797**			**575**	**231**	**3015**	**542**	**78**		**700**	**17086**
Total NISP by Group		**10600**					**2151**					**3635**			**700**	**17086**
Average Percentage NISP by Group		**62**					**13**					**21**			**4**	**100**
mid-12th - mid-14th Religious																
Dominican Friary, Chester	210	67	148				5			2	12	1			12	457
St Gregory's Priory, Kent	67	117	83	3	1	1			1	1	28	7	3		8	320
Austin Friars, Leicestershire	388	1925	717	2		115			3	5	246	232	17		21	3671
Dean Court, Oxfordshire	187	278	43								15	6			1	530
Witney, Oxfordshire	293	787	435	12	3	31			4	58	96	25			16	1760
Dominican Priory, Oxfordshire	213	194	42		1					2	65	72				589
Osney Abbey, Oxfordshire	32	4	5								1	1				43
Eynsham Abbey, Oxfordshire	471	466	541	15	33	1				43	133	68	6		21	1798
Shrewsbury Abbey, Shropshire	973	321	546	16	24	12	8			4	73	45	2		5	2029
Wells Museum, Somerset	17	4	26	1	1						56	10	1		5	121
Wells Museum, Somerset	326	191	281	2		21			4	2	182	116	33		37	1195
Battle Abbey, Sussex	83	20	31						1	1	4	1			2	143
Lurk Lane, Yorkshire	1068	1339	500			75		2	2	8	220	100	2		30	3346
Fishergate, Yorkshire (Phase 6)	863	476	137								136	36	1		19	1668
Total NISP	**5191**	**6189**	**3535**	**51**	**63**	**256**	**13**	**2**	**15**	**126**	**1267**	**720**	**65**		**177**	**17670**
Total NISP by Group		**14915**					**526**					**2052**			**177**	**17670**
Average Percentage NISP by Group		**84.4**					**3**					**11.6**			**1**	**100**
French Sites																
Site	Domestic Mammals			Wild Mammals							Domestic Birds				Wild	Total
	cattle	sheep	pig	red	roe	fallow	deer'	boar	rabbit	hare	fowl	goose	duck	bird	Bird	
Village																
Dury, Somme	241	161	85								16	1	1			505
Noyon, Picardie, 12-13th century	32	27	13								9		1			82
Noyon, Picardie, 14th century	73	60	27								31	2	1			194
Coquelles, Nord, 14th century	51	15	18								3	1				88
Total NISP	**397**	**263**	**143**								**59**	**4**	**3**			**869**
Total NISP by Group		**803**										**66**				**869**
Average Percentage NISP by Group		**96**										**4**				**100**
Town																
Rue de Lutece, Paris, 9th century	22	35	63								6	1			10	137
Rue de Lutece, Paris, 10th century	230	745	497	2						5				140		1619
Saint-Denis, Paris, layer 953, 6-10th	16	129	88								8	71	3		2	317
Saint-Denis, Paris, layer 837, 8-11th	3	40	25								12		1		5	86
Saint-Denis, Paris, layer 937, 9-11th	42	100	111								21		4		10	288
Saint-Denis, Paris, layer 14779, 10th	44	123	196							8	30	8	2		2	413
Saint-Denis, Paris, layer 960, 10-11th	44	374	70								42	30	10		19	589
Saint-Denis, Paris, layer 856, 10-12th	59	120	172								17	3	3			374
Saint-Germain, Amiens 10-12th	63	92	70							1	48	2	1	3		280
Saint-Denis, Paris, layer 828, 11-12th	130	82	171								23	2	5			413
Saint-Denis, Paris, layer 902, 11-12th	129	62	136								8	13	7			355

Appendix Ia: Taxa representation (NISP) by site type and date. See Table 2 for references.

Site	Domestic Mammals			Wild Mammals							Domestic Birds				Wild	Total
	cattle	sheep	pig	red	roe	fallow	deer'	boar	rabbi	hare	fowl	goose	duck	bird	Bird	
Saint-Denis, Paris, layer 783, 12-13th	73	139	167						40		33	5	2		14	473
Saint-Denis, Paris, layer 956, 12-13th	41	77	124							1	29	7	4		3	286
Rue de Lutece, Paris. 12th century	162	522	121								147	120		194	26	1292
Rue de Lutece, Paris. 12th century	269	497	597		1				2	9	274	65		151	20	1885
Rue de Lutece, Paris. 12th/13th	46	35	149						1		22	2		7	5	267
Rue de Lutece, Paris. 12th/13th	35	92	58							1	45	9		24	8	272
Rue de Lutece, Paris. 13th	194	389	353							1	28	6	1	11		983
Rue de Collegiale, Paris, 13th century	56	119	118							8	730	75	8		24	1138
l'Hoptial general, Compiegne, 9th century	31	269	92								46	9				447
Compeigne, Picardie, 9th	163	86	129	6				7			2	5				398
Compeigne, Picardie, 9th-10th	83	119	297	6				1		2	49	21			6	584
Compeigne, Picardie, 10th	385	372	1056	25	1			14		4	24	20			8	1909
Compeigne, Picardie, 11th	121	285	1069	16	3			42		3	160	15	2		3	1719
Compeigne, Picardie, 11th/12th	33	75	412					4		1	197	13	3		10	748
Compeigne, Picardie12th	116	42	249	1				1			80	3	6		5	503
Abbville, Picardie, 14th century	48	85	96							2	73	6	3		2	315
Saint Michel, Paris 14th century	112	221	60						2		47	14			6	462
l'Hoptial General, Picardie, 14th	462	320	225	5				1	5	3	17	5				1043
ZAC Cathederal, Amiens 14th century	182	156	81						8	1	227	3	2			660
Saint-Germain, Amiens 14th	92	130	15						1		19	13	4			274
Total NISP	**3486**	**5932**	**7067**	**61**	**5**			**70**	**59**	**50**	**2464**	**546**	**71**	**530**	**188**	**20529**
Total NISP by Group		**16485**				**245**						**3611**			**188**	**20529**
Average Percentage NISP by Group		**80.3**				**1.2**						**17.5**			**1**	**100**
High Status																
Chateau de Mayenne 10th	6	7	52				26							7		98
Chateau de Mayenne 11-12th	1145	632	1832				361							676		4646
Grand Besle, Normandy 11-12th	34	35	208	1					1	4	62	4			12	361
Compeigne, Picardie, Phase 11-12th	73	75	347	6	1			12		1	4	4			2	525
Douai, Nord 10th-12th	473	866	2136	8	12					10	667	34	47		62	4315
Compiegne, Picardie US 12th	53	110	413	5	5			17			32	2	2		3	642
Vatteville-la-Rue, Seine Maritime 11-12th	103	72	221	59	21			*	1	3	164	7			10	661
Vatteville-la-Rue, Seine Maritime 13th	52	19	28	15				*			18		1		2	135
Vatteville-la-Rue, Seine Maritime 14th	41	17	16	40	2			*			23	4			4	147
Vatteville-la-Rue, Seine Maritime 12-15th	476	271	824	311	40	17		*	7	30	589	36	3		64	2668
Cour Carree, Paris 13th century	75	127	128	2					11	24	177	25		71	24	664
Chateaude Mayenne 13-14th	29	42	159				26							160		416
Cour Carree, Paris 14th century	367	163	297	1	1			2	11	2	171	6	2	23	10	1056
Vincennes 14th	80	176	134						41	1	142	6			20	600
Chateau de Courtrai 14th century	229	1180	310						8	17	197	22	4		38	2005
Total NISP	**3236**	**3792**	**7105**	**448**	**82**	**17**	**413**	**31**	**80**	**92**	**2246**	**150**	**59**	**937**	**251**	**18939**
Total NISP by Group		**14133**					**1163**					**3392**			**251**	**18939**
Average Percentage NISP by Group		**74.6**					**6.1**					**18**			**1.3**	**100**
Religious House																
Saint Avit-Seinuer																
Moreaucourt late 14th century	49	95	213						80		206	48			6	697
Charite-sur-Loire, Nievre 11-12th	842	1092	1460				23		30	150	1518	589	71		200	5975
Gensac-sur-Garonne, 8th-9th	159	160	205							45	53				2	624
Compeigne, Oise US1505	144	119	302	5	6			1		10	88	13	2		7	697
Total NISP	**1194**	**1466**	**2180**	**5**	**6**		**23**	**1**	**110**	**205**	**1865**	**650**	**73**		**215**	**7993**
Total NISP by Group		**4840**					**350**					**2588**			**215**	**7993**
Average Percentage NISP by Group		**61**					**4**					**32**			**3**	**100**

Appendix Ib: Cattle, caprine and pig NISP and relative frequencies by site type and date

Site	Cattle		Caprine		Pig		Total
	nisp	%	nisp	%	nisp	%	nisp
5th - mid-9th Rural (n=37)							
Lot's Hole, Berkshire	196	51	111	29	78	20	385
Pennylands, Buckinghamshire	1170	49	894	37	330	14	2394
Hartigans, Buckinghamshire	167	86	28	14			195
Wolverton Turn, Buckinghamshire	217	32	344	50	121	18	682
Latimer, Buckinghamshire	196	82	18	8	24	10	238
Chichley, Buckinghamshire	70	37	106	56	13	7	189
Eynesbury, Cambridgeshire	188	35	218	41	126	24	532
Poundbury, Dorset	57	33	89	52	25	15	171
Mucking, Essex	322	62	66	12	135	26	523
Lechdale Gloucestershire	266	60	154	35	22	5	442
Northbrook, Hampshire	43	38	67	59	4	43	114
Old Down Farm, Hampshire	125	43	140	48	25	9	290
Bonners lane, Leicestershire	45	27	47	29	73	44	165
Quarrington, Lincolnshire	1229	61	598	30	173	9	2000
Riby Cross Roads, Lincolnshire	902	60	546	35	70	5	1518
Nettleton, Lincolnshire	461	84	52	10	33	6	546
Sedgeford, Norfolk	199	29	430	63	55	8	684
Marefair, Northampton	193	39	176	32	130	29	499
Mid-Saxon Yarnton, Oxfordshire	200	56	110	31	48	13	358
Eynsham, Oxfordshire	230	36	291	45	121	19	642
Barton Court Farm, Oxfordshire	377	29	573	44	354	27	1304
Yarnton (Cresswell Fields), Oxfordshire	311	57	156	29	78	14	545
Yarnton (Worton), Oxfordshire	54	52	32	31	18	17	104
Audlett Drive, Oxfordshire	91	53	67	39	15	8	173
Cassington, Oxfordshire		51		14		35	
Shaken-Oak, Oxfordshire	86	83	17	16	1	1	104
Shaken-Oak, Oxfordshire	802	29	1347	48	641	23	2790
New Wintles, Oxfordshire	435	53	281	35	104	12	820
Cadbury Congresbury, Somerset	612	39	151	25	533	36	1296
Melford Meadows, Suffolk	286	62	130	28	46	10	462
Spong Hill, Norfolk	634	87	75	10	22	3	731
West Stow, Suffolk	7873	34	11148	49	3903	17	22824
Old Erringham, Sussex	96	41	107	46	30	13	233
Bishopstone, Sussex	35	31	55	48	24	21	114
Cadley Road, Wiltshire	377	37	573	56	66	7	1016
Wharram South Manor Area, Yorkshire	2319	35	3747	56	573	9	6639
Crossgates, Yorkshire		86				14	
Sum of Percentages		**1859**		**1290**		**591**	**51722**
Average Percentage		**50**		**35**		**15**	
5th - mid-9th Urban (n=17)							
Bantham, Devon	400	43	337	37	183	20	920
Abbots Worthy, Hampshire	502	48	443	42	108	10	1053
Hamwic, Hampshire	23896	53	14606	32	6953	15	45455
Sandtun, Kent	71	39	96	52	16	9	183
Sandtun, Kent	91	36	109	43	51	21	251
Maiden Lane, London	2898	55	861	16	1547	29	5306
Jubilee Hall, London	843	55	336	22	365	23	1544
Peaody Buildings, London	63	30	76	36	71	34	210
National Gallery, London	475	30	661	41	470	29	1606
Brandon Road, Norfolk	1427	66	1050	26	483	8	2960
Caister-On-Sea, Norfolk	305	62	108	22	77	16	490
Red Castle Furze, Norfolk	407	50	301	37	105	13	813
Marefair, Northamptonshire	193	39	176	35	130	26	499
St Aldates, Oxford	91	42	109	52	18	9	218
Ipswich, Suffolk	4282	45	2206	23	3130	32	9618
Ipswich, Suffolk	3408	48	1809	25	1973	27	7190
Fishergate, Yorkshire	8296	64	3421	26	1295	10	13012
Sum of Percentages		**805**		**567**		**331**	
Average Percentage		**48**		**33**		**19**	

Appendix Ib: Cattle, caprine and pig NISP and relative frequencies by site type and date

Site	Cattle		Caprine		Pig		Total
	nisp	%	nisp	%	nisp	%	nisp
5th - mid-9th Elite (n=11)							
Wicken Bonhunt, Essex	5138	17	3858	13	20954	70	29950
Lake End Road, Essex	2108	51	704	17	1288	31	4100
Portchester Castle, Hampshire	5627	54	2892	28	1857	18	10376
Eynsham 2b, Oxfordshire	350	20	1011	57	407	23	1768
Eynsham 2c, Oxfordshire	33	31	43	40	32	29	108
Saxon Palaces, Northampton	194	24	476	58	151	18	821
Caister-on-Sea, Norfolk	305	62	108	22	77	16	490
Wroxeter, Shropshire	10106	63	2212	14	3793	24	16111
Brandon, Suffolk	13441	29	24562	52	9121	19	47124
Boltophs, Sussex	113	42	55	21	98	37	266
Ramsbury, Wiltshire	339	40	247	30	249	30	835
Sum of Percentages		**433**		**352**		**315**	
Average Percentage		**39**		**32**		**29**	
5th - mid-9th Religious (n=4)							
St Albans, Hertfordshire	136	18	84	12	510	70	730
North Elmham, Norfolk	2424	33	2808	38	2182	29	7414
Hartlepool, Cleeveland	36	19	146	73	16	8	198
Canterbury Cathedral, Kent	601	70	92	11	163	19	856
Sum of Percentages		**140**		**134**		**126**	
Average Percentage		**35**		**34**		**32**	
5th - mid-9th Sum of Percentages (n=69)		**3237**		**2343**		**1363**	
5th - mid-9th Average Percentage		**47**		**34**		**19**	
mid-9th - mid-11th Rural (n=11)							
Walton Vicarage, Buckinghamshire	603	42	497	35	331	23	1431
Walton Lodge, Buckinghamshire	141	37	175	46	65	17	381
Mawgan Porth, Cornwall	506	45	577	53	37	3	1120
Goltho, Lincoln	103	71	32	22	10	7	145
Whitehall, London		42		39		19	100
Brixworth, Northamptonshire	177	39	224	49	56	12	457
Maxey, Northamptonshire	194	44	191	43	59	13	444
West Cotton, Northamptonshire	46	42	35	32	28	26	109
Steyning, Sussex	624	24	1768	66	260	10	2652
Wharram Percy, Yorkshire	299	39	401	52	69	9	769
Wharram South Manor Area, Yorkshire	646	38	917	53	158	9	1721
Sum of Percentages		**463**		**490**		**148**	
Average Percentage		**42**		**44**		**14**	
mid-9th - mid-11th Urban (n=29)							
Bath, Avon	143	38	183	48	51	13	377
Newbury, Berkshire	485	37	623	48	202	15	1310
Westgate, Gloucestershire	62	35	68	38	49	27	179
Winchcombe, Gloucestershire	343	37	512	55	76	8	931
Winchester, Hampshire	1411	43	1176	36	670	21	3257
Victoria Road, Hampshire	152	29	301	58	63	12	516
Hereford City, Herefordshire	1648	64	433	17	492	19	2573
Friars Street, Droitwich, Hereforeshire	140	49	103	30	93	21	336
Flaxengate, Lincoln	11301	58	6106	31	2174	11	19581
Lincoln City, Lincolnshire	1037	61	449	27	203	12	1689
Upwich, Hereford and Worcester	56	20	212	74	18	6	286
St Peters Street, Northamptonshire	162	34	228	48	88	18	478
Woolmonger Street, Northamptonshire		39		52		9	
Mairfair, Northamptonshire	130	42	122	40	55	18	307
Brandon Road, Norfolk	1757	48	1574	36	687	16	4018
Fishergate, Norfolk	893	67	222	17	216	16	1331
Thetford 1092, Norfolk	919	47	650	33	394	20	1963
St Nicholas Street, Norfolk	112	58	58	30	21	11	191
Mill Lane, Thetford, Norfolk	548	46	336	28	318	26	1202
White Friars, Norfolk	144	39	137	37	89	24	370
St Martin, Norfolk	1524	41	1102	29	1140	30	3766
Red Castle Furze, Norfolk	143	46	106	34	61	20	310
St Aldates, Oxfordshire	143	41	159	45	51	14	353

Appendix Ib: Cattle, caprine and pig NISP and relative frequencies by site type and date

Site	Cattle		Caprine		Pig		Total
	nisp	%	nisp	%	nisp	%	nisp
Ipswich, Suffolk	398	63	114	18	119	19	631
Steyning, Sussex	192	23	550	67	84	10	826
Lurk Lane, Yorkshire	4083	60	1533	22	1263	18	6879
Eastgate, Yorkshire	234	61	104	27	47	12	385
Coppergate, Yorkshire	17832	61	7024	24	4588	16	29444
Skeldergate, Yorkshire	814	74	201	18	88	8	1103
Sum of Percentages		**1361**		**1067**		**470**	
Average Percentage		**47**		**37**		**16**	
mid-9th - mid-11th Elite (n=13)							
Bristol Castle, Avon	150	46	86	26	92	28	328
Wraysbury, Berkshire	1937	41	1330	28	1453	31	
Morison Hall, Cleeveland	36	18	146	74	16	8	198
Faccombe Netherton, Hampshire	284	25	449	39	405	36	1138
Portchester Castle, Hampshire	1029	50	626	31	385	19	2040
Canterbury Castle, Kent	78	33	91	39	65	28	234
Goltho, Lincolnshire	234	42	152	27	177	31	563
Castle Mall, Norfolk	541	51	236	22	277	26	1054
Saxon Palaces, Northampton	733	39	785	42	349	19	1867
Cheddar Palaces, Somerset	1806	43	1064	26	1288	31	4158
Bishopstone, Sussex	511	22	1057	46	748	32	2316
Trowbridge, Wiltshire	718	49	415	28	342	23	1475
Ramsbury, Wiltshire	447	40	307	29	353	29	1107
Sum of Percentages		**499**		**457**		**341**	
Average Percentage		**38**		**36**		**26**	
mid-9th - mid-11th Religious (n=9)							
St Albans, Hertfordshire	104	15	102	14	506	71	712
Canterbury Cathedral, Kent	333	51	163	25	153	24	649
Westminster Abbey, London	769	40	535	28	615	32	1919
Jarrow, Northumberland	319	65	106	22	63	13	488
North Elmham, Norfolk	1046	31	1503	45	827	24	3376
Eynsham 2d, Oxfordshire	294	31	444	46	226	23	964
Eynsham 2e, Oxfordshire	118	28	166	40	133	32	417
Eynsham 2f, Oxfordshire	260	31	322	38	261	31	843
St Ebbes, Oxford	536	25	1251	58	380	18	2268
Sum of Percentages		**317**		**316**		**268**	
Average Percentage		**35**		**35**		**30**	
mid-9th - mid-11th Total Sum of Percentages (n=62)		**2640**		**2330**		**1227**	
mid-9th - mid-11th Average Percentage		**43**		**37**		**20**	
mid-11th - mid-12th Rural (n=6)							
Walton Road, Buckinghamshire	132	33	214	53	57	14	403
Walton Vicarcage, Buckinghamshire	726	36	871	44	396	20	1993
Pepperhill Lane, Kent	91	64	24	17	27	19	142
Brighton Hill South, Hampshire	198	45	140	32	98	23	536
West Cotton, Northampton	726	36	871	44	396	20	2093
Postern Mill, Wiltshire		53		20		27	
Sum of Percentages		**267**		**210**		**123**	
Average Percentage		**44**		**35**		**21**	
mid-11th - mid-12th Urban (n=34)							
Bath, Avon	1023	42	1115	46	292	12	2430
Bartholomew Street, Newbury, Berkshire	59	35	69	40	42	25	179
Newbury, Berkshire	82	56	39	27	25	17	146
Exeter, Devon	735	45	591	37	274	17	1600
Trickhay, Devon	821	46	683	38	295	16	1799
High St, Exeter, Devon		51		28		21	
Culver Street, Colchester, Essex	125	51	53	21	68	28	246
Winchester, Hampshire		36		50		14	
Winchester, Hampshire		41		35		24	
Winchester, Hampshire		31		59		10	
Friars Street, Droitwich, Herefordshire	257	49	159	30	110	21	526
Lincoln City, Lincolnshire	88	39	110	49	26	12	224

Appendix Ib: Cattle, caprine and pig NISP and relative frequencies by site type and date

Site	Cattle		Caprine		Pig		Total
	nisp	%	nisp	%	nisp	%	nisp
Flaxengate, Lincolnshire	8521	47	7593	42	2028	11	18142
St Magnus, London	496	45	345	31	264	24	1105
Fishergate, Norfolk	952	61	309	20	289	19	1550
White Friars, Norfolk	412	46	269	30	216	24	897
St Nicholas Street, Norfolk	838	39	937	43	382	18	2157
St Martin, Norfolk	953	41	702	30	660	29	2315
Red Castle Furze, Norfolk	313	37	386	45	149	18	848
Kings Lynn, Norfolk	603	36	715	43	350	21	1668
Brandon Road, Norfolk	1757	44	1577	39	687	17	4021
Woolmongers Street, Northamptonshire		44		48		7	
Marefair, Northamptonshire	110	49	63	28	50	22	223
St Aldates, Oxfordshire	53	31	100	58	18	11	171
Codrington Library, Oxfordshire	186	33	278	49	102	18	566
Benhams Garage, Somerset	374	58	242	38	27	4	643
Ipswich, Suffolk	295	54	121	22	132	24	548
Ilchester, Somerset	1483	44	1614	48	250	8	3347
North Street, Sussex		27		53		20	
Skeldergate SkK, Yorkshire	1223	65	509	27	159	8	1891
General Accident, Yorkshire	139	65	41	19	33	15	213
Coppergate 5Cf 1 and 2, Yorkshire	1096	63	384	22	255	15	1735
Fishergate, Yorkshire	1025	53	660	34	237	12	1922
Eastgate, Yorkshire	947	65	248	17	256	18	1451
Sum of Percentages		**1569**		**1246**		**580**	
Average Percentage		**46**		**37**		**17**	
mid-11th - mid-12th Elite (23)							
Bristol Castle, Avon	1237	39	972	31	959	30	3168
Launceston Castle, Cornwall	54	31	59	34	62	35	175
Faccombe Netherton, Hampshire	333	28	372	32	471	40	1176
Portchester Castle, Hampshire	136	47	78	27	77	27	291
Carisbrooke Castle, Isle of Wight	321	14	882	39	1030	46	2233
Goltho, Lincolnshire	146	24	328	55	127	21	601
Castle Rising Castle, Norfolk	68	35	66	34	59	31	193
North Elmham, Norfolk	290	32	291	32	321	36	902
Castle Mall, Norfolk	446	51	208	24	216	25	870
Saxon Palaces, Northampton	48	39	60	42	18	19	126
Prudhoe Castle, Northumberland	192	54	147	41	16	5	355
Middleton Stony, Oxfordshire	78	24	123	37	129	39	330
Oxford Castle, Oxfordshire	165	27	328	54	108	18	601
Banbury Castle, Oxordshire	158	50	79	25	78	25	315
Cheddar Palaces, Somerset	274	64	95	22	57	13	426
Wirral Park, Somerset	175	30	236	40	180	30	591
Guildford Castle, Surrey	79	30	126	47	62	23	267
Pevensey Castle, Sussex	291	29	366	37	341	34	998
Bramber Castle, Sussex	274	39	182	25	254	36	710
Sandal Castle, Yorkshire	31	28	16	15	64	57	111
Dudley Castle, West Midlands	108	43	67	26	78	31	253
Yatesbury, Wiltshire		36		44		20	
Trowbridge, Wiltshire	619	36	771	44	342	20	1732
Sum of Percentages		**830**		**807**		**661**	
Average Percentage		**36**		**35**		**29**	
mid-11th - mid-12th Religious (n=5)							
Canterbury Cathedral, Kent	1227	49	823	33	365	18	2415
Jarrow, Northumberland	218	40	167	31	159	29	544
Eynsham Abbey, Oxfordshire	1432	39	908	25	1325	36	3665
St Ebbes, Oxfordshire	1370	23	3468	59	1094	18	5932
Lurk Lane, Yorkshire	2535	46	1887	34	1061	19	5483
Sum of Percentages		**197**		**182**		**120**	
Average Percentage		**40**		**36**		**24**	
mid-11th - mid-12th Total Sum of Percentages (n=68)		**2863**		**2451**		**1470**	
mid-11th - mid-12th Average Percentage		**42**		**36**		**22**	

Appendix Ib: Cattle, caprine and pig NISP and relative frequencies by site type and date

Site	Cattle		Caprine		Pig		Total
	nisp	%	nisp	%	nisp	%	nisp
mid-12th - mid-14th Rural (n=15)							
Walton Vicarage, Buckinghamshire	645	37	827	47	292	16	1764
Walton Road, Buckinghamshire	22	38	22	38	14	24	58
Thrislington Tofts, County Durham	163	47	132	38	51	15	346
Rivenhall, Chelmsford, Essex	110	40	117	43	48	17	275
Upton, Gloucestershire	106	18	452	78	23	4	581
Brighton Hill South, Hampshire	296	36	404	49	126	15	826
Botelers Castle, Alcester, Leicestershire	122	37	167	51	41	12	330
West Cotton, Northamptonshire	760	44	658	38	318	18	1736
Lyvden, Northamptonshire	122	33	161	44	82	23	365
Marefair, Northamptonshire	193	25	484	64	80	11	757
Brixworth, Northamptonshire	371	28	773	59	162	12	1306
Castle Lane, Northamptonshire	455	31	904	61	123	8	1482
Seacourt, Oxfordshire	87	46	76	41	24	13	187
Eckweek, Avon	72	22	230	71	24	7	326
Postern Mill, Wiltshire		51		29		20	
Sum of Percentages		**533**		**751**		**215**	
Average Percentage		**36**		**50**		**14**	
mid-12th - mid-14th Urban (n=43)							
Mary-Le-Port, Avon	660	48	715	43	133	9	1508
Bath, Avon	166	27	369	60	76	12	611
Newbury, Berkshire	229	44	168	32	125	24	522
Bartholomew Street, Newbury Berkshire	299	39	368	48	106	14	773
Cheap Sreet, Berkshire	96	27	205	58	52	15	353
George Street, Buckinghamshire	313	45	277	40	106	15	696
The Lanes, Cumbria	337	73	66	14	60	13	463
Exeter, Devon	114	48	124	42	59	20	297
Exeter, Devon	257	35	349	47	135	18	741
West Gate Gloucestershire	162	48	91	27	84	25	337
Southampton, Hampshire	145	45	73	23	104	32	322
Quilters Vault, Hampshire	412	43	442	45	118	12	972
Victoria Road 700, Hampshire	89	30	109	37	97	33	295
Victoria Road 975, Hampshire	116	36	104	32	102	32	322
St Johns Street, Winchester, Hampshire	95	34	111	40	72	26	278
Eastern Suburbs 53, Winchester, Hampshire	335	40	293	35	214	25	842
Hereford City, Hereford	1006	63	387	25	185	12	1578
Friars Street, Droitwich, Herefordshire	90	44	64	32	48	24	202
St Peters Lane, Leicestershire	490	43	468	41	174	15	1132
Little Lane, Leicester	281	43	271	42	101	15	653
Flaxengate, Lincoln	1022	49	813	39	240	12	2075
Lincoln city, Lincolnshire	218	54	143	36	42	10	403
Closegate I and II, Newcastle	39	32	71	58	13	10	123
Heigham Street, Norwich, Norfolk	63	54	24	21	29	25	116
Castle Mall, Norfolk	171	47	133	36	62	17	366
St Martin, Norfolk	2040	39	1801	34	1433	27	5274
White Friars, Norfolk	59	39	57	37	37	24	153
Alms Lane, Norfolk	80	44	77	42	25	14	182
St Peters St, Northamptonshire	1042	30	2006	59	377	11	3425
Marefair, Northamptonshire	193	25	484	64	80	11	757
St Aldates, Oxford	260	38	382	55	46	7	688
Old Gaol sites, Oxford	84	39	107	50	24	11	215
The Hamel, Oxford	257	30	435	51	158	19	850
Stert Street, Oxford	37	22	96	56	38	22	171
Queens Street, Oxford	63	40	69	44	26	16	158
Benhams Garage, Somerset	1500	49	1436	47	131	4	3067
Priory Barn, Somerset		32		59		9	
Bell Street, Surrey	75	36	99	47	35	17	209
Steyning, Sussex	186	52	112	31	62	17	360
North Street, Sussex	161	27	316	52	123	20	600
Brandon Road, Suffolk	1051	42	1066	42	411	16	2528
Eastgate, Yorkshire	607	30	1245	63	146	7	1998
General Accident, Yorkshire	4640	69	1337	20	732	11	6709
Sum of Percentages		**1774**		**1806**		**728**	
Average Percentage		**41**		**42**		**17**	

Appendix Ib: Cattle, caprine and pig NISP and relative frequencies by site type and date

mid-12th - mid-14th Elite (n=14)							
Launceston Castle, Cornwall	503	33	553	36	490	31	1546
Oakhampton Castle, Devon	382	28	422	31	538	40	1342
Thrislington, County Durham	76	41	63	34	48	25	187
Portchester Castle, Hampshire	385	39	348	35	262	26	995
North Elmham, Norfolk	1025	31	1063	32	1225	37	3313
Castle Rising, Norfolk	402	33	297	24	521	43	1220
Banbury Castle, Oxfordshire	48	30	67	43	42	27	157
Guildford Castle, Surrey	99	27	196	54	70	19	365
Bramber Castle, Sussex	274	39	182	25	254	36	710
Lewes Castle, Sussex	81	14	240	40	278	46	599
Dudley Castle, West Midlands	1213	49	404	16	861	35	2478
Trowbridge, Wiltshire	305	36	339	39	213	25	857
Ramsbury, Wiltshire	99	42	96	40	44	18	239
Sandal Castle, Yorkshire	99	55	49	27	33	18	181
Sum of Percentages		**497**		**476**		**426**	
Average Percentage		**36**		**34**		**30**	

Site	**Cattle**		**Caprine**		**Pig**		**Total**
	nisp	**%**	**nisp**	**%**	**nisp**	**%**	**nisp**
mid-12th - mid-14th Religious (n=16)							
Greyfriars, Avon	168	44	166	44	47	12	381
Dominican Friary, Chester, Cheshire	331	45	217	30	182	25	730
Wells Museum, Hereforsdhire	343	41	195	23	307	36	845
St Gregorys Priory, Kent	67	25	117	44	83	31	267
Canterbury Cathedral, Kent	474	40	482	40	235	20	1191
Canterbury Cathedral, Kent	1258	47	938	35	495	18	2691
Austin Friars, Leicestershire	388	13	1925	64	717	24	3030
Dominican Priory, Oxford	213	47	194	43	42	9	449
Dean Court, Oxford	187	37	278	55	43	8	508
Eynsham Abbey, Oxford	471	32	466	32	541	37	1478
Witney, Oxfordshire	293	19	787	52	435	29	1515
Lewes Friary, Sussex	385	26	754	51	314	21	1453
Battle Abbey, Sussex	83	62	20	15	31	23	134
Shrewsbury Abbey, Shropshire	973	53	321	17	546	30	1840
Lurk Lane, Yorkshire	1068	37	1339	46	500	17	2907
Fishergate, Yorkshire (Phase 6)	863	59	476	32	137	9	1476
Sum of Percentages		**627**		**623**		**349**	
Average Percentage		**39**		**39**		**22**	
mid-12th - mid-14th Total Sum of Percentages (n=88)		**3308**		**3511**		**1686**	
mid-12th - mid-14th Average Percentage		**39**		**41**		**19**	

French Sites	**Cattle**		**Caprine**		**Pig**		**Total**
	nisp	**%**	**nisp**	**%**	**nisp**	**%**	**NISP**
Rural (n=9)							
Saleux, Somme HMA 7th-8th	404	54	225	30	124	16	753
Villiers-le-sec 7th-8th		39		26		35	
Villiers-le-sec 8th-10th		45		27		28	
Saleux, Somme IX-X 9th-10th	69	48	56	39	18	13	143
Villiers-le-sec 10th-11th		51		25		24	
Saleux, Somme XI 11th century	89	49	70	38	23	13	182
Dury, Somme 11th-12th century	297	50	198	33	105	17	600
Noyon 14th century	73	46	60	37	27	17	160
Coquelles 14th century	51	61	15	18	18	21	84
Sum of Percentages		**443**		**218**		**184**	
Average Percentage		**49**		**30**		**21**	

Appendix Ib: Cattle, caprine and pig NISP and relative frequencies by site type and date

French Sites	Cattle		Caprine		Pig		Total
	nisp	%	nisp	%	nisp	%	NISP
Urban (n=28)							
Saint-Denis, layer 14779 (10th)	44	12	123	34	196	54	363
Saint-Denis, layer 960 (10-11th)	44	9	374	77	70	14	488
Saint-Denis 11-12th	259	36	144	20	307	43	710
Saint-Denis 12-13th	114	18	216	35	291	47	621
Rue de Lutece 9th century	22	18	35	29	63	53	120
L'Hopital general, Compiegne 9th century	31	8	269	69	92	24	392
Compeigne, Oise Phase I-III 9th century	163	43	86	23	129	34	378
Compeigne, Oise Phase IV 9th-10th	83	17	119	24	297	59	499
Rue de Lutece 10th century	230	15	745	51	497	34	1472
Compeigne, Oise Phase V 10th century	385	21	372	21	1056	58	1813
Amiens, Somme LHZ2 9th-11th	630	43	289	19	566	38	1485
Amiens, Somme ALHZ4 9th-11th	45	40	50	44	18	16	113
Compiegne, Oise Phase VI 11th century	121	8	285	19	1069	73	1475
Saint-Germain, Amiens 10th-12th century	63	28	92	41	70	31	225
Compeigne, Oise US 1622 11th-12th century	33	6	75	14	412	79	520
Compeigne, Oise US 1460 12th century	116	29	42	10	249	61	407
Rue de Lutece 12th century	162	20	522	65	121	15	805
Rue de Lutece 12th century	269	20	497	36	597	44	1363
Rue de Lutece 12th-13th century	46	20	35	15	149	65	230
Rue de Lutece, Paris. 13th	194	21	389	41	353	38	936
Parc St Julien, Nord 12th-13th		44		29		27	
Parc St Julien, Nord 12th-13th		17		56		27	
Site	**Cattle**		**Caprine**		**Pig**		**Total**
	nisp	**%**	**nisp**	**%**	**nisp**	**%**	**nisp**
Rue de Collegiale 13th century	56	19	119	41	118	40	293
Saint Michel, Paris 14th century	112	29	221	56	60	15	393
Abbville 14th century	48	21	85	37	96	42	229
L'Hopital general 14th	462	46	320	32	225	22	1007
ZAC Cathederal, Amiens 14th century	182	43	156	37	81	19	419
Saint-Germain, Amiens 14th century	92	39	130	55	15	6	237
Sum of Percentages		**596**		**823**		**880**	
Average Percentage		**26**		**36**		**38**	
Elite (n=13)							
Chateau de Mayenne 10th	6	9	7	11	52	80	65
Comble, Somme PII 7th-8th century		32		30		38	
Compeigne, Oise US 1056 12th	53	9	110	19	413	72	576
Cour Carree, Paris 13th century	75	23	127	38	128	39	330
Cour Carree, Paris 14th century	367	44	163	20	297	36	827
Chateau de Courtrai 14th century	229	13	1180	69	310	18	1719
Douai, Nord 10-12th	473	14	866	25	2136	61	3475
Chateau de Mayenne 11-12th	1145	32	632	18	1832	50	3609
Vatteville, Seine Maritime 12th	103	26	72	18	221	56	396
Vatteville, Seine Maritime 13-14th	93	54	36	21	44	25	173
Chateaude Mayenne 13-14th	29	13	42	18	159	69	230
Vincennes 14th	80	21	176	45	134	34	390
Compeigne, Oise US1505 12th century	144	25	119	21	302	53	565
Sum of Percentages		**315**		**353**		**631**	
Average Percentage		**24**		**27**		**49**	
Religious (n=5)							
Compeigne, Oise Phase VII 11th-12th	73	**15**	75	**15**	347	**70**	495
Charite-sur-Loire 11th-12th		**24**		**31**		**46**	
Abbey de Saint-avit-senieur 11-13th		**11**		**24**		**65**	
Moreaucourt late 14th century	49	**14**	95	**27**	213	**59**	357
Gensac-sur-Garonne, 8th-9th	159	30	160	30	205	40	524
Sum of Percentages		**19**		**26**		**56**	
Average Percentage		19		26		56	

Appendix Ic: Domestic bird (fowl, goose and duck) NISP and relative frequencies by site type and date.

Site	Domestic Birds			Total
	fowl	goose	duck	NISP
5th - mid-9th Rural				
Lot's Hole, Berkshire	10	4	1	15
Chichley, Buckinghamshire	1	1		2
Pennylands, Buckinghamshire	68	28	1	97
Wolverton Turn, Buckinghamshire	13	6		
Eynesbury, Cambridgeshire	9	9		18
Lechdale Gloucestershire	16	11	1	28
Northbrook, Hampshire	7			7
Old Down Farm, Hampshire	14			14
Bonners lane, Leicestershire	8	2	2	12
Riby Cross Roads, Lincolnshire	14	30		44
Quarrington, Lincolnshire	33	14	2	49
Audlett Drive, Oxfordshire	2	1	1	4
Wantage, Oxfordshire	2	1		3
Eynsham, Oxfordshire	16	18		34
Yarnton, Oxfordshire	8	13		21
Yarnton (Cresswell Fields), Oxfordshire	4	11		15
Yarnton (Worton), Oxfordshire	4	1		5
New Wintles, Oxfordshire	4	3		7
West Stow, Suffolk	521	285		806
Melford Meadows, Suffolk	2	6		8
Bishopstone, Sussex	6	3		9
Cadley Road, Wiltshire	74	21		95
Wharram South Manor Area, Yorkshire	74	20		94
Total NISP	**910**	**488**	**8**	**1150**
Relative Percentage	**65**	**34.5**	**0.5**	**100**
5th - mid-9th Urban				
Abbots Worthy, Hampshire	70	17		87
Hamwic, Hampshire	800	353		1153
Maiden Lane, London	80	61	4	145
Jubilee Hall, London	24	14		38
Peaody Buildings, London	33	15	1	49
National Gallery, London	45	30	2	77
Redcastle Furze, Norfolk	19	5		24
Marefair, Northamptonshire	12	21	5	38
St Aldates, Oxford	6	1		7
Ipswich, Suffolk	513	129	9	651
Ipswich, Suffolk	397	176	11	584
Fishergate, Yorkshire	603	222		825
Total NISP	**2602**	**1044**	**32**	**3678**
Relative Percentage	**71**	**28**	**1**	**100**
5th - mid-9th Elite				
Lake End Road, Essex	198	64	67	329
Wicken Bonhunt, Essex	3082	2038	147	5267
Portchester Castle, Hampshire	31	33	2	66
Caister-on-Sea, Norfolk	52	16	18	86
Saxon Palaces, Northamptonshire	30	29	11	70
Eynsham 2b, Oxfordshire	271	118	37	426
Eynsham 2c, Oxfordshire	9	4	1	14
Wroxeter, Shropshire	391	6	23	420
Brandon, Suffolk	1306	964	755	3025
Boltophs, Sussex	2	1		3
Total NISP	**5372**	**3272**	**1061**	**9705**
Relative Percentage	**55**	**33**	**11**	**100**

Appendix Ic: Domestic bird (fowl, goose and duck) NISP and relative frequencies by site type and date.

Site	Domestic Birds			Total
	fowl	goose	duck	NISP
5th - mid-9th Religious				
Hartlepool, Cleeveland	9	15	1	25
North Elmham, Norfolk	37	18	5	
St Albans, Hertfordshire	23	11	3	37
Total NISP	**69**	**44**	**9**	**122**
Relative Percentage	**56**	**36**	**8**	**100**
5th - mid-9th Average Percentage	*62*	*33*	*5*	*100*
mid-9th - mid-11th Rural				
Walton Vicarage, Buckinghamshire	20	44		64
Goltho, Lincoln	2	3	1	6
West Cotton, Northamptonshire	5	4	1	10
Steyning, Sussex	62	19	23	104
Wharram Percy, Yorkshire	96	8		104
Wharram South Manor Area, Yorkshire	16	7		23
Total NISP	**201**	**85**	**25**	**311**
Relative Percentage	**65**	**27**	**8**	**100**
mid-9th - mid-11th Urban				
Winchester, Hampshire	161	6	14	181
Victoria Road, Hampshire	11		1	12
Winchcombe, Gloucestershire	36	5		41
Westgate, Gloucestershire	3	3	1	7
Newbury, Berkshire	40	14	1	55
Hereford City, Herefordshire	58	4	1	63
Ipswich, Suffolk	24	5	3	32
Brandon Road, Norfolk	100	35	15	150
St Nicholas Street, Norfolk		1		1
Thetford 1092, Norfolk	29	13	3	45
Mill Lane, Thetford, Norfolk	69	11	11	91
Shrewsbury, Shropshire	6			6
Marefair, Northamptonshire	7			7
Fishergate, Norfolk	13	10	1	24
St Peters Street, Northamptonshire	2	4		6
Lurk Lane, Yorkshire	88	109	3	200
Coppergate, Yorkshire	1217	348	1	1566
Lincoln City, Lincolnshire	63	13		76
Total NISP	**1927**	**581**	**55**	**2563**
Relative Percentage	**75**	**23**	**2**	**100**
mid-9th - mid-11th Elite				
Portchester Castle, Hampshire	511	147	62	720
Faccombe Netherton, Hampshire	61	15		76
Goltho, Lincolnshire	36	15	5	56
Castle Mall, Norfolk	191	22	9	222
Saxon Palaces, Northampton	113	68	19	200
Bishopstone, Sussex	388	76	1	465
Trowbridge, Wiltshire	20	7		27
Total NISP	**1320**	**350**	**96**	**1766**
Relative Percentage	**75**	**20**	**5**	**100**

Appendix Ic: Domestic bird (fowl, goose and duck) NISP and relative frequencies by site type and date.

Site	Domestic Birds			Total
	fowl	goose	duck	NISP
mid-9th - mid-11th Religious				
St Albans, Hertfordshire	22	8	4	34
Westminster Abbey, London	205	12	17	234
Jarrow, Northumberland	94	56	1	151
Eynsham 2d, Oxfordshire	45	16	9	70
Eynsham 2e, Oxfordshire	9	7	1	17
Eynsham 2f, Oxfordshire	29	11	9	49
Lewes Priory, Sussex	9			9
Total NISP	**413**	**110**	**41**	**564**
Relative Percentage	**73**	**20**	**7**	**100**
mid-9th - mid-11th Average Percentage	***72***	***23***	***5***	
mid-11th - mid-12th Rural				
Walton Vicarage, Buckinghamshire	56	19	2	77
Brighton Hill South, Hampshire	6			6
Pepperhill Lane, Kent	1			1
Total NISP	**63**	**19**	**2**	**84**
Relative Percentage	**75**	**23**	**2**	**100**
mid-11th - mid-12th Urban				
Bartholomew Street, Newbury, Berkshire	24	11		35
Newbury, Berkshire	37	8	2	47
Exeter, Devon	204	42	1	247
Flaxengate, Lincolnshire	44	19		63
Lincoln City, Lincolnshire	8	13		21
Brandon Road, Norfolk	166	69	8	243
St Nicholas Street, Norfolk	111	49	7	167
Fishergate, Norfolk	113	33	4	150
Kings Lynn, Norfolk	77	42	5	124
St Aldates, Oxfordshire	66	15		81
Ipswich, Suffolk	32	13	3	48
Benhams Garage, Somerset	1	1		2
North Street, Sussex	200	13		213
Coppergate 5Cf 1 and 2, Yorkshire	50	15		65
General Accident, Yorkshire	11	2		13
Fishergate, Yorkshire	159	40		199
Eastgate, Yorkshire	855	662	131	1648
Total NISP	**2158**	**1047**	**161**	**3366**
Relative Percentage	**64**	**31**	**5**	**100**
mid-11th - mid-12th Elite				
Bristol Castle, Avon	252	555		807
Launceston Castle, Cornwall	5	1		6
Portchester Castle, Hampshire	10			10
Faccombe Netherton, Hampshire	629	164		793
Castle Rising Castle, Norfolk	41	7	1	49
Castle Mall, Norfolk	99	30	9	138
Oxford Castle, Oxfordshire	29	2	1	32
Middleton Stony, Oxfordshire	1055	98		1153
Wirral Park, Somerset	14	10	1	25
Guildford Castle, Surrey	11			11
Pevensey Castle, Sussex	49	28	10	87
Bramber Castle, Sussex	87	7		94
Dudley Castle, West Midlands	20	6	3	29
Trowbridge, Wiltshire	96	33		129
Total NISP	**2397**	**941**	**25**	**3363**
Relative Percentage	**71**	**28**	**1**	**100**

Appendix Ic: Domestic bird (fowl, goose and duck) NISP and relative frequencies by site type and date.

Site	Domestic Birds			Total
	fowl	goose	duck	NISP
mid-11th - mid-12th Religious				
Eynsham Abbey, Oxfordshire	217	47	11	275
Jarrow, Northumberland	129	27	1	157
Lurk Lane, Yorkshire	198	61	2	261
Total NISP	**544**	**135**	**14**	**693**
Relative Percentage	**78**	**19**	**2**	**100**
mid-11th - mid-12th Average Percentage	*72*	*25*	*3*	*100*
mid-12th - mid-14th Rural				
Walton Vicarage, Buckinghamshire	31	18	2	51
Thrislington Tofts, County Durham	117	6	2	
Rivenhall, Chelmsford, Essex	12	5	1	18
Upton, Gloucestershire	8			8
Botelers Castle, Alcester, Leicestershire	8	1		9
West Cotton, Northamptonshire	68	40	21	129
Lyvden, Northamptonshire	13			13
Seacourt, Oxfordshire	10	1		11
Eckweek, Somerset	30	11	1	42
Total NISP	**121**	**52**	**22**	**195**
Relative Percentage	**73**	**20**	**7**	**100**
mid-12th - mid-14th Urban				
Newbury, Berkshire	80	37	3	120
Bartholomew Street, Newbury Berkshire	30	17		47
George Streeet, Buckinghamshire	168	39	2	209
The Lanes, Cumbria	17	5		22
Old Grapes Lane, Cumbria	13	35		48
Exeter, Devon	352	135	3	490
West Gate Gloucestershire	29	9		38
Southampton, Hampshire	6	2	2	10
Quilter's vault, Hampshire	53	9	2	64
Victoria Road 700, Hampshire	57	7	1	65
Victoria Road 975, Hampshire	39	37	2	78
St Johns Street, Winchester, Hampshire	61	14	1	76
Eastern Suburbs 53, Winchester, Hampshire	73	26	2	101
Hereford city, Herefordshire	13	7	1	21
Little Lane, Leicester	56	25		81
Little Lane, Leicester	10	5		15
Lincoln city, Lincolnshire	9	31		40
Closegate I and II, Newcastle	8	1	2	11
Castle Mall, Norfolk	83	18	3	104
Heigham Street, Norwich, Norfolk			1	1
Coslany Street, Norfolk	4	4		8
Alms Lane, Norfolk	93	86	5	184
Marefair, Northamptonshire	77	63	1	141
Benhams Garage, Somerset	1	11	1	13
Brandon Road, Suffolk	98	33	5	136
Marlipins, Sussex	27	9	1	37
General Accident, Yorkshire	192	52	4	248
Total NISP	**960**	**440**	**34**	**1434**
Relative Percentage	**68**	**30**	**2**	**100**

Appendix Ic: Domestic bird (fowl, goose and duck) NISP and relative frequencies by site type and date.

Site	Domestic Birds			Total
	fowl	goose	duck	NISP
mid-12th - mid-14th Elite				
Launceston Castle, Cornwall	504	60	5	569
Thrislington, County Durham	110	9		119
Oakhampton, Devon	178	15	4	197
Portchester, Hampshire	139	12	4	155
Portchester, Hampshire	91	48	14	153
Castle Rising, Norfolk	952	291	5	1248
Banbury Castle, Oxfordshire	41	6	1	48
Guildford Castle, Surrey	600	30	12	642
Dudley Castle, West Midlands	367	62	33	462
Trowbridge, Wiltshire	33	9		42
Total NISP	**2190**	**449**	**69**	**2708**
Relative Percentage	**83**	**15**	**2**	**100**
mid-12th - mid-14th Religious				
Dominican Friary, Chester	12	1		13
St Gregory's Priory, Kent	28	7	3	38
Austin Friars, Leicestershire	246	232	17	495
Wells Museum, Herefordshire	56	10	1	67
Wells Museum, Herefordshire	182	116	33	331
Witney, Oxfordshire	96	25		121
Dominican Priory, Oxfordshire	65	72		137
Osney Abbey, Oxfordshire	1	1		2
Dean Court, Oxfordshire	15	6		21
Eynsham Abbey, Oxfordshire	133	68	6	207
Shrewsbury Abbey, Shropshire	62	27	1	90
Shrewsbury Abbey, Shropshire	11	18	1	30
Battle Abbey, Sussex	4	1		5
Lurk Lane, Yorkshire	220	100	2	322
Fishergate, Yorkshire (Phase 6)	136	36	1	173
Total NISP	**1227**	**712**	**62**	**2001**
Relative Percentage	**62**	**35**	**3**	**100**
mid-12th - mid-14th Average Percentage	***72***	***25***	***3***	*100*
French Sites	**Domestic Birds**			
	fowl	**goose**	**duck**	
Rural				
Dury, Somme	16	1	1	18
Noyon 12-13th century	9		1	10
Noyon 14th century	31	2	1	34
Coquelles 14th century	3	1		4
Total NISP	**59**	**4**	**3**	**66**
Relative Percentage	**89**	**6**	**5**	**100**
Urban				
Saint-Denis, layer 953 (6-10th)	8	71	3	82
Saint-Denis, layer 837 (8-11th)	12		1	13
l'Hoptial general 9th century	46	9		55
Rue de Lutece, Paris. 9th century	6	1		7
Saint-Denis, layer 937 (9-11th)	21		4	25
Saint-Denis, layer 14779 (10th)	30	8	2	40
Saint-Denis, layer 960 (10-11th)	42	30	10	82
Compeigne 9th	2	5		7
Compeigne, 9th-10th	49	21		70
Compeigne, Oise Phase 10th	24	20		44

Appendix Ic: Domestic bird (fowl, goose and duck) NISP and relative frequencies by site type and date.

Site	Domestic Birds			Total
	fowl	goose	duck	NISP
Saint-Denis, layer 856 (10-12th)	17	3	3	23
Saint-Germain, Amiens 10-12th	48	2	1	51
Saint-Denis, layer 828 (11-12th)	23	2	5	30
Saint-Denis, layer 902 (11-12th)	8	13	7	28
Saint-Denis, layer 783 (12-13th)	33	5	2	40
Saint-Denis, layer 956 (12-13th)	29	7	4	40
Rue de Lutece, Paris. 12th century	147	120		267
Rue de Lutece, Paris. 12th century	274	65		339
Rue de Lutece, Paris. 12th/13th	22	2		24
Rue de Lutece, Paris. 12th/13th	45	9		54
Rue de Lutece, Paris. 13th	28	6	1	35
Rue de Collegiale 13th century	730	75	8	813
Compeigne, Oise Phase 11th	160	15	2	177
Compeigne, Oise 11th/12th	197	13	3	213
Abbville 14th century	73	6	3	82
Saint Michel, Paris 14th century	47	14		61
l'Hoptial general 14th	17	5		22
ZAC Cathederal, Amiens 14th century	227	3	2	232
Saint-Germain, Amiens 14th	19	13	4	36
Compeigne, 12th	80	3	6	89
Total NISP	**2464**	**546**	**71**	**3081**
Relative Percentage	**80**	**18**	**2**	**100**
Elite				
Grand Besle, Normandy 11-12th	62	4		66
Compeigne, Oise Phase 11-12th	4	4		8
Douai, Nord10th-12th	667	34	47	748
Compeigne, Oise US 12th	32	2	2	36
Vatteville-la-Rue, Seine Maritime 11-12th	164	7		171
Vatteville-la-Rue, Seine Maritime 13th	18		1	19
Vatteville-la-Rue, Seine Maritime 14th	23	4		27
Vatteville-la-Rue, Seine Maritime 12-15th	589	36	3	628
Cour Carree, Paris 13th century	177	25		202
Cour Carree, Paris 14th century	171	6	2	179
Vincennes 14th	142	6		148
Chateau de Courtrai 14th century	197	22	4	223
Total NISP	**2246**	**150**	**59**	**2455**
Relative Percentage	**91**	**6**	**3**	**100**
Religious House				
Moreaucourt late 14th century	206	48		254
Charite-sur-Loire, Nievre 11-12th	1518	589	71	2178
Compeigne, Oise US1505	88	13	2	103
Total NISP	**1812**	**650**	**73**	**2535**
Relative Percentage	**71**	**26**	**3**	**100**

Appendix Idi: Representation and presence/diversity of fish taxa by site type and period

Site	eel	herring	smelt	sprat	mackrel	cod	haddock	gadid	ling	hake	whiting	plaice	flounder	flatfish	roach	cyprinid	salmon	perch	trout	pike	Presence	P/D Index
5th - mid-9th Rural (n=4)																						
Wolverton Turn, Buckinghamshire (Sykes 2004a)	*																					
Riby Cross Roads, Lincolnshire (Scott 1994)	*																					
Yarnton, Oxfordshire (Mulville and Ayres 2004)	*																					
West Stow, Suffolk (Crabtree 1996)												*						*		*		
% of sites on which represented	**75**											**25**						**25**		**25**	6 of 80	7.5
5th - mid-9th Urban (n=8)																						
Bantham, Devon					*	*											*					
Hamwic, Hampshire (Bourdillon and Coy 1980)	*	*			*						*		*				*					
Sandtun, Kent (Murray and Hamilton-Dyer 2001)	*	*			*	*	*	*	*	*	*	*	*	*								
Maiden Lane, London (Locker 1993a)	*	*				*	*		*	*	*	*	*	*	*	*	*			*		
Jubilee Hall, London (Locker 1993a)	*	*										*			*	*	*			*		
Peabody Buildings, London (Locker 1993b)	*	*				*				*	*	*	*	*		*	*			*		
Redcastle Furze, Norfolk (Nicholson 1995)	*	*																		*		
Fishergate, Yorkshire (O'Connor 1991)	*	*				*	*							*		*	*		*	*		
% of sites on which represented	**88**	**88**			**37**	**62**	**37**	**12**	**17**	**37**	**50**	**50**	**50**	**50**	**17**	**50**	**75**		**12**	**62**	65 of 160	40.6
5th - mid-9th Elite																						
Lake End Road, Essex (Powell 2002)	*											*						*		*		
Saxon Palaces, Northamptonshire (Harman 1985b)	*					*																
% of sites on which represented	**100**					**50**						**50**						**50**		**50**	6 of 40	15
5th - mid-9th Religious	**Insufficient Data**																					
Total 5th-mid-9th (n=14)	*86*	*50*			*21*	*42*	*21*	*7*	*14*	*21*	*29*	*42*	*29*	*29*	*14*	*29*	*42*	*14*	*7*	*40*	*77 of 280*	*27.5*
mid-9th - mid-11th Rural	**Insufficient Data**																					
mid-9th - mid-11th Urban (n=9)																						
Newbury, Berkshire (Coy 1997)	*	*																				
Brandon Road, Norfolk (Jones 1993)						*		*														
Thetford 1092, Norfolk (Jones 1984)	*	*				*	*									*						
St Martins-at palace plain (Locker 1987)	*	*	*		*	*	*	*			*	*	*	*		*			*			
Alms Lane, Norfolk (Jones and Scott 1985)	*	*			*	*		*			*											
Fishergate Norfolk (Locker 1994)	*	*			*	*		*			*	*		*								
Redcastle Furze, Norfolk (Nicholson 1995)	*	*																		*		
Fishergate, Yorkshire (O'Connor 1991)	*	*	*			*	*							*		*	*			*		
Lincoln city, Lincolnshire (Dobney*et al.* 1995)	*		*				*								*	*	*	*		*		
% of sites on which represented	**89**	77	**33**		**33**	**66**	**44**	**44**			**33**	**22**	**11**	**33**	**11**	**44**	**22**	**11**	**11**	**33**	56 of 180	31.1

Appendix Idi: Representation and presence/diversity of fish taxa by site type and period

Site	eel	herring	smelt	sprat	mackrel	cod	haddock	gadid	ling	hake	whiting	plaice	flounder	flatfish	roach	cyprinid	salmon	perch	trout	pike	Presence	P/D Index
mid-9th - mid-11th Elite (n=4)																						
Faccombe Netherton, Hampshire (Sadler 1990)																			*			
Wraysbury, Berkshire (Coy 1989)	*	*												*	*	*	*	*	*	*		
Castle Mall, Norfolk (Locker 1997)	*	*	*	*	*	*	*	*	*		*	*		*	*	*	*			*		
Saxon Palaces, Northamptonshire (Harman, 1985b)	*	*				*								*								
% of sites on which represented	**75**	**75**	**25**	**25**	**25**	**50**	**25**	**25**	**25**		**25**	**25**		**75**	**50**	**50**	**50**	**25**	**50**	**50**	30 of 80	37.5
mid-9th - mid-11th Religious (n=3)																						
Westminster Abbey, London (Rackham 1994)	*	*	*			*	*		*		*	*	*	*	*	*	*			*		
Eynsham Abbey, Oxfordshire (Ayres*et al*. 2003)								*										*		*		
Lewes Priory, Sussex (Stevens1997)	*	*				*					*	*					*					
% of sites on which represented	**66**	**66**	**33**			**66**	**33**	**33**	**33**		**66**	**66**	**33**	**33**	**33**	**33**	**66**	**33**		**66**	23 of 60	38
Tota mid-9th-mid-11th (n=16)	***81***	***75***	***31***	***6***	***25***	***63***	***37***	***37***	***13***		***37***	***31***	***13***	***44***	***25***	***44***	***37***	***19***	***19***	***44***	*109 of 320*	*34*
mid-11th - mid-12th Rural (n=1)																						
Pepperhill Lane, Kent (Charles and Ingrem 2001)		*						*						*								
% of sites on which represented		**100**						**100**						**100**							3 of 20	15
mid-11th - mid-12th Urban (n=10)																						
Exeter, Devon (Maltby 1979)	*					*		*		*	*			*								
Brandon Road, Norfolk (Jones 1993)						*		*						*								
St Martins-at-Palace Plain, Norfolk (Locker 1987)	*	*			*	*	*	*			*	*		*					*			
Fishergate Norfolk (Locker 1994)	*	*	*		*	*	*	*			*	*	*	*		*						
Redcastle Furze, Norfolk (Nicholson 1995)	*	*																				
Alms Lane, Norfolk (Jones and Scott 1995)		*			*	*		*														
General Accident, Yorkshire (O'Connor 1988)	*	*			*	*		*								*	*	*				
Fishergate, Yorkshire (O'Connor 1991)	*	*	*			*	*						*	*		*	*		*	*		
Eastgate, Yorkshire (Scott 1992)	*	*			*	*	*	*	*		*	*	*	*	*	*	*	*	*	*		
Lincoln city, Lincolnshire (Dobney*et al.* 1995)	*														*	*						
% of sites on which represented	**80**	**70**	**20**		**50**	**80**	**40**	**70**	**10**	**10**	**40**	**30**	**30**	**60**	**20**	**50**	**30**	**20**	**30**	**20**	76 of 200	38
mid-11th - mid-12th Elite (n=6)																						
Trowbridge, Wiltshire (Bourdillon 1994)	*	*														*						
Castle Acre, Norfolk (Lawrance 1982)	*																			*		
Ludgershall Castle (Poole n.d.)		*				*				*		*		*								
Boltophs Bramber, Sussex (Stevens 1990)						*																
Castle Mall, Norfolk (Locker 1997)	*	*	*		*	*	*	*			*	*		*		*	*			*		
Launceston Castle, Cornwall (Smith 1995)										*												
% of sites on which represented	**50**	**50**	**17**		**17**	**50**	**17**	**17**		**33**	**17**	**33**		**33**		**33**	**17**			**33**	25 of 120	20.8

Appendix Idi: Representation and presence/diversity of fish taxa by site type and period

Site	eel	herring	smelt	sprat	mackrel	cod	haddock	gadid	ling	hake	whiting	plaice	flounder	flatfish	roach	cyprinid	salmon	perch	trout	pike	Presence	P/D Index
mid-11th - mid-12th Religious (n=4)																						
St Albans Abbey, Hertfordshire (Crabtree 1993)	*	*			*	*	*	*				*		*			*			*		
Eynsham Abbey, Oxfordshire (Ayres *et al*. 2003)	*	*			*	*	*	*	*		*	*		*	*	*	*	*	*	*		
Jarrow, Northumberland (Noddle 1987)						*					*											
Battle Abbey, Sussex (Locker 1985)	*					*	*		*		*	*		*								
% of sites on which represented	**75**	**50**			**50**	**100**	**75**	**50**	**50**		**75**	**50**		**75**	**25**	**25**	**50**	**25**	**25**	**50**	35 of 80	44
Total mid-11th - mid-12th (n=21)	***67***	***62***	***14***		***38***	***71***	***40***	***52***	***14***	***14***	***38***	***38***	***14***	***57***	***14***	***38***	***29***	***14***	***19***	***29***	*139 of 420*	*33*
mid-12th - mid-14th Rural (n = 2)																						
Thrislington Tofts, Durham (Rackham 1989)	*	*					*															
West Cotton, Northamptonshire (Albarella and Davis 1994)	*															*						
% of sites on which represented	**100**	**50**					**50**									**50**					5 of 40	12.5
mid-12th - mid-14th Urban (n=9)																						
Exeter, Devon (Maltby 1979)	*					*	*	*	*	*	*			*			*					
Newbury, Berkshire (Coy 1997)	*					*		*	*			*										
Stert Street Oxford (Wheeler 1979)	*			*	*		*								*			*		*		
Brandon Road, Norfolk (Jones 1993)						*	*	*	*													
St Martins-at-Palace Plain, Norfolk (Locker 1987)	*	*	*		*	*	*	*			*	*	*	*			*					
George Street, Aylesbury (Jones 1983)					*	*			*		*								*			
Alms Lane, Norfolk (Jones and Scott 1995)																						
General Accident, Yorkshire (O'Connor 1988)	*	*	*			*	*	*			*		*	*	*	*	*	*	*	*		
Lincoln city, Lincolnshire (Dobney *et al.* 1995)	*							*						*	*	*	*	*		*		
% of sites on which represented	**67**	**22**	**22**	**11**	**33**	**67**	**55**	**67**	**44**	**11**	**44**	**22**	**22**	**44**	**33**	**22**	**44**	**33**	**22**	**33**	65 of 180	36.1
mid-12th - mid-14th Elite (n=7)																						
Guildford Castle, Surrey (Sykes et al. 2005)	*	*			*	*	*	*			*	*		*		*	*					
Thrislington, Durham (Rackham 1989)	*	*					*	*	*							*						
Castle Mall, Norfolk (Locker 1997)	*	*	*		*	*	*	*	*		*	*		*		*	*			*		
Middleton Stoney (Levitan 1984)	*	*			*											*						
Saxon Palaces, Northamptonshire (Harman, 1985b)					*	*	*	*	*			*					*			*		
Trowbridge, Wiltshire (Bourdillon 1994)	*	*			*											*	*					
Launceston Castle, Cornwall (Smith 1995	*					*	*		*	*												
% of sites on which represented	**86**	**71**	**14**		**71**	**57**	**71**	**57**	**57**	**14**	**28**	**42**		**28**		**71**	**57**			**33**	53 of 140	37.8

Appendix Idi: Representation and presence/diversity of fish taxa by site type and period

Site	eel	herring	smelt	sprat	mackrel	cod	haddock	gadid	ling	hake	whiting	plaice	flounder	flatfish	roach	cyprinid	salmon	perch	trout	pike	Presence	P/D Index
mid-12th - mid-14th Religious (n=9)																						
Battle Abbey, Sussex (Locker 1985)						*	*	*				*										
Eynsham Abbey, Oxfordshire (Ayres *et al*. 2003)	*	*				*	*	*	*		*	*		*	*	*		*		*		
St Mary's, Clerkwell (Locker 1996)	*	*				*	*				*	*	*	*								
Cleeve Abbey, Somerset (Locker 1989)	*	*			*		*	*		*	*	*										
Dean Court, Oxfordshire (Jones 1994)	*	*						*			*	*		*	*	*				*		
Shrewsbury Abbey (Jones 2002)						*	*	*						*						*		
Witney, Ocfordshire (Ayres and Serjeantson 2002)	*	*				*								*						*		
Austin Friars, Leicester (Thawley 1981)	*					*	*		*			*		*			*					
Fishergate, Yorkshire 6a (O'Connor 1991)	*	*	*			*	*	*			*		*	*		*	*	*	*	*		
% of sites on which represented	**78**	**67**	**11**		**11**	**78**	**78**	**67**	**22**	**11**	**55**	**67**	**22**	**78**	**22**	**33**	**22**	**22**	**11**	**55**	73 of 180	41
Total mid-12th - mid-14th (n=27)	*78*	*52*	*15*	*4*	*33*	*63*	*67*	*59*	*37*	*11*	*41*	*41*	*11*	*48*	*19*	*41*	*37*	*19*	*11*	*37*	*196 of 540*	*36*

Appendix Idii: Presence/diversity of freshwater, migratory and marine fish by site type and date.

	Freshwater				Migratory							Marine								
Site	roach	cyprinid	perch	pike	eel	salmon	trout	smelt	flatfish	plaice	flounder	herring	sprat	mackrel	cod	haddock	gadid	ling	hake	whiting
5th - mid-9th Rural																				
Wolverton Turn, Buckinghamshire (Sykes 2004a)					*															
Riby Cross Roads, Lincolnshire (Scott 1994)					*															
Yarnton, Oxfordshire (Mulville and Ayres 2004)					*															
West Stow, Suffolk (Crabtree 1996)			*	*						*										
Presence/Diversity Index	12.5				14							**0**								
5th - mid-9th Urban (n=8)																				
Bantham, Devon						*								*	*					
Hamwic, Hampshire (Bourdillon and Coy 1980)					*	*					*	*		*						*
Sandtun, Kent (Murray and Hamilton-Dyer 2001)					*				*	*	*	*		*	*	*	*	*	*	*
Maiden Lane, London (Locker 1993a)	*	*		*	*	*			*	*	*	*			*	*		*	*	*
Jubilee Hall, London (Locker 1993a)	*	*		*	*	*				*		*								
Peabody Buildings, London (Locker 1993b)		*		*	*	*			*	*	*	*			*				*	*
Redcastle Furze, Norfolk (Nicholson 1995)				*	*							*								
Fishergate, Yorkshire (O'Connor 1991)		*		*	*	*	*		*			*			*	*				
Presence/Diversity Index	**34**				**46**							**39**								
5th - mid-9th Elite																				
Lake End Road, Essex (Powell 2002)			*	*	*					*										
Saxon Palaces, Northamptonshire (Harman 1985b)					*										*					
Presence/Diversity Index	25				21							5								
5th - mid-9th Religious	**Insufficient Data**				**Insufficient Data**							**Insufficient Data**								
Total 5th-mid-9 Presence/Diversity Index	*26.7*				*33.6*							*23*								
mid-9th - mid-11th Rural					**Insufficient Data**															
mid-9th - mid-11th Urban (n=9)																				
Newbury, Berkshire (Coy 1997)					*							*								
Brandon Road, Norfolk (Jones 1993)															*		*			
Thetford 1092, Norfolk (Jones 1984)		*			*							*			*	*				
St Martins-at palace plain (Locker 1987)		*			*		*	*	*	*	*	*		*	*	*	*			*
Alms Lane, Norfolk (Jones and Scott 1985)					*							*		*	*		*			*
Fishergate Norfolk (Locker 1994)					*				*	*		*		*	*		*			*
Redcastle Furze, Norfolk (Nicholson 1995)				*	*							*								
Fishergate, Yorkshire (O'Connor 1991)		*		*	*	*		*	*			*			*	*				
Lincoln city, Lincolnshire (Dobney *et al.* 1995)	*	*	*	*	*	*		*								*				
Presence/Diversity Index	**25**				**31**							**33**								

Appendix Idii: Presence/diversity of freshwater, migratory and marine fish by site type and date.

	Freshwater				Migratory							Marine								
Site	roach	cyprinid	perch	pike	eel	salmon	trout	smelt	flatfish	plaice	flounder	herring	sprat	mackrel	cod	haddock	gadid	ling	hake	whiting
mid-9th - mid-11th Elite (n=4)																				
Faccombe Netherton, Hampshire (Sadler 1990)							*													
Wraysbury, Berkshire (Coy 1989)	*	*	*	*	*	*	*		*			*								
Castle Mall, Norfolk (Locker 1997)	*	*		*	*	*		*	*	*		*	*	*	*	*	*	*		*
Saxon Palaces, Northamptonshire (Harman, 1985b)					*				*			*			*					
Presence/Diversity Index	**44**				**43**							**31**								
mid-9th - mid-11th Religious (n=3)																				
Westminster Abbey, London (Rackham 1994)	*	*		*	*	*		*	*	*	*	*			*	*		*		*
Eynsham Abbey, Oxfordshire (Ayres *et al*. 2003)			*	*													*			
Lewes Priory, Sussex (Stevens1997)					*	*				*		*			*					*
Presence/Diversity Index	**42**				**43**							**33**								
Tota mid-9th-mid-11th Presenve/Diversity Index	*32.8*				*37*							*32*								
mid-11th - mid-12th Rural																				
Pepperhill Lane, Kent (Charles and Ingrem 2001)									*			*					*			
Presence/Diversity Index	**0**				**14**							**22**								
mid-11th - mid-12th Urban (n=10)																				
Exeter, Devon (Maltby 1979)					*				*						*		*		*	*
Brandon Road, Norfolk (Jones 1993)									*						*		*			
St Martins-at-Palace Plain, Norfolk (Locker 1987)					*		*		*	*		*		*	*	*	*			*
Fishergate Norfolk (Locker 1994)		*			*			*	*	*	*	*		*	*	*	*			*
Redcastle Furze, Norfolk (Nicholson 1995)					*							*								
Alms Lane, Norfolk (Jones and Scott 1995)												*		*	*		*			
General Accident, Yorkshire (O'Connor 1988)		*	*		*	*						*		*	*		*			
Fishergate, Yorkshire (O'Connor 1991)		*		*	*	*	*	*	*		*	*			*	*				
Eastgate, Yorkshire (Scott 1992)	*	*	*	*	*	*	*		*	*	*	*		*	*	*	*	*		*
Lincoln city, Lincolnshire (Dobney *et al.* 1995)	*	*			*															
Presence/Diversity Index	**27.5**				**40**							**41**								
mid-11th - mid-12th Elite (n=6)																				
Trowbridge, Wiltshire (Bourdillon 1994)		*			*							*								
Castle Acre, Norfolk (Lawrance 1982)				*	*															
Ludgershall Castle (Poole n.d.)									*	*		*			*				*	
Boltophs Bramber, Sussex (Stevens 1990)															*					
Castle Mall, Norfolk (Locker 1997)		*		*	*	*		*	*	*		*		*	*	*	*			*
Launceston Castle, Cornwall (Smith 1995)																			*	
Presence/Diversity Index	**17**				**21**							**22**								

Appendix Idii: Presence/diversity of freshwater, migratory and marine fish by site type and date.

	Freshwater				Migratory							Marine								
Site	roach	cyprinid	perch	pike	eel	salmon	trout	smelt	flatfish	plaice	flounder	herring	sprat	mackrel	cod	haddock	gadid	ling	hake	whiting
mid-11th - mid-12th Religious (n=4)																				
St Albans Abbey, Hertfordshire (Crabtree 1993)				*	*	*			*	*		*		*	*	*	*			
Eynsham Abbey, Oxfordshire (Ayres *et al*. 2003)	*	*	*	*	*	*	*		*	*		*		*	*	*	*	*		*
Jarrow, Northumberland (Noddle 1987)															*					*
Battle Abbey, Sussex (Locker 1985)					*				*	*					*	*		*		*
Presence/Diversity Index	**25**				**43**							**50**								
Total mid-11th - mid-12th Presence/Diversity Index	*25*				*34*							*37*								
mid-12th - mid-14th Rural																				
Thrislington Tofts, Durham (Rackham 1989)					*							*				*				
West Cotton, Northamptonshire (Albarella and Davis 1994)		*			*															
Presence/Diversity Index	**13**				**14**							**11**								
mid-12th - mid-14th Urban (n=9)																				
Exeter, Devon (Maltby 1979)					*	*			*						*	*	*	*	*	*
Newbury, Berkshire (Coy 1997)					*					*					*		*	*		
Stert Street Oxford (Wheeler 1979)	*		*	*	*								*	*		*				
Brandon Road, Norfolk (Jones 1993)															*	*	*	*		
St Martins-at-Palace Plain, Norfolk (Locker 1987)					*	*		*	*	*	*	*		*	*	*	*			*
George Street, Aylesbury (Jones 1983)							*							*	*			*		*
Alms Lane, Norfolk (Jones and Scott 1995)																				
General Accident, Yorkshire (O'Connor 1988)	*	*	*	*	*	*	*	*	*		*	*			*	*	*			*
Lincoln city, Lincolnshire (Dobney *et al.* 1995)	*	*	*	*	*	*			*								*			
Presence/Diversity Index	**31**				**34.9**							**39.5**								
mid-12th - mid-14th Elite (n=7)																				
Guildford Castle, Surrey (Sykes et al. 2005)		*			*	*			*	*		*		*	*	*	*			*
Thrislington, Durham (Rackham 1989)		*			*							*				*	*	*		
Castle Mall, Norfolk (Locker 1997)		*		*	*	*		*	*	*		*		*	*	*	*	*		*
Middleton Stoney (Levitan 1984)		*			*							*		*						
Saxon Palaces, Northamptonshire (Harman, 1985b)				*		*				*				*	*	*	*	*		
Trowbridge, Wiltshire (Bourdillon 1994)		*			*	*						*		*						
Launceston Castle, Cornwall (Smith 1995)					*										*	*		*	*	
Presence/Diversity Index	**25**				**33**							**47.6**								

Appendix Idii: Presence/diversity of freshwater, migratory and marine fish by site type and date.

	Freshwater				Migratory							Marine								
Site	roach	cyprinid	perch	pike	eel	salmon	trout	smelt	flatfish	plaice	flounder	herring	sprat	mackrel	cod	haddock	gadid	ling	hake	whiting
mid-12th - mid-14th Religious (n=9)																				
Battle Abbey, Sussex (Locker 1985)										*					*	*	*			
Eynsham Abbey, Oxfordshire (Ayres *et al*. 2003)	*	*	*	*	*				*	*		*			*	*	*	*		*
St Mary's, Clerkwell (Locker 1996)					*				*	*	*	*			*	*				*
Cleeve Abbey, Somerset (Locker 1989)					*					*		*		*		*	*		*	*
Dean Court, Oxfordshire (Jones 1994)	*	*		*	*				*	*		*					*			*
Shrewsbury Abbey (Jones 2002)				*					*						*	*	*			
Witney, Oxfordshire (Ayres and Serjeantson 2002)				*	*				*			*			*					
Austin Friars, Leicester (Thawley 1981)					*	*			*	*					*	*		*		
Fishergate, Yorkshire 6a (O'Connor 1991)		*	*	*	*	*	*	*	*		*	*			*	*	*			*
Presence/Diversity Index	**33**				**41.2**							**43**								
Total mid-12th - mid-14th Presence DiversityIndex	*28.8*				*35.7*							*41.4*								

Appendix Idiii: Representation and presence/diversity of fish taxa in assemblages from northern France (See Table 3 for references).

	eel	herring	mackrel	cod	haddock	gadid	whiting	plaice	flounder	flatfish	sturgeon	roach	dace	chub	tench	cyprinid	carp	salmon	perch	trout	pike	Presence	P/D Index
Rural (n=2)																							
Dury, Somme, 11th-12th										*													
Noyon 12-13th century																							
% Representation										**100**												**1 of 21**	**4.7**
Urban (n=9)																							
l'Hoptial general 9th	*	*					*			*		*									*		
Compiegne 9th																							
Compiegne, 9th-10th										*													
Compiegne, Oise Phase 10th											*												
Saint-Germain, Amiens 10-12th	*	*					*	*		*										*			
Saint-Denis, layer 856 (10-13th)	*												*	*					*	*	*		
Compeigne, Oise Phase 11th						*				*	*												
Compeigne, Oise 11th/12th						*				*	*												
Compeigne, 12th										*	*												
% Representation	**33**	**22**				**22**	**22**	**11**		**67**	**44**	**11**	**11**	**11**					**11**	**22**	**22**	**28 of 189**	**15**
Elite (n=5)																							
Compiegne, Oise, 2453, 9th-10th											*												
Grand Besle, Normandy 11-12th				*												*							
Compiegne, Oise Phase 11-12th										*	*												
L'eveche de Senlis 11th-12th	*					*	*	*			*					*					*		
Compeigne, Oise US 12th										*	*												
% By Catergory	**20**			**20**		**20**	**20**	**20**		**40**	**80**					**40**					**20**	**14 of 105**	**13**
Religious House (n=3)																							
Charite-sur-Loire, Nievre 11-12th	*															*				*	*		
Saint Avit-Seineur														*	*								
Compeigne, Oise US1505						*				*													
% Representation	**33**					**33**				**33**				**33**	**33**	**33**				**33**	**33**	**8 of 63**	**13**
Rural (n=3)																							
Noyon 12-13th century																							
Noyon 14th century																							
Coquelles 14th century				*		*				*													
% of sites on which represented				**25**		**25**				**50**												**3 of 63**	**4.7**
Urban (n=8)																							
Restaurant Universitaire, Amiens 13th	*	*	*				*	*	*			*				*							
Rue de Collegiale 13th century	*	*	*																*				
Rank Xeros, Amiens 14th	*	*	*				*	*	*	*						*			*				
Abbville 14th century	*	*	*	*	*	*	*	*	*	*		*		*		*			*		*		
Saint Michel, Paris 14th century			*	*	*		*	*		*						*	*				*		

Appendix Idiii: Representation and presence/diversity of fish taxa in assemblages from northern France (See Table 3 for references).

	eel	herring	mackrel	cod	haddock	gadid	whiting	plaice	flounder	flatfish	sturgeon	roach	dace	chub	tench	cyprinid	carp	salmon	perch	trout	pike	**Presence**	**P/D Index**
l'Hoptial general 14th		*					*	*				*				*							
ZAC Cathederal, Amiens 14th century	*	*		*		*	*	*		*						*							
Saint-Germain, Amiens 14th	*	*	*				*	*	*	*											*		
% of sites on which represented	**53**	**53**	**35**	**18**	**12**	**24**	**53**	**47**	**24**	**65**	**24**	**24**	**6**	**12**		**35**	**6**		**24**	**12**	**29**	**66 of 168**	**39**
Elite (n=4)																							
Cour Carree, Paris 13th century	*	*	*	*			*	*		*					*	*	*		*	*	*		
Cour Carree, Paris 14th century		*		*			*			*							*						
Vincennes 14th	*	*		*	*	*	*	*	*	*	*	*			*	*	*	*	*		*		
Chateau de Courtrai 14th century	*	*	*	*	*	*	*	*	*	*		*				*	*		*	*	*		
% of sites on which represented	**44**	**44**	**22**	**55**	**22**	**33**	**55**	**44**	**22**	**66**	**55**	**22**			**22**	**55**	**44**	**11**	**33**	**22**	**44**	**51 of 84**	**61**
Religious House (n=1)																							
Moreaucourt late 14th century	*	*	*	*	*		*	*	*							*	*		*		*		
% of sites on which represented	**40**	**20**	**20**	**20**	**20**	**20**	**20**	**20**	**20**	**20**				**20**	**20**	**40**	**20**		**20**	**20**	**40**	**12 of 21**	**57**

Appendix Ie: Representation of wild bird taxa by site type and date (French and British sites).

Site	Wild Birds																
	duck	goose	swan	heron	crane	bittern	curlew	pigeon	peafowl	woodcock	partridge	pheasant	plover	lapwing	snipe	godwit	small bird
5th - mid-9th Rural																	
Lot's Hole, Berkshire								*									
Pennylands, Buckinghamshire								*									*
Eynesbury, Cambridgeshire					*								*				
Mucking, Essex													*				
Lechdale Gloucestershire					*												*
Barton Court Farm, Oxfordshire	*	*						*					*				*
West Stow, Suffolk	*	*	*	*	*					*			*	*	*		*
Cadley Road, Wiltshire					*												*
% of sites on which represented	**25**	**25**	**13**	**13**	**50**			**38**		**13**			**50**	**13**	**13**		**63**
5th - mid-9th Urban																	
Hamwic, Hampshire	*									*							*
Sandtun, Kent			*		*			*					*	*			
Maiden Lane, London	*	*															
Ipswich, Suffolk			*					*									
Ipswich, Suffolk	*				*			*									*
Fishergate, Yorkshire								*				*					*
% of sites on which represented	**50**	**17**	**33**		**33**			**67**		**17**		**17**	**17**	**17**			**50**
5th - mid-9th Elite																	
Lake End Road, Essex								*		*			*				*
Wicken Bonhunt, Essex	*	*	*		*	*		*	*				*				
Portchester Castle, Hampshire						*	*	*			*						
Flixborough, Lincolnshire			*	*	*												
Caister-on-Sea, Norfolk	*	*	*					*									
Saxon Palaces, Northampton		*															
Eynsham, Oxfordshire					*			*						*			*
Wroxeter, Shropshire	*	*	*	*				*		*	*		*				*
Brandon, Suffolk	*		*		*	*		*									*
Ramsbury, Wiltshire															*		
% of sites on which represented	**40**	**40**	**50**	**20**	**40**	**30**	**10**	**70**	**10**	**20**	**20**		**30**	**10**	**10**		**40**
5th - mid-9th Religious																	
Hartlepool, Cleeveland								*		*							*
Morrison Hall, Cleeveland								*									
St Albans, Hertfordshire					*								*				
Holy Island					*												
North Elmham, Norfolk	*	*			*		*	*					*				
% of sites on which represented	**20**	**20**			**60**		**20**	**60**		**20**			**40**				**20**

Appendix Ie: Representation of wild bird taxa by site type and date (French and British sites).

Site	Wild Birds																
	duck	goose	swan	heron	crane	bittern	curlew	pigeon	peafowl	woodcock	partridge	pheasant	plover	lapwing	snipe	godwit	small bird
mid-9th - mid-11th Rural																	
Walton Vicarage, Buckinghamshire					*								*				
West Cotton, Northamptonshire								*									
% of sites on which represented					**50**			**50**					**50**				
mid-9th - mid-11th Urban																	
Newbury, Berkshire																	
Winchester, Hampshire		*								*							*
Lincoln City, Lincolnshire	*	*								*		*					
Thetford 1092, Norfolk	*				*				*								
Castle Mall, Nofolk								*									
Mill Lane, Thetford, Norfolk	*																
St Peters Street, Northamptonshire					*												
Chalk Lane, Northamptonshire		*													*		
Shrewsbury, Shropshire										*							
Ipswich, Suffolk																	
Steyning, Sussex					*			*									
Coppergate, Yorkshire		*	*		*			*		*			*				
Lurk Lane, Yorkshire																	
% of sites on which represented	**23**	**31**	**8**		**30**			**23**	**8**	**23**		**8**	**8**		**8**		**8**
mid-9th - mid-11th Elite																	
Wraysbury, Berkshire	*							*		*	*		*	*			*
Portchester Castle, Hampshire	*						*	*					*			*	
Faccombe Netherton, Hampshire								*			*						*
Saxon Palaces, Northampton	*		*					*									
Trowbridge, Wiltshire										*							
% of sites on which represented	**60**		**20**				**20**	**80**		**40**	**40**		**40**	**20**		**20**	**40**
mid-9th - mid-11th Religious																	
Westminster Abbey, London	*	*			*			*		*					*		*
St Albans, Hertfordshire					*					*							
North Elmham, Norfolk	*				*			*									*
Jarrow, Northumberland																	*
Eynsham Abbey, Oxfordshire							*	*		*				*	*		
Lewes Priory, Sussex			*														
% of sites on which represented	**33**	**17**	**17**		**50**		**17**	**50**		**50**				**17**	**33**		**50**
mid-11th - mid-12th Rural																	
Walton Vicarcage, Buckinghamshire	*										*		*				
Brighton Hill South, Hampshire																	
West Cotton, Northampton								*									*
Boltophs Bramber, Sussex								*									
% of sites on which represented	**25**							**50**			**25**		**25**				**25**

Appendix Ie: Representation of wild bird taxa by site type and date (French and British sites).

Site	Wild Birds																
	duck	goose	swan	heron	crane	bittern	curlew	pigeon	peafowl	woodcock	partridge	pheasant	plover	lapwing	snipe	godwit	small bird
mid-11th - mid-12th Urban																	
Newbury, Berkshire										*							
Bartholomew Street, Newbury, Berkshire								*									
Exeter, Devon							*	*		*							*
Lincoln City, Lincolnshire	*	*						*									
Causeway Lane, Lincolnshire	*									*			*				*
Brandon Road, Norfolk		*			*			*									
Kings Lynn, Norfolk	*	*					*	*			*		*	*		*	*
St Nicholas Street, Norfolk	*										*	*				*	*
Fishergate, Norfolk	*									*							
North Street, Sussex					*			*		*							*
Coppergate 5Cf 1 and 2, Yorkshire		*															
General Accident, Yorkshire	*									*			*				
Fishergate, Yorkshire								*					*				*
Eastgate, Yorkshire	*	*	*	*	*			*	*	*		*					*
% of sites on which represented	**50**	**36**	**7**	**7**	**21**		**14**	**57**	**7**	**50**	**14**	**14**	**4**	**7**		**14**	**50**
mid-11th - mid-12th Elite																	
Launceston Castle, Cornwall										*	*						*
Portchester Castle, Hampshire	*																
Faccombe Netherton, Hamphire			*	*				*		*	*		*		*		*
Carisbrooke Castle, I.O.W	*							*	*								
North Elmham, Norfolk	*							*									
Castle Mall, Norfolk	*		*	*				*									*
Castle Mall, Norfolk				*				*	*		*		*		*		*
Castle Rising Castle, Norfolk	*		*	*	*			*					*				*
Saxon Palaces, Northampton													*				
Wirral Park, Somerset	*		*														
Guildford Castle, Surrey										*							*
Wharram South Manor, Yorkshire			*														
Dudley Castle, West Midlands	*			*						*							
Ludgershall Castle, Wiltshire	*							*	*	*	*		*		*		*
Trowbridge, Wiltshire	*							*		*					*		
% of sites on which represented	**60**		**33**	**33**	**7**			**53**	**20**	**36**	**13**		**33**		**13**		**47**

Appendix Ie: Representation of wild bird taxa by site type and date (French and British sites).

Site	Wild Birds																
	duck	goose	swan	heron	crane	bittern	curlew	pigeon	peafowl	woodcock	partridge	pheasant	plover	lapwing	snipe	godwit	small bird
mid-11th - mid-12th Religious																	
St Albans, Herefordshire	*	*		*	*			*		*							*
Jarrow, Northumberland								*						*			*
Eynsham Abbey, Oxfordshire	*	*			*			*		*			*		*		*
Osney Abbey, Oxfordshire			*														
Lurk Lane, Yorkshire			*	*				*		*			*				*
% of sites on which represented	**40**	**40**	**40**	**40**	**40**			**80**		**60**			**40**	**20**	**20**		**80**
mid-12th - mid-14th Rural																	
Eckweek, Avon																	
Walton Vicarage, Buckinghamshire					*			*									
Rivenhall, Chelmsford, Essex								*						*			
Grenstein, Norfolk	*																
West Cotton, Northamptonshire								*						*	*		*
Boetlers Castle, Leicestershire								*									
% of sites on which represented	**17**				**17**			**67**						**33**	**17**		**17**
mid-12th - mid-14th Urban																	
Newbury, Berkshire	*																
Bartholomew Street, Newbury, Berkshire										*							
The Lanes, Cumbria			*														
Exeter, Devon	*							*		*							*
Greyhounds Yard, Dorchester, Dorest	*							*		*			*				
West Gate, Gloucestershire								*									
Winchcombe, Gloucestershire																	
Southampton, Hampshire	*				*		*		*	*				*			
Hereford city, Herefordshire	*								*	*			*				*
Little Lane, Leicester										*							
Lincoln city, Lincolnshire	*																
Closegate I and II, Newcastle										*							*
Coslany Street, Norfolk	*									*							
Alms Lane, Norwich, Norfolk	*		*								*		*				*
Brandon Road, Norfolk					*												
Kings Lynn, Norfolk	*	*	*	*			*						*			*	
Marefair, Northamptonshire	*			*						*							
Benhams Garage, Somerset			*														
Marlipins, Sussex								*			*						
General Accident, Yorkshire	*	*			*									*			*
Lurk Lane, Yorkshire		*	*	*						*							
% of sites on which represented	**52**	**14**	**24**	**14**	**14**		**9**	**19**	**9**	**48**	**9**		**19**	**9**		**5**	**24**

Appendix Ie: Representation of wild bird taxa by site type and date (French and British sites).

Site	Wild Birds																
	duck	goose	swan	heron	crane	bittern	curlew	pigeon	peafowl	woodcock	partridge	pheasant	plover	lapwing	snipe	godwit	small bird
mid-12th - mid-14th Elite																	
Launceston Castle, Cornwall			*		*			*		*	*		*	*	*		
Oakhampton, Devon	*				*		*	*					*		*		*
Thrislington, County Durham								*			*		*				*
Hadleigh Castle, Essex								*			*	*					
Brighton Hill South, Hampshire			*								*						
Portchester Castle, Hampshire	*						*	*		*	*			*			*
Baynard's Castle, London	*		*					*		*	*				*		*
North Elmham, Norfolk	*							*					*				*
Castle Mall, Norfolk	*						*				*						
Castle Rising, Norfolk	*		*	*			*	*	*	*			*				*
Stafford Castle, Staffordshire	*		*	*	*	*		*		*	*	*	*	*	*	*	*
Guildford Castle, Surrey			*					*		*	*		*	*	*		
Dudley Castle, West Midlands	*		*	*				*		*							
% of sites on which represented	**62**		**54**	**23**	**23**	**7**	**31**	**85**	**7**	**54**	**70**	**15**	**54**	**31**	**39**	**7**	**54**
mid-12th - mid-14th Religious																	
Dominican Friary, Chester					*					*							*
Cristchurch, Dorset	*		*	*													
St Gregory's Priory, Kent	*							*		*							*
Austin Friars, Leicestershire				*	*			*			*						*
Jarrow, Northumberland								*		*			*				
Eynsham Abbey, Oxfordshire	*		*					*		*	*		*	*			
Witney, Oxfordshire			*														*
Dean Court, Oxfordshire								*									
Shrewsbury Abbey, Shropshire										*	*						
Wells Museum, Somerset	*		*					*	*	*						*	
Battle Abbey, Sussex								*									
Fishergate, Yorkshire	*							*					*				*
% of sites on which represented	**42**		**33**	**17**	**17**			**67**	**8**	**50**	**25**		**25**	**8**		**8**	**42**

Appendix Ie: Representation of wild bird taxa by site type and date (French and British sites).

Site	Wild Birds																
	duck	goose	swan	heron	crane	bittern	curlew	pigeon	peafowl	woodcock	partridge	pheasant	plover	lapwing	snipe	godwit	small bird
French Rural (n=1)																	
Argentan											*						
French Urban (n= 26)																	
Saint-Denis, layer 953 (6-10th)												*					
Saint-Denis, layer 837 (8-11th)												*					
l'Hoptial general 9th century																	
Rue de Lutece, Paris. 9th century		*		*				*									
Saint-Denis, layer 937 (9-11th)												*					
Saint-Denis, layer 14779 (10th)											*	*					
Saint-Denis, layer 960 (10-11th)												*					
Saint-Denis, layer 856 (10-12th)												*					
Saint-Germain, Amiens 10-12th																	*
Saint-Denis, layer 828 (11-12th)												*					
Saint-Denis, layer 783 (12-13th)			*					*			*	*					
Saint-Denis, layer 956 (12-13th)								*			*						
Rue de Lutece, Paris. 12th century				*		*		*			*						
Rue de Lutece, Paris. 12th/13th	*							*			*						
Rue de Collegiale 13th century											*						
Abbville 14th century	*									*	*						*
Saint Michel, Paris 14th century																	
l'Hoptial general 14th																	
ZAC Cathederal, Amiens 14th century																	
Saint-Germain, Amiens 14th																	
Compeigne, Oise Phase IV	*	*								*							*
Compeigne, Oise Phase V 10th		*	*		*												*
Compeigne, Oise Phase VI 11th	*										*						
Compeigne, Oise US 1622									*	*							
Compeigne, Oise US 1460			*								*						*
% of sites on which represented	**15**	**12**	**12**	**8**	**4**	**4**		**19**	**4**	**12**	**35**	**31**					**19**
French Elite (n=16)																	
l'eveche de Senlis, 11th-12th	*		*	*	*	*		*	*	*	*		*	*			*
Vatteville-la-Rue, Seine Maritime 11-12th	*		*	*				*		*	*						
Vatteville-la-Rue, Seine Maritime 13th										*							
Vatteville-la-Rue, Seine Maritime 14th												*					
Vatteville-la-Rue, Seine Maritime 12-15th	*		*	*			*	*	*	*	*	*	*	*		*	*
Cour Carree, Paris 13th century										*	*						*
Chateaude Mayenne 13-14th																	
Cour Carree, Paris 14th century											*						*
Rue Fromenteau, Paris 14th century	*			*	*			*			*						
Vincennes 14th											*						*

Appendix Ie: Representation of wild bird taxa by site type and date (French and British sites).

Site	Wild Birds																
	duck	goose	swan	heron	crane	bittern	curlew	pigeon	peafowl	woodcock	partridge	pheasant	plover	lapwing	snipe	godwit	small bird
Chateau de Courtrai 14th century			*								*						*
Grand Besle, Normandy				*						*	*						
Compeigne, Oise U2453 9-10th					*				*								*
Compeigne, Oise Phase VII nivellement					*												
Compeigne, Oise US 1056					*				*								
Douai, Nord	*			*	*	*		*		*	*			*	*		*
% of sites on which represented	**31**		**31**	**38**	**38**	**13**	**6**	**29**	**25**	**44**	**63**		**13**	**19**	**6**	**6**	**50**
French Religious House (n= 5)																	
Gensac-sur-Garonne, 8th-9th											*						
Moreaucourt late 14th century								*									
Compiegne, Oise US1505					*			*			*						*
Saint Avit-Senieur, Dordogne								*		*	*						
Charite-sur-Loire, Nievre	*			*	*			*		*	*	*					*
% of sites on which represented	**20**			**20**	**40**			**80**		**20**	**80**	**20**					**40**

Appendix IIa: Dental ageing for cattle, by site type and period.

N.B. The data for some sites (highlighted by *) are shown as percentages.
For wear stage correlations see Table 5.

	0-6 mths	6-12 mths	1-2 yrs	2-3 yrs	3-6 yrs	6-8 yrs	8-10 yrs +	Total
5th - mid-9th Rural								
Chichley, Buckinghamshire		3						3
Wolverton Turn, Buckinghamshire	1	2	1	1	2	1		8
Pennyland, Buckinghamshire	6	6	15	2	2	2	1	34
Hartigans, Buckinghamshire			1	1	2	1		5
Eynesbury, Cambridgeshire		1	2	1	1		1	6
Mucking, Essex			2	2	1			5
Riby Cross Roads, Lincolnshire		3		1	1			5
Netleton, Lincolnshire		1	1	7				9
West Cotton, Northamptonshire		3						3
Barton Court Farm, Oxfordshire	3	4	2	2				11
Eynsham, Oxfordshire	1	4	1	1		1	1	9
Yarnton, Oxfordshire	3	3	1	2	1		1	11
Yarnton (Cresswell Field), Oxfordshire	1			1	2			4
Yarnton (Worton), Oxforshire		3	1		1	1	1	7
West Stow, Suffolk	13	14	26	19	10	15	3	100*
Wharram South Manor, Yorkshire	1		4	3		6	5	19
Total	**29**	**47**	**57**	**43**	**23**	**27**	**13**	**239**
%	12	20	24	18	10	11	5	100
5th - mid-9th Urban								
Hamwic, Hampshire	4	12	24	39	19	1	1	100*
Marefair, Northamptonshire		1		3	1	4		9
Middle Saxon Southampton, Hampshire			4	8	3	3		18
Ipswich, Suffolk	5	8	18	9	10	4	1	55
Ipwsich, Suffolk		15	27	13	30	10	5	100*
Fishergate, Yorkshire	2	2	6	1	5	3	2	21
Total	**11**	**38**	**79**	**73**	**68**	**25**	**9**	**303**
%	4	13	26	24	22	8	3	100
5th - mid-9th Elite								
Wicken Bonhunt, Essex	1	6	6	12	37	28	10	100*
Lake End Road, Essex	7	14	29	3	26	11	10	100*
Eynsham 2b, Oxfordshire		1			5	1	1	8
Brandon, Suffolk	1	12	7	12	30	28	10	100*
Boltophs Bramber, Sussex		1	4	4	1			10
Total	**9**	**33**	**42**	**27**	**98**	**68**	**31**	**308**
%	3	11	14	10	31	21	10	100
L9th-M11th century Rural								
West Cotton, Northamptonshire	2	3	1			2		8
Wharram Percy, Yorkshire	3	5	1			1	1	11
Wharram South Manor, Yorkshire		1	2	2	1	2	3	11
Total	**5**	**9**	**4**	**2**	**1**	**4**	**3**	**28**
%	18	32	14	7	4	14	11	100
L9th-M11th century Urban								
Winchester, Hampshire			4	6	3	3		16
Quilters Vault, Hampshire					2			2
Flaxengate, Lincolnshire	1	2	3	17	25	10	5	63
Thetford 1092, Norfolk								
Mill Lane, Thetford, Norfolk			4	8	8	8	2	30
Brandon Road, Norfolk		1	9	6	5	12	2	35
Castle Mall, Norfolk	2		5	7	5	7	1	27
Fishergate, Norfolk	1	1	6	9	1	11		29
Marefair, Northamptonshire				1		4		5
St Peters Street, Northamptonshire	2	1	4	9	6	11		33
St Ebbes, Oxfordshire	2	1	3	1	3			10
Codrington, Oxfordshire			1					1
Ipswich, Suffolk		1	1		1	1		4
Steyning, Sussex								
Malmsbury, Wiltshire		1	1					2
Eastgate, Yorkshire	1		1	2				4
Coppergate, Yorkshire	4	11	44	90	11	10	2	172
Total	**13**	**19**	**86**	**156**	**70**	**77**	**12**	**433**
%	3	4	20	36	16	18	3	100

Appendix IIa: Dental ageing for cattle, by site type and period.

N.B. The data for some sites (highlighted by *) are shown as percentages.
For wear stage correlations see Table 5.

	0-6 mths	6-12 mths	1-2 yrs	2-3 yrs	3-6 yrs	6-8 yrs	8-10 yrs +	Total
L9th-M11th century Elite								
Portchester Castle, Hampshire	1	1	18	15	10	21	16	82
Faccombe Netherton, Hampshire	1		1	3	1	3		9
Eynsham, Oxfordshire	1			5		2	2	10
Eynsham 2d, Oxfordshire	1		1		4	1		7
Eynsham 2e, Oxfordshire			1		2	2	1	6
Trowbridge, Wiltshire			5	1	1	1	1	9
Total	**4**	**1**	**26**	**24**	**18**	**30**	**20**	**123**
%	3	1	21	20	15	24	16	100
mid-11th - mid-12th Rural								
Walton Road, Buckinghamshire			4	4	1	2		
Pepperhill Lane, Kent							1	
Total			**4**	**4**	**1**	**2**	**1**	**12**
%			33	34	8	17	8	100
mid-11th - mid-12th Urban								
Flaxengate, Lincolnshire	1	1	1	6	18	8		35
St Nicholas Street, Norfolk								
Brandon Road, Norfolk		2	4	5	6	8		25
Redcastle Furze, Norfolk								
Marefair, Northamptonshire			1	2		4	1	8
St Ebbes, Oxfordshire		2	4	2	1			9
Ipswich, Suffolk		1	1		1			3
Steyning, Sussex								
Malmesbury, Wiltshire		1	1					2
Coppergate, Yorkshie			3	3	1	1		8
General Accident, Yorkshire			1	2				3
Fishergate, Yorkshire								
Eastgate, Yorkshire				1				1
Total	**1**	**7**	**16**	**21**	**27**	**21**	**1**	**94**
%	1	7	17	23	29	22	1	100
mid-11th - mid-12th Elite								
Portchester Castle, Hampshire				1	5			6
Faccombe Netherton, Hampshire		1	1	9	6	10	2	29
Castle Rising castle, Norfolk					6	1		7
Castle Mall, Norfolk			11	6	7	16	3	43
Eynsham, Oxford		1	3	11	22	18	42	97
Dudley Castle, West Midlands			1					1
Trowbridge, Wiltshire	1	1	1	3	1			7
Yatesbury, Wiltshire								
Wharram South Manor, Yorkshire	2	1	3	1		5		12
Total	**3**	**4**	**20**	**31**	**47**	**50**	**49**	**204**
%	1	2	10	15	23	25	24	100
mid-12th - mid-14th Rural								
West Cotton, Northamptonshire	3	17	4	12	12	29	1	78
Boetlers Castle, Leicestershire	2	4	3				2	11
Total	**5**	**21**	**7**	**12**	**12**	**29**	**3**	**89**
%	6	24	8	13	13	33	3	100

Appendix IIa: Dental ageing for cattle, by site type and period.

N.B. The data for some sites (highlighted by *) are shown as percentages.
For wear stage correlations see Table 5.

mid-12th - mid-14th Urban								
Quilters Vault, Hampshire			2	3	6	1		12
Droitwich, Hereford					3			3
Flaxengate, Lincolnshire	2		1	2	4	1		10
Heigham St, Norfolk					1	1	1	3
Brandon Road, Norfolk	1	1	6	5		5		18
St Peters Street, Northampton	3	2	2	2	4	11		24
St Ebbes, Oxfordshire	4		9	3	5			21
Codrington Library, Oxfordshire				1				1
George Street, Oxfordshire	2	3	1	1	2	1		10
The Hamel, Oxfordshire								
Dean Court, Oxfordshire	1	2			2	1		6
Bell St Reigate, Surrey			1	2		2		5
General Accident, Yorkshire	1	1	4	13	12	11	4	46
Eastgate, Yorkshire	4	2	1	2				9
Total	**12**	**8**	**16**	**22**	**21**	**15**	**4**	**98**
%	11	7	16	20	23	20	3	100
	0-6 mths	**6-12 mths**	**1-2 yrs**	**2-3 yrs**	**3-6 yrs**	**6-8 yrs**	**8-10 yrs +**	**Total**
mid-12th - mid-14th Elite								
Faccombe Netherton, Hampshire	16	1	2	6	4	12	2	43
Portchester Castle, Hampshire	1		1	8	10	5	2	27
Canterbury Cathedral, Kent								
Castle Mall, Norfolk				1	4	1	2	8
Castle Rising Castle, Norfolk	1				2	3	2	8
Eynsham, Oxfordshire	1	2	9	18	7	5	7	49
Witney Palace, Oxfordhsire	2							2
Guildford Castle, Surrey								
Dudley Castle, West Midlands	4		1	6	9	17	3	40
Trowbridge, Wiltshire					1			1
Total	**8**	**2**	**10**	**25**	**23**	**26**	**14**	**108**
%	14	2	7	22	21	24	10	100
French Cattle								
Dury (Rural)			1	4	1	3		9
Compeigne 9-10th (Urban)				3	2	3	4	12
Compeigne 11-12th (Urban)		1	1	6	1	4	5	18
Vatteville (Elite)	3		1	1	4	1	1	11
Total	**3**	**1**	**3**	**14**	**8**	**11**	**10**	50
%	6	2	6	28	16	22	20	100

Appendix IIb: Dental ageing data for caprines by site type and date.

N.B. the data for some assemblages (highlighted *) are shown as percentages.
For wear stage correlations see Table 5

	0-6mths	6-12mths	1-2 yrs	2-3 yrs	3-4 yrs	4-6 yrs	6-8 yrs	8-10yrs	Total
5th - mid-9th Rural									
Wolverton Turn, Buckinghamshire	3	4	15	10	5	7		1	45
Hartigans, Buckinghamshire					1	1			2
Pennylands, Buckinghamshire	4	30	23	9	11	8	8		93
Chichley, Buckinghamshire			3	1	1	2			7
Eynesbury, Cambridgeshire	2	3	3	1	2	4	2	2	19
Mucking, Essex	1	3			1	2	1		8
Old Downe Farm, Hampshire	3	1	1	1					6
Riby Cross Roads, Lincolnshire	3	2	1		2			1	9
Riby Cross Roads, Lincolnshire	2	3	2	1	1	1			10
Nettleton, Lincolnshire		1	2		2	1			6
West Cotton, Northamptonshire			1	1					2
Eynsham, Oxfordshire	2	4	5	3	4				18
Barton Court Farm, Oxforshire	4	15	20	7	12	5	1	1	65
Yarnton, Oxfordshire		3	4	5	3	4	1		20
Yanton (Cresswell Field), Oxfordshire			1			3	1		5
Yarnton (Worton), Oxforshire					1				1
West Stow, Suffolk	7	37	18	7	11	17	1	2	100*
Cadley Road, Wiltshire	3	18	7	1	2		6	6	43
Wharram South Manor, Yorkshire	11	17	12	12	19	10	8		89
Total	**45**	**141**	**118**	**59**	**78**	**65**	**29**	**13**	**548**
%	8	26	22	11	14	12	5.0	2	100
5th - mid-9th Urban									
Hamwic, Hampshire	3	12	15	36	16	16	1	1	100*
Middle Saxon Southampton, Hampshire		3	5	18	5	7	2		40
Redcastle Furze, Norfolk	3	12	12	1	12	4			44
Marefair, Northamptonshire			4	3	2	4			13
St Peters Street, Northamptonshire		2	1	1	2	3	1		10
Ipswich, Suffolk		15	25	6	7	30	14	3	100*
Total	**6**	**17**	**30**	**10**	**11**	**37**	**15**	**3**	129
%	2	14	20	22	14	21	6	1	**100**
5th - mid-9th Elite									
Wicken Bonhunt, Essex	1	6	6	6	6	50	13	10	98
Lake End Road, Essex	1	16	16	14	20	7	6	6	86
Brandon, Norfolk	1	10	7	6	6	50	15	5	100*
Eynsham, Oxfordshire	3	13	6	1	1				24
Boltophs Bramber, Sussex	1	1	1	1				1	5
Total	**4**	**23**	**13**	**7**	**7**	**50**	**15**	**5**	**124**
%	2	15	12	9	10	34	11	7	100
L9th-M11th century Rural									
West Cotton, Northamptonshire		1		2	2				5
Wharram South Manor, Yorkshire	5		4	13	13	5	1	1	42
Wharram Percy, Yorkshire	4	9	9	4	5	6			37
Total	**9**	**10**	**13**	**19**	**20**	**11**	**1**	**1**	**84**
%	11	12	15	23	24	13	1	1	100
L9th-M11th century Urban									
Winchester, Hampshire	9	1	4	14	2	10	5	2	47
Quilters Vault, Hampshire	2		2	2	1	2			9
Marefair, Northamptonshire						6			6
Lincoln, Lincolnshire			4	8	7	5	2		26
St Magnus, London		3	1	1	3	1			9
Mill Lane, Thetford, Norfolk	2		5	11	6	6			30
Fishergate, Norfolk	1		4		3	1			9
Castle Mall, Norfolk	4	2	8	11	2	4	1		32
Brandon Road, Norfolk	2	2	16	15	15	11	2		63
Alms Lane, Norfolk							2		2
St Peters Street, Northamptonshire		2	1	8	12	23	2		48
St Ebbes, Oxford	9	3	41	20	28	3			104
Codrington Library, Oxfordshire					1				1
Ipswich, Suffolk	10	38	23	10	11	5	2	1	100*
Flaxengate, Yorkshire	3	2	19	20	10	10	2		66
Coppergate, Yorkshire	4	8	21	25	28	12	1		99
Eastgate, Yorkshire			2	1	1				4
Malmsbury, Wiltshire	1		2	2	3	1			9
Total	**47**	**61**	**153**	**148**	**132**	**100**	**19**	**3**	**663**

Appendix IIb: Dental ageing data for caprines by site type and date.

%	7	9	23	22	20	15	3	1	100
	0-6mths	**6-12mths**	**1-2 yrs**	**2-3 yrs**	**3-4 yrs**	**4-6 yrs**	**6-8 yrs**	**8-10yrs**	**Total**
L9th-M11th century Elite									
Portchester Castle, Hampshire	3	12	16	14	22	31	1	1	100
Faccombe Netherton, Hampshire	3	8	2	11	4	8	1		37
Eynsham 2d, Oxfordshire	1		3	3	4	3	3	1	18
Eynsham 2e, Oxfordshire		1	3		3		1		8
Eynsham, Oxfordshire	1	5	2	1	5		3	1	18
Trowbridge, Wiltshire		3	2	4	4	2	2		17
Total	**8**	**9**	**10**	**8**	**16**	**5**	**9**	**2**	**67**
%	4	15	14	17	21	22	5	2	100
mid-11th - mid-12th Rural									
Walton Road, Buckinghamshire	1	3	6	9	2	2	1		24
Marefair, Northamptonshire					2	1			3
Boltophs Bramber, Sussex				1	2	1	1	2	7
Total	**1**	**0**	**0**	**1**	**4**	**2**	**1**	**2**	**11**
%	3	9	18	29	18	12	6	6	100
mid-11th - mid-12th Urban									
Bath, Avon	6	3	7	9	5	1			31
Exeter, Devon	4	5	16	8	1	1			35
St Magnus, London			2	1	1	3			7
Ipswich, Norfolk		1	1	1		1	1		5
St Nicholas Street, Norfolk		1	7		3	6	8	2	27
Brandon Road, Norfolk	1	1	20	18	18	21	5	1	85
Redcastle Furze, Norfolk	1	1	14	5	16	15	14		66
St Ebbes, Oxfordshire	4	4	20	16	13	2			59
The Hamel, Oxfordshire		8	12	15	16	7	3		61
Codrington Library, Oxfordshire			3	5	1	6	2		17
Malmesbury, Wiltshire	1	1	1	1		3	1		8
Flaxengate, Yorkshire	12	6	53	17	17	60	30	3	198
Eastgate, Yorkshire	2	2	5	8	6	5	2	1	31
Coppergate, Yorkshire			3	4	1				8
Total	**21**	**25**	**136**	**86**	**90**	**126**	**66**	**7**	**557**
%	5	5	26	17	15	21	10	1	100
mid-11th - mid-12th Elite									
Portchester Castle, Hampshire	1		1			1			3
Faccombe Netherton, Hampshire	2	2	1	5	3	2	1		16
Castle Rising Castle, Norfolk				1	3	1	1		6
Castle Mall, Norfolk	3	2	8	14	7	8	1	1	44
Eynsham, Oxfordshire	2	4	17	6	9		8	11	57
Middleton Stony, Oxfordshire			1						1
Guildford Castle, Surrey									
Trowbridge, Wiltshire		2	5	5					12
Yatesbury, Wiltshire	2	4	10	3	5	12	2	4	42
Wharram South Manor, Yorshire	8	4	6	8	14	10	5	3	58
Total	**8**	**8**	**28**	**26**	**22**	**12**	**11**	**12**	**127**
%	8	8	20	18	17	14	8	8	100
mid-12th - mid-14th Rural									
Walton Road, Buckinghamshire				1					
Boetlers Castle, Leicestershire	1	2	15	12	2	2	2		
West Cotton, Northamptonshire	1	16	22	14	16	18	11	1	
Total	**2**	**18**	**37**	**27**	**18**	**20**	**13**	**1**	**136**
%	1	13	27	20	13	15	10	1	100
mid-12th - mid-14th Urban									
Exeter, Devon	9	7	32	23	20	7	2		100*
Quilters Vault, Hampshire	1	11	8	7	6	12	1		46
Droitwich, Herefordshire			3	1	5	1	1		11
Lincoln City, Lincolnshire			3	9	14	4			30
St Magnus, London	3		9	8	12	11	1		44
Brandon Road, Norfolk	4	3	16	14	17	11	13	3	81
Redcastle Furze, Norfolk	2	15	12	1	24	32	12	3	101

Appendix IIb: Dental ageing data for caprines by site type and date.

	0-6mths	**6-12mths**	**1-2 yrs**	**2-3 yrs**	**3-4 yrs**	**4-6 yrs**	**6-8 yrs**	**8-10yrs**	**Total**
Alms Lane, Norfolk				1	1	1			3
Heigham St, Norfolk							1		1
St Peters Street, Northamptonshire	6	1	3	21	22	44	3		100
St Ebbes, Oxfordshire	7	7	21	31	26	5	2	1	100
The Hamel, Oxfordshire	1	1	1	5	7	10	2	2	29
Codrington Library, Oxfordshhire				1	1				2
Dean Court, Oxfordshire		1	3	1	7	10	2	2	26
George Street, Oxfordshire	1		4	2	2	2			11
Ipswich, Suffolk				1	1	1	1		4
Bell St, Reigate, Surrey				2	1	3	2		8
Eastgate, Yorkshire	4	1	20	26	16	12	6	3	88
General Accident, Yorkshire	1	2	10	4	10	7	1	1	36
Total	**21**	**28**	**60**	**77**	**107**	**115**	**35**	**11**	**454**
%	5	6	18	19	23	21	7	2	101
mid-12th - mid-14th Elite									
Faccombe Netherton, Hampshire	2	4	1	10	9	24	3	4	57
Portchester Castle, Hampshire	2		2	3	5	2			14
Canterbury Cathedral, Kent									
Castle Rising, Norfolk		1	1	3	2	3			10
Castle Mall, Norfolk	1		5	5	1	7	2	1	22
Prudhoe Castle, Northumberland	4			1	1	1			7
Eynsham, Oxfordshire	2		1	7	8	1	3	1	23
Middleton Stony, Oxfordshire		2		1		4	5		12
Witney Palace, Oxfordshire	4	1				2			7
Guildford Castle, Surrey									
Dudley Castle, West Midlands	2		3	2	2	3	1		13
Trowbridge, Wiltshire		2		1					3
Ludgershall Castle, Wiltshire	5	1	2	3	1	1	1		14
Total	**22**	**11**	**15**	**36**	**29**	**48**	**15**	**6**	**111**
%	12	6	8	20	16	26	8	3	99
French Caprines									
Dury (Rural)	1	3	8	2	2	9	8	2	**35**
Compeigne 9-10th (Urban)		2	9	4	4	4	3		**26**
Compeigne 11-12th (Urban)		1	3	1					**5**
Douai (Elite)	55	11	12	8	4	2	1	1	**94**
Vatteville (Elite)	3	2				4	1	1	**11**
Montsoreau (Elite)	14	3	19						**36**
Total	**73**	**22**	**51**	**15**	**10**	**19**	**13**	**4**	**207**
%	35	11	25	7	5	9	6	2	100

Appendix IIc: Dental ageing data for pigs by site type and date.

N.B. The data for some assemblages (highlighted *) are shown as percentages.
For wear stage correlations see Table 5.

	0-6 mths	6-12 mths	1-2 yrs	2-3 yrs	3 yrs +	Total
5th - mid-9th Rural						
West Stow, Suffolk	13	14	28	43	2	100*
Wolverton Turn, Buckinghamshire	4	3	5	1	3	16
Pennyland, Buckinghamshire	2	1	6	10	1	20
Yarnton, Oxfordshire	2	1		1	1	5
Yarnton (Cresswell Field), Oxfordshire			4			4
Yarnton (Worton), Oxfordshire				1		1
Mucking, Essex	1	7		1		9
Marefair, Northamptonshire		1	2	3	1	7
Total	**22**	**27**	**45**	**60**	**8**	**162**
%	13	17	28	37	5	100
5th - mid-9th Urban						
Hamwih, Hampshire	1	21	31	27	19	99
Ipswich, Suffolk	14	22	26	20	3	85
Ipswich, Suffolk	3	18	30	52	0	103
Fishergate, Yorkshire	0	8	0	10	0	18
Total	**18**	**69**	**87**	**109**	**22**	**305**
%	6	22	29	36	7	100
5th - mid-9th Elite						
Lake End Road, Essex	7	37	41	44	2	131
Wicken Bonhunt, Essex	6	19	29	36	10	100
St Albans, Hertforshire	1	5	11	21	10	48
Brandon, Suffolk	31	20	21	25	2	99
Total	**14**	**61**	**81**	**101**	**22**	**279**
%	12	21	27	34	6	100
L9th-M11th century Rural						
Marefair, Northamptonshire			1	4		
Wharram Percy, Yorkshire		1	4	1		
Total		**1**	**5**	**5**		**11**
%	0	10	45	45	0	100
L9th-M11th century Urban						
Winchester, Hampshire	2	6	17	8	6	39
Lincoln City, Lincolnshire		3	2	4		9
St Magnus, London				4	5	9
Brandon Road, Norfolk		8	11	9	3	31
Thetford 1092, Norfolk	1	7	12	6	5	31
St Ebbes, Oxfordshire	1	7	11	8		27
Ipswich, Suffolk		1	3			4
Steyning, Sussex	1	3	4	1		9
Coppergate, Yorkshire, period 3	2	7	13	16		38
Coppergate, Yorkshire, period 4		8	15	9		32
Coppergate, Yorkshire, period 5a	1	7	19	14		41
Coppergate, Yorkshire, period 5b		7	28	15		50
Total	**8**	**64**	**135**	**94**	**19**	**320**
%	3	20	42	29	6	100
L9th-M11th century Elite						
Faccombe Netherton, Hampshire	1	3	9	5	5	23
Portchester Castle, Hampshire	5	13	53	39	14	124
Castle Mall, Norfolk	3	10	16	6	2	37
Eynsham Abbey, Oxford	1	7	11	8		27
Trowbridge, Wiltshire		2	17	6	3	28
Total	**10**	**35**	**106**	**64**	**24**	**239**
%	4	15	44	27	10	100
mid-11th - mid-12th Rural						
West Cotton, Northamptonshire	3	11	16	6	0	36
Marefair, Northamptonshire	0	0	0	2	0	2
Total	**3**	**11**	**16**	**8**	**0**	**38**
%	8	29	42	21	0	100

Appendix IIc: Dental ageing data for pigs by site type and date.

	0-6 mths	6-12 mths	1-2 yrs	2-3 yrs	3 yrs +	Total
mid-11th - mid-12th Urban						
Exeter, Devon	5	6	38	13	8	70
Lincoln City, Lincolnshire				1		1
St Magnus, London			1	3	6	10
St Nicholas Street, Norfolk	2	5	1			8
Fishergate, Norfolk		5	4	9		18
Red Castle Furze, Norfolk	9		14	2		25
Brandon Road, Norfolk		9	17	12	4	42
St Ebbes, Oxfordshire	2	1	7	3		13
Ipswich, Suffolk		3	5	1		9
Coppergate, Yorkshire			2	4		6
Fishergate, Yorkshire		1		1		2
General Accident, Yorkshire		1				1
Eastgate, Yorkshire	3	3	6			12
Total	**21**	**34**	**95**	**49**	**18**	**217**
%	10	16	44	22	8	100
mid-11th - mid-12th Elite						
Portchester Castle, Hampshire	1	2	4	3	1	11
Faccombe Netherton, Hampshire	9	2	7	4	4	26
Castle Mall, Norfolk	4	13	21	5		43
Eynsham Abbey, Oxfordshire	10	7	44	41		102
Guildford Castle, Surrey	1		1	1		3
Trowbridge, Wiltshire	2	6	12	1		21
Total	**27**	**30**	**89**	**55**	**5**	**206**
%	13	15	43	27	2	100
mid-12th - mid-14th Rural						
Boteler's Castle, Leicestershire	4	4	1	1		10
Marefair, Northamptonshire		1	3	3	3	10
Total	**4**	**5**	**4**	**4**	**3**	**20**
%	20	25	20	20	15	100
mid-12th - mid-14th Urban						
Exeter, Devon	2	4	17	3	5	31
Lincoln City, Lincolnshire			1	2		3
St Magnus, London			3	5	1	9
St Ebbes, Oxfordshire	1	9	8	5		23
General Accident, Yorkshire			14	9		23
General Accident, Yorkshire		2	1	2		5
Eastgate, Yorkshire	3	7	1			11
Total	**6**	**22**	**45**	**26**	**6**	**105**
%	6	21	43	25	5	100
mid-12th - mid-14th Elite						
Faccombe Netherton, Hampshire	3	5	12	21	7	48
Portchester Castle, Hampshire	1	4	4	6	1	16
Castle Mall, Norfolk	2	2	4	1	1	10
Eynsham Abbey, Oxfordshire		2	14	15	1	32
Guildford Castle, Surrey	6	17	24	6	3	56
Trowbridge, Wiltshire			6	1		7
Total	**12**	**30**	**64**	**50**	**13**	**169**
%	7	18	38	30	7	100
French Pigs						
Dury, Somme (Village)		3	4		1	8
Compeigne, Oise (Town) 9-10th	26	20	50	6	1	103
Compeigne, Oise (Town) 11-12th	50	40	20			110
Compeigne, Oise (High Status)	25	14	12	1	1	53
Vatteville, Seine Maritime (High Status)		10	44	19	1	74
Douai, Nord (High Status)	3	55	8	17	17	100
Compeigne (Monastic)	5	20	1			26
Total	**109**	**162**	**139**	**43**	**21**	**474**
%	23	34	29	10	4	100

Appendix IIIa: Anatomical representation data for cattle from pre- and post-Conquest rural, urban and elite sites.

N.B. the totals for each element have been standardised

Mid-ninth to mid-eleventh century Rural	Horn	Mandible	Scapula	Humerus	Radius	Pelvis	Femur	Tibia	Metapodia	Phalanges
Maxey, Northamptonshire	1	13	4	3	7	5	5	3	4	1
Goltho, Lincolnshire	6	10	2	2	6	1	2	1	11	0
Wharram Percy, Yorkshire	0	4	6	4	3	9	4	3	10	3
Total	7	27	12	9	16	15	11	7	25	4
% MNI	**26**	**100**	**44**	**33**	**59**	**56**	**41**	**26**	**93**	**15**
Mid-eleventh to mid-twelfth century Rural										
Brighton Hill South, Hampshire	1	7	4	5	3	7	6	3	2	2
Walton Road, Buckinghamshire	2	8	4	4	4	1	3	2	2	1
West Cotton, Northamptonshire		29	19	13	18	30	7	27	20	9
Total	3	44	27	22	25	38	16	32	24	12
% MNI	**7**	**100**	**61**	**50**	**57**	**86**	**36**	**73**	**55**	**27**
Mid-ninth to mid-eleventh century Urban a)										
Thetford 1092, Norfolk	188	27	18	14	19	12	9	13	25	3
% MNI	**100**	**14**	**9**	**7**	**10**	**6**	**5**	**7**	**13**	**1**
Mid-ninth to mid-eleventh century Urban b)										
Winchester, Hampshire	17	40	45	42	40	39	38	38	23	4
Ipsich, Suffolk	6	24	10	14	17	23	9	12	9	1
Redcastle Furze, Norfolk	5	25	20	9	8	11	6	8	11	3
Redcastle Furze, Norfolk	1	20	5	3	4	5	1	3	4	2
St Peters Street, Northamptonshire		4	11	9	8	10	7	12	5	1
Waterfront, Lincolnshire	4	8	3	4	6	6	3	7	8	1
Wigford, Lincolnshire	33	62	9	9	16	17	8	10	31	2
Flaxengate, Lincolnshire	7	36	30	13	54		12	24	30	12
Fishergate, Yorkshire 3c	7	11	6	8	7	8	6	7	8	4
Fishergate, Yorkshire 3c	3	6	5	4	4	4	3	3	2	2
Coppergate, Yorkshire 15311	4	8	6	12	7	7	7	7	10	
Total	87	**244**	150	127	171	130	100	131	141	32
% MNI	**36**	**100**	**61**	**52**	**70**	**53**	**40**	**54**	**58**	**13**
Mid-eleventh to mid-twelfth century Urban										
Bath, Avon	13	8	9	11	14	24	20	22	14	4
Westgate, Gloucestershire	0	11	7	5	5	5	7	7	5	1
Brandon Road, Norfolk	12	24	22	15	23	16	9	17	26	6
Hereford City, Herefordshire	6	3	2	3	2		3	2	4	1
North Street, Sussex	4	7	3	14	10	6	11	18	9	2
Wigford, Lincolnshire	17	13	1	9	15	7	9	5	11	1
Flaxengate, Lincolnshire	7	17	35	22	38		11	14	11	8
Coppergate, Yorkshire 5981	4	6	6	11	8	14	12	8	6	
Coppergate, Yorkshire 16734	8	11	12	11	13	20	19	14	10	
Total	71	100	97	101	**128**	92	101	107	96	23
% MNI	**55**	**78**	**76**	**79**	**100**	**72**	**78**	**84**	**75**	**18**
Mid-ninth to mid-eleventh century Elite										
Canterbury Castle, Sussex	3	3	3	3	5	4	3	2	5	1
North Elmham, Norfolk		108	34	49	36	34	44	45	28	2
Faccombe Netherton, Hampshire	5	9	7	15	6	5	12	13	7	1
Portchester Castle, Hampshire	99	139	65	80	77	96	55	83	98	12
Goltho, Lincolnshire	6	18	6	10	7	9	4	8	13	1
Eynsham Abbey, Oxfordshire (2f)	1	7	11	7	4	5	8	8	8	2
Castle Mall, Norfolk		17	14	19	12	14	5	28	19	9
St Albans, Hertfordshire	0	2	6	5	5	7	5	5	1	1
Total	114	**303**	146	188	152	174	136	192	179	29
% MNI	**38**	**100**	**48**	**62**	**50**	**58**	**45**	**64**	**59**	**10**
Mid-eleventh to mid-twelfth century Elite										
Prudhoe Castle, Northumberland		1	2	1	3		1	5	1	
Sandal Castle, Yorkshire		5		2	1		1	6	2	
Goltho, Lincolnshire	2	9	18	12	19	10	30	10	26	8
North Elmham, Norfolk		8	8	17	9	10	10	10	12	1
Castle Mall, Norfolk period 2		3	15	12	8	11	4	13	16	6
Castle Mall, Norfolk period 3		6	4	1	1	2	1	2	3	1
Faccombe Netherton, Hampshire	2	21	13	9	10	7	15	8	11	7
Pevensey Castle, Sussex	2	6	5	5	4	6	4	8	4	2
Carisbrooke Castle, Isle of Wight		1	6	5	4	5	5	5	6	
Carisbrooke Castle, Isle of Wight		3	7	10	8	14	10	10	2	
Carisbrooke Castle, Isle of Wight		0	1	6	2	2	2	3	2	
Eynsham Abbey, Oxford (3a)	2	23	13	10	10	11	8	9	23	2
Eynsham Abbey, Oxford (3b)	6	34	22	10	10	10	21	10	20	2
Total	14	120	114	100	89	88	112	99	**128**	29
% MNI	**11**	**94**	**89**	**78**	**69**	**68**	**87**	**77**	**100**	**22**

Appendix IIIb: Anatomical representation data for caprines from pre- and post-Conquest rural, urban and elite sites.

N.B. the totals for each element have been standardised

Mid-ninth to mid-eleventh century Rural	Horn	Mandible	Scapula	Humerus	Radius	Pelvis	Femur	Tibia	Metapodia
Maxey, Northamptonshire	1	13	4	7	4	5	13	4	1
Wharram Percy, Yorkshire		25	8	13	4	7	18	7	2
Total	1	**38**	12	20	8	12	31	11	3
% MNI	**3**	**100**	**32**	**53**	**21**	**32**	**82**	**29**	**8**
Mid-eleventh to mid-twelfth century Rural									
Brighton Hill South, Hampshire	1	8	2	4	7	2	10	6	1
Walton Road, Buckinghamshire	2	8	1	8	6	5	12	8	1
West Cotton, Northamptonshire		43	9	15	13	5	24	10	3
Total	3	**59**	12	27	26	12	46	24	5
% MNI	**5**	**100**	**20**	**46**	**44**	**20**	**78**	**41**	**8**
Mid-ninth to mid-eleventh century Urban a)									
Wigford, Lincolnshire	2	18	10	8	15	6	16	20	1
% MNI	**10**	**90**	**50**	**40**	**75**	**30**	**80**	**100**	**5**
Mid-ninth to mid-eleventh century Urban b)									
Bath, Avon	3	3	3	3	4	1	2	2	1
Westgate, Gloucestershire		5	2	2	3	1	6	4	
Thetford 1092, Norfolk	22	25	39	20	28	12	25	12	2
Hereford city, Herefordshire	3	6	4	3	5	2	6	4	1
Winchester, Hampshire	10	42	27	19	33	10	45	30	1
Ipswich, Suffolk,	6	5	1	4	10	3	6	4	1
St Peters Street, Northamptonshire		4	11	9	8	7	12	5	1
Flaxengate, Lincolnshire	4	33	34	32	31	5	29	33	2
Fishergate, Yorkshire	5	10	5	4	5	5	5	6	1
Total	53	133	126	96	127	46	**136**	100	10
% MNI	**39**	**98**	**93**	**71**	**93**	**34**	**100**	**74**	**7**
Mid-eleventh to mid-twelfth century Urban									
Bath, Avon	25	25	17	24	22	7	9	18	1
Brandon Road, Norfolk	19	54	29	20	22	9	4	36	1
Redcastle Furze, Norfolk	14	59	8	15	14	3	13	12	2
Hereford city, Herefordshire	2	4	2	3	5	3	8	5	1
North Street, Sussex	33	44	13	28	23	32	51	44	3
Wigford, Lincolnshire	5	15	9	7	13	1	7	9	1
Flaxengate, Lincolnshire	21	88	52	59	77	16	72	48	4
Copperghate, Yorkshire 16612	4	6	5	16	9	7	12	7	
Total	123	**295**	135	172	185	78	176	179	13
% MNI	**42**	**100**	**46**	**58**	**63**	**26**	**60**	**61**	**4**
Mid-ninth to mid-eleventh century Elite									
North Elmham, Norfolk		170	54	54	78	25	128	22	7
Castle Mall, Norfolk (Phase 1)		21	7	9	8	3	17	8	2
Canterbury Castle, Kent	2	6	5	2	6	1	9	4	
Faccombe Netherton, Hampshire	3	16	8	12	34	9	34	12	1
Portchester Castle, Hampshire	52	268	42	56	50	23	50	45	1
Morrison Hall, Cleeveland	0	4	13	7	14	1	13	1	1
Goltho, Lincolnshire	5	10	5	10	10	1	24	2	0
Eynsham Abbey, Oxfordshire (2f)	1	9	9	11	20	6	23	4	1
St Albans, Hertfordshire	1	10	4	4	10	3	14	4	1
Total	64	**514**	147	165	230	72	312	102	14
% MNI	**12**	**100**	**29**	**32**	**45**	**14**	**61**	**20**	**3**
Mid-eleventh to mid-twelfth century Elite									
Faccombe Netherton, Hampshire	1	11	17	16	35	6	32	6	1
Portchester Castle, Hampshire		3	2	1	2	1	1	3	1
Goltho, Lincolnshire	3	15	12	29	29	7	38	3	
North Elmham, Norfolk		22	5	12	13	9	28	9	1
Castle Mall, Norfolk (Phase 2)		9	8	9	6	2	14	7	1
Castle Mall, Norfolk (Phase 3)		3	2	3	2	1	4	2	1
Eynsham Abbey, Oxfordshire (3a)	2	15	13	15	29	13	38	5	1
Eynsham Abbey, Oxfordshire (3b)	4	19	13	16	21	8	48	7	1
Guildford Castle, Surrey	0	0	0	2	6	1	4	4	1
Pevensey Castle, Sussex	2	9	6	10	19	4	18	8	1
Carisbrooke Castle, Isle of Wight		1	34	22	37	18	32	2	
Carisbrooke Castle, Isle of Wight			5	3	7	2	4	1	
Carisbrooke Castle, Isle of Wight		1	17	35	36	16	18		
Total	12	108	134	173	242	88	**279**	57	9
% MNI	**4**	**39**	**48**	**62**	**87**	**32**	**100**	**20**	**3**

Appendix IIIc: Anatomical representation data for pigs from pre- and post-Conquest rural, urban and elite sites.

N.B. The totals for each element have been standardised

Mid-ninth to mid-eleventh century Rural	Mandible	Scapula	Humerus	Radius	Pelvis	Femur	Tibia	Metapodia	Phalanges
Maxey, Northamptonshire	3	4	1	1		1	1	1	1
Wharram Percy, Yorkshire	3	2	2	2	2		1	2	1
Total	**6**	**6**	3	3	2	1	2	3	**2**
% MNI	**100**	**100**	**50**	**50**	**33**	**16**	**33**	**50**	**33**
Mid-eleventh to mid-twelfth century Rural									
Brighton Hill South, Hampshire	8	3	4	2	2	2	3	1	1
Walton Road, Buckinghamshire	2	2	1			2	1	1	1
West Cotton, Northamptonshire	14	5	4		4	1	5	2	2
Total	**24**	10	9	2	6	5	9	4	4
% MNI	**100**	**42**	**38**	**8**	**25**	**21**	**38**	**17**	**17**
Mid-ninth to mid-eleventh century Urban									
Thetford 1092, Norfolk	50	10	7	9	7	5	14	7	1
Red Castle, Norfolk	9	4	1	2	3	1	3	1	1
Winchester, Hampshire	38	23	16	12	12	12	16	9	1
Ipswich, Suffolk	11	5	5	5	5	1	8	2	1
St Peters Street, Northamptonshire	8	7	5	1	3	3	4	2	1
Lincoln City, Lincolnshire	15	6	7	12	8	5	8	3	1
Flaxengate, Lincolnshire	27	13	13	29		2	5	9	1
Fishergate, Yorkshire 3c	3	2	2	3	5	2	3	1	1
Total	**161**	70	56	73	43	31	61	34	8
% MNI	**100**	**43**	**35**	**45**	**27**	**19**	**38**	**21**	**5**
Mid-eleventh to mid-twelfth century Urban									
Bath, Avon	10	8	6	7	4	2	2	4	1
Brandon Road, Norfolk	33	11	12	12	4	8	16	7	1
Red Castle, Norfolk	25	5	3	1	3		8	4	
Westgate, Gloucestershire	2	2	5	4	5	4	3	1	
Brighton Hill South, Hampshire	8	3	4	2	2	2	3	1	1
Noth Street, Sussex	8	3	15	12	7	8	12	6	1
Lincoln City, Lincolnshire	10	3	7	7	2	3	5	1	1
Coppergate, Yorkshire 16612	9	5	6	6	5	6	10		
Total	**105**	40	58	51	32	33	59	24	5
% MNI	**100**	**38**	**55**	**48**	**30**	**31**	**56**	**23**	**5**
Mid-ninth to mid-eleventh century Elite									
St Albans, Hertfordshire	35	22	12	4	4	9	8	3	1
Canterbury Castle, Kent	7	3	2	3	1	3	6	1	
Faccombe Netherton, Hampshire	23	8	13	4	5	10	10	4	1
Portchester Castle, Hampshire	248	30	27	21	33	8	23	12	1
North Elmham, Norfolk	67	33	39	10	33	17	25	7	1
Eynsham Abbey, Oxfordshire (2f)	12	10	5	6	9	3	9	4	1
Castle Mall, Norfolk	17	8	9	6	7	3	16	11	2
Goltho, Lincolnshire	37	2	8	4	3	1	6	2	1
Total	**446**	116	115	58	95	54	103	44	8
% MNI	**100**	**26**	**25**	**13**	**21**	**12**	**23**	**2**	**2**
Mid-eleventh to mid-twelfth century Elite									
Goltho, Lincolnshire	30	4	6	7	3		9	3	
Faccombe Netherton, Hampshire	24	13	17	11	14	11	14	7	2
Sandal Castle, Yorkshire	5		8	5		2	6	4	
North Elmham. Norfolk	19	8	16	6	6	5	14	3	1
Castle Mall, Norfolk	12	4	6	5	7	3	11	9	2
Pevensey Castle, Sussex	10	6	8	5	4	6	9	4	1
Eynsham Abbey, Oxfordshire (3a)	25	14	13	18	20	8	29	17	5
Eynsham Abbey, Oxfordshire (3b)	87	36	41	28	33	22	47	27	3
Carisbrooke Castle, Isle of Wight	25	9	9	16	9	27	9	1	
Carisbrooke Castle, Isle of Wight	12	8	12	3	3	3	8		
Carisbrooke Castle, Isle of Wight		7	4	7		4		1	
Carisbrooke Castle, Isle of Wight	61	24	38	25	13	13	18		
Total	**310**	133	178	136	112	104	174	76	14
% MNI	**100**	**43**	**57**	**44**	**36**	**34**	**56**	**25**	**5**

Appendix IIId: Anatomical representation data for deer from elite sites in Britain, from Vatteville-la-Rue and sites in Germany.

Site	Antler	Mandible	Scapula	Humerus	Radius	Metacarpal	Pelvis	Femur	Tibia	Astragalus	Calcaneum	Metatarsal
7th/mid9th-century Roe deer												
Ramsbury, Wiltshire	5	23	5	4	4	3	1		1			3
St Albans Abbey, Hertfordshire	2	12	3	4	5	5	6	4	5	1	1	5
Eynsham Abbey, Oxfordshire	2		9	6	3	3	4	2	11			7
Goltho, Lincolnshire		1			1							
Faccombe Netherton, Hampshire	1	4	1	5	3	1	2	4	12		1	9
Portchester Castle, Hampshire	6	27	18	16	14	11	2	5	21	3	1	18
Cheddar Palaces, Somerset			3	5	12	13			1			14
7th/mid9th-century Red deer												
Ramsbury, Wiltshire	1	7		2		1						1
Eynsham Abbey, Oxfordshire		3		1	3			2	1			2
Goltho, Lincolnshire											1	
Faccombe Netherton, Hampshire	1		2	4	2	3	2		4	2	1	
Portchester Castle, Hampshire	11	10	7	9	13	39	6	11	35	8	6	39
Cheddar Palaces, Somerset				1		1	1		3			
Total	29	87	48	57	60	80	24	28	94	14	11	**98**
%		**90**	**49**	**59**	**61**	**82**	**24**	**29**	**96**	**14**	**12**	**100**
mid11th/mid-12th century Roe deer												
Eynsham Abbey, Oxfordshire	3	3	4	2	1	13	2	2	14		3	23
Goltho 1000-1080, Lincolnshire		6	2	3	10	8			3		3	8
Goltho 1080-1150, Lincolnshire	1		2	1	2	2	3		10			2
Faccombe Netherton, Hampshire	7	10			1	13			2	1		18
Portchester Castle, Hampshire	3	3	1	5	2	1			3	1		7
mid11th/mid-12th century Red deer												
Eynsham Abbey, Oxfordshire	1				1			4	3		1	
Goltho 1000-1080, Lincolnshire	3							1	2	1	1	
Goltho 1080-1150, Lincolnshire	1							1		4	2	
Faccombe Netherton, Hampshire	5	11	1	4	7	45	3	7	9	13	9	43
Portchester Castle, Hampshire	6	3	1	1	5	10			7	3		15
mid11th/mid-12th century Fallow												
Guildford Castle, Surrey								1	2			
Goltho 1000-1080, Lincolnshire		6	1	1	2	1			6		2	6
Goltho 1080-1150, Lincolnshire		1							1		1	1
Faccombe Netherton, Hampshire	2	2				7		1	1	1	1	3
Castle Rising Castle, Norfolk		1							2	1		
Cheddar Palaces, Somerset												1
Total	32	46	12	17	31	100	8	17	65	25	23	**127**
%		**36**	**9**	**13**	**24**	**79**	**6**	**13**	**51**	**20**	**18**	**100**
mid-12th/mid-14th century Roe deer												
Guildford Castle, Surrey		2	3		1	3		1	2	3		3
Eynsham Abbey, Oxfordshire	2	3	2		7	9	1	4	17			8
Faccombe Netherton, Hampshire		2				1			3			4
Launceston Castle *, Cornwall			4	4		4						6
mid-12th/mid-14th century Red deer												
Guildford Castle, Surrey				1	1	1	1	6	9	10	5	2
Eynsham Abbey, Oxfordshire	2	1	3		7		1	10	4		1	1
Faccombe Netherton, Hampshire	1	1				1	1				2	
Launceston Castle, Cornwall									2	6	6	
Sandal Castle, Yorkshire					2	5		1	1			2
mid-12th/mid-14th century Fallow												
Guildford Castle, Surrey				3	2		13	10	29	20	12	16
Eynsham Abbey, Oxfordshire	1				1				2			3
Faccombe Netherton, Hampshire	5	6	1	3	5	6	4	6	18	5	2	14
Launceston Castle, Cornwall				2			2	2	2	20	6	4
Sandal Castle, Yorkshire		4		4	9	13		7	11			67
Castle Rising, Norfolk	4	7	3	3	8		7	12	26	10	8	45
Okehampton Castle, Devon	8		2	1	4	2	3	6	7			7
Total	23	26	18	21	47	45	33	65	133	74	42	**182**
%		**14**	**10**	**12**	**25**	**25**	**18**	**36**	**73**	**41**	**23**	**100**

Appendix IIId: Anatomical representation data for deer from elite sites in Britain, from Vatteville-la-Rue and sites in Germany.

Vatteville Castle	Antler	Mandible	Scapula	Humerus	Radius	Metacarpal	Pelvis	Femur	Tibia	Astragalus	Calcaneum	Metatarsal
12th century	4		4	5	4	2	3	6	9	3	2	9
13th-15th century	45	7		1	6	4	1	4	8	5	6	6
German Deer												
Munsterberg	1	1		4	2	3	1	2	2		1	
Hochstetten	4		1	2	1	1		1	7		2	1
Unterregenbach (Red deer)	22	20	25	35	41	18	18					13
Unterregenbach (Roe deer)	2	7		4	2	3	1					2
Schiedberg (Red deer)	8	5	1	7	12	9	5	1	4	7	3	10
Schiedburg (Roe deer)	1											
Heuneburg (Red deer)	3			1			1	1			2	
Heuneburg (Roe deer)												1

www.ingramcontent.com/pod-product-compliance
Lightning Source LLC
LaVergne TN
LVHW072256300926
849160LV00018B/1677